Advances in Intelligent Systems and Computing

Volume 402

Series editor

Janusz Kacprzyk, Polish Academy of Sciences, Warsaw, Poland
e-mail: kacprzyk@ibspan.waw.pl

About this Series

The series "Advances in Intelligent Systems and Computing" contains publications on theory, applications, and design methods of Intelligent Systems and Intelligent Computing. Virtually all disciplines such as engineering, natural sciences, computer and information science, ICT, economics, business, e-commerce, environment, healthcare, life science are covered. The list of topics spans all the areas of modern intelligent systems and computing.

The publications within "Advances in Intelligent Systems and Computing" are primarily textbooks and proceedings of important conferences, symposia and congresses. They cover significant recent developments in the field, both of a foundational and applicable character. An important characteristic feature of the series is the short publication time and world-wide distribution. This permits a rapid and broad dissemination of research results.

Advisory Board

Chairman

Nikhil R. Pal, Indian Statistical Institute, Kolkata, India
e-mail: nikhil@isical.ac.in

Members

Rafael Bello, Universidad Central "Marta Abreu" de Las Villas, Santa Clara, Cuba
e-mail: rbellop@uclv.edu.cu

Emilio S. Corchado, University of Salamanca, Salamanca, Spain
e-mail: escorchado@usal.es

Hani Hagras, University of Essex, Colchester, UK
e-mail: hani@essex.ac.uk

László T. Kóczy, Széchenyi István University, Győr, Hungary
e-mail: koczy@sze.hu

Vladik Kreinovich, University of Texas at El Paso, El Paso, USA
e-mail: vladik@utep.edu

Chin-Teng Lin, National Chiao Tung University, Hsinchu, Taiwan
e-mail: ctlin@mail.nctu.edu.tw

Jie Lu, University of Technology, Sydney, Australia
e-mail: Jie.Lu@uts.edu.au

Patricia Melin, Tijuana Institute of Technology, Tijuana, Mexico
e-mail: epmelin@hafsamx.org

Nadia Nedjah, State University of Rio de Janeiro, Rio de Janeiro, Brazil
e-mail: nadia@eng.uerj.br

Ngoc Thanh Nguyen, Wroclaw University of Technology, Wroclaw, Poland
e-mail: Ngoc-Thanh.Nguyen@pwr.edu.pl

Jun Wang, The Chinese University of Hong Kong, Shatin, Hong Kong
e-mail: jwang@mae.cuhk.edu.hk

More information about this series at http://www.springer.com/series/11156

Mohammad S. Obaidat · Tuncer Ören
Janusz Kacprzyk · Joaquim Filipe
Editors

Simulation and Modeling Methodologies, Technologies and Applications

International Conference, SIMULTECH 2014
Vienna, Austria, August 28–30, 2014 Revised
Selected Papers

 Springer

Editors

Mohammad S. Obaidat
Department of Computer and Information
  Science
Fordham University
Bronx, NY
USA

Tuncer Ören
Faculty of Engineering, School of ECE
University of Ottawa
Ottawa, ON
Canada

Janusz Kacprzyk
Intelligent Systems Laboratory,
  Systems Research Institute
Polish Academy of Sciences
Warsaw
Poland

Joaquim Filipe
INSTICC
Setúbal
Portugal

ISSN 2194-5357   ISSN 2194-5365   (electronic)
Advances in Intelligent Systems and Computing
ISBN 978-3-319-26469-1   ISBN 978-3-319-26470-7   (eBook)
DOI 10.1007/978-3-319-26470-7

Library of Congress Control Number: 2015955364

Springer Cham Heidelberg New York Dordrecht London

Springer International Publishing AG Switzerland is part of Springer Science+Business Media
(www.springer.com)

Preface

This book includes extended and revised versions of a set of selected papers from the 2014 International Conference on Simulation and Modeling Methodologies, Technologies and Applications (SIMULTECH 2014), which was sponsored by the Institute for Systems and Technologies of Information, Control and Communication (INSTICC) and co-organized by Austrian Computer Society and Vienna University of Technology—TU Wien (TUW). SIMULTECH 2014 was held in cooperation with the ACM SIGSIM—Special Interest Group (SIG) on SImulation and Modeling (SIM), ACM SIGMIS—ACM Special Interest Group on Management Information Systems, Movimento Italiano Modellazione e Simulazione (MIMOS), Japan Society for Simulation Technology, Federation of Asia Simulation Societies and AIS Special Interest Group on Modeling and Simulation (AIS SIGMAS) and technically co-sponsored by the Society for Modeling & Simulation International (SCS), IEEE Computer Society's Technical Committee on Simulation (TCSIM), and Liophant Simulation, Simulation Team.

This conference brings together researchers, engineers, and practitioners interested in methodologies and applications of modeling and simulation. The main topics covered in the papers accepted in the conference are: Methodologies and Technologies, and Applications and Tools of Modeling and simulation. We believe the accepted papers demonstrate new and innovative solutions. They also highlight technical issues and challenges in this field.

SIMULTECH 2014 received 167 paper submissions, including special sessions, from 45 countries in all continents, of which 23 % were presented as full papers. A double blind paper review was performed by the International Program Committee members, all of them recognized in at least one of the main conference topic areas.

The papers included in this book were selected from those with the best reviews also taking into account the quality of their presentation at the conference, assessed by the session chairs. Therefore, we hope that you find the papers included in this book interesting, and we trust they will represent as helpful references.

We wish to thank all those who supported and helped to organize the conference. On behalf of the conference Organizing Committee, we would like to thank the authors, whose work mostly contributed to a very successful conference, the members of the Program Committee, whose expertise and diligence were instrumental to ensure the quality of final contributions, and the invited distinguished keynote speakers for their invaluable contribution and for taking the time to synthesize and deliver their talks. Special thanks to the plenary panel organizer and panelists for their outstanding contribution. We also wish to thank all the members of the Organizing Committee whose work and commitment were invaluable. Thanks also are due to the organizations that technically co-sponsored the conference. Special thanks go to the Staff members of SIMULTECH 2014 for their outstanding support and dedicated work. Last but not least, we would like to thank INSTICC for sponsoring and organizing the conference.

April 2015
Mohammad S. Obaidat
Tuncer Ören
Janusz Kacprzyk
Joaquim Filipe

Organization

Conference Chair

Mohammad S. Obaidat, Fordham University, USA

Program Co-chairs

Tuncer Ören, University of Ottawa, Canada
Janusz Kacprzyk, Polish Academy of Sciences, Poland

Organizing Committee

Marina Carvalho, INSTICC, Portugal
Helder Coelhas, INSTICC, Portugal
Bruno Encarnação, INSTICC, Portugal
João Francisco, INSTICC, Portugal
Lucia Gomes, INSTICC, Portugal
Rúben Gonçalves, INSTICC, Portugal
André Lista, INSTICC, Portugal
Ana Guerreiro, INSTICC, Portugal
Filipe Mariano, INSTICC, Portugal
Andreia Moita, INSTICC, Portugal
Raquel Pedrosa, INSTICC, Portugal
Vitor Pedrosa, INSTICC, Portugal
Cláudia Pinto, INSTICC, Portugal
Cátia Pires, INSTICC, Portugal
Carolina Ribeiro, INSTICC, Portugal

João Ribeiro, INSTICC, Portugal
Susana Ribeiro, INSTICC, Portugal
Sara Santiago, INSTICC, Portugal
Fábio Santos, INSTICC, Portugal
Mara Silva, INSTICC, Portugal
José Varela, INSTICC, Portugal
Pedro Varela, INSTICC, Portugal

Program Committee

Magdiel Ablan, Universidad de Los Andes, Venezuela
Erika Ábrahám, RWTH Aachen University, Germany
Nael Abu-Ghazaleh, State University of New York at Binghamton, USA
Lyuba Alboul, Sheffield Hallam University, UK
Marco Aldinucci, University of Torino, Italy
Mikulas Alexik, University of Zilina, Slovak Republic
Manuel Alfonseca, Universidad Autonoma de Madrid, Spain
Jan Awrejcewicz, The Technical University of Lódz (TUL), Poland
Gianfranco Balbo, University of Torino, Italy
Simonetta Balsamo, University of Venezia Ca' Foscari, Italy
M. Gisela Bardossy, University of Baltimore, USA
Ana Isabel Barros, TNO, The Netherlands
Ildar Batyrshin, Mexican Petroleum Institute, Mexico
Nicolas Belloir, LIUPPA, France
Lucian Bentea, University of Oslo, Norway
Louis Birta, University of Ottawa, Canada
Gennaro Boggia, Politecnico di Bari, Italy
Wolfgang Borutzky, Bonn-Rhein-Sieg University of Applied Sciences, Germany
Christos Bouras, University of Patras and CTI&P Diophantus, Greece
Felix Breitenecker, Vienna University of Technology, Austria
Christian Callegari, University of Pisa, Italy
Jesus Carretero, Computer Architecture Group, University Carlos III of Madrid,
Spain
Emiliano Casalicchio, University of Tor Vergata, Italy
Srinivas Chakravarthy, Kettering University, USA
Chun-Hung Chen, George Mason University, USA
Dan Chen, China University of Geosciences, China
Jiangzhuo Chen, Virginia Polytechnic Institute and State University, USA
Lawrence Chung, The University of Texas at Dallas, USA
Franco Cicirelli, Università della Calabria, Italy
Tanja Clees, Fraunhofer Institute for Algorithms and Scientific Computing (SCAI),
Germany
Priami Corrado, CoSBi, Italy

Andrea D'Ambrogio, Università di Roma "Tor Vergata", Italy
Eugen Dedu, University of Franche-Comté, France
Gabriella Dellino, University of Foggia, Italy
Atakan Dogan, Anadolu University, Turkey
Susanna Donatelli, Università degli Studi di Torino, Italy
Zhihui Du, Tsinghua University, China
Julie Dugdale, Laboratoire d'Informatique de Grenoble, France
Mostafa El-Said, Grand Valley State University, USA
Paul Fishwick, University of Texas at Dallas, USA
Jason Friedman, Tel-Aviv University, Israel
Claudia Frydman, Aix Marseille University, France
Richard Fujimoto, Georgia Institute of Technology, USA
Marco Furini, Università di Modena e Reggio Emilia, Italy
José Manuel Galán, Universidad de Burgos, Spain
Petia Georgieva, University of Aveiro, Portugal
Nikolaos Geroliminis, Ecole Polytechnique Federal De Lausanne, Switzerland
Charlotte Gerritsen, Vrije Universiteit Amsterdam, The Netherlands
Nigel Gilbert, University of Surrey, UK
David Goldsman, Georgia Institute of Technology, USA
John (Yannis) Goulermas, The University of Liverpool, UK
Jan Tommy Gravdahl, Norwegian University of Science and Technology, Norway
Feng Gu, The City University of New York, USA
Murat Gunal, Turkish Naval Academy, Turkey
Mykola Gusti, International Institute for Applied Systems Analysis, Austria
Sigurdur Hafstein, Reykajvik University, Iceland
Scott Y. Harmon, Zetetix, USA
Monika Heiner, Brandenburg University of Technology Cottbus, Germany
Herbert Hoeger, Universidad de los Andes, Venezuela
Tsan-sheng Hsu, Institute of Information Science, Academia Sinica, Taiwan
Xiaolin Hu, Georgia State University, USA
Zbigniew Hulicki, AGH University of Science and Technology, Poland
Eric Innocenti, IUT DE CORSE—University of Corsica, France
Nobuaki Ishii, Bunkyo University, Japan
Mhamed Itmi, INSA, Rouen, France, France
Yumi Iwashita, Kyushu University, Japan
Luis Izquierdo, University of Burgos, Spain
Segismundo Samuel Izquierdo, University of Valladolid, Spain
Shafagh Jafer, Embry-Riddle University, USA
András Jávor, Budapest University of Technology and Economics, Hungary
Catholijn Jonker, Delft University of Technology, The Netherlands
Cara Kahl, Hamburg University of Technology, Germany
Rihard Karba, University of Ljubljana, Slovenia
Peter Kemper, College of William and Mary, USA
Matthias Klumpp, University of Duisburg-Essen, Germany
Juš Kocijan, Jozef Stefan Institute, Slovenia

James J. Nutaro, Oak Ridge National Laboratory, USA
Mohammad S. Obaidat, Monmouth University, USA
Paulo Moura Oliveira, Universidade de Tras os Montes e Alto Douro, Portugal
Peter Csaba Ölveczky, University of Oslo, Norway
Stephan Onggo, Lancaster University, UK
Tuncer Ören, University of Ottawa, Canada
Paul Ormerod, Volterra Partners, UK
C. Michael Overstreet, Old Dominion University, USA
Ioannis Paraskevopoulos, Brunel University, UK
George Pavlidis, "Athena" Research Centre, Greece
Krzysztof Pawlikowski, University of Canterbury, New Zealand
Ana Peleteiro, Universidade de Vigo, Spain
L. Felipe Perrone, Bucknell University, USA
Tomas Potuzak, University of West Bohemia, Czech Republic
Francesco Quaglia, Sapienza Università di Roma, Italy
Jacinto A. Dávila Quintero, Universidad de Los Andes, Venezuela
Stanislaw Raczynski, Universidad Panamericana, Mexico
Urvashi Rathod, Symbiosis Centre for Information Technology (SCIT), India
Manuel Resinas, Universidad de Sevilla, Spain
M.R. Riazi, Kuwait University, Kuwait
José Risco-Martín, Universidad Complutense de Madrid, Spain
Theresa Roeder, San Francisco State University, USA
Paolo Romano, IST/INESC-ID, Portugal
Ella E. Roubtsova, Open University of the Netherlands, The Netherlands
Willem Hermanus le Roux, CSIR, South Africa
Maarouf Saad, Ecole de technologie superieure, Université du Québec, Canada
Janos Sallai, Vanderbilt University, USA
Jean-François Santucci, SPE UMR CNRS 6134—University of Corsica, France
Massimo La Scala, Politecnico di Bari, Italy
Philippe Serré, Supméca, France
Tielong Shen, Sophia University, Japan
Jaime Sichman, Usp, Brazil
Peer-Olaf Siebers, University of Nottingham, UK
Flavio S. Correa Da Silva, University of Sao Paulo, Brazil
Yuri Skiba, Centro de Ciencias de la Atmosfera, Universidad Nacional Autonoma
de Mexico, Mexico
Jaroslav Sklenar, University of Malta, Malta
Andrzej Sluzek, Khalifa University, UAE
Jefrey Smith, Auburn University, USA
Yuri Sotskov, United Institute of Informatics Problems of the National Academy of
Belarus, UIIP, Minsk, Belarus
James C. Spall, The Johns Hopkins University, USA
Giovanni Stea, University of Pisa, Italy
Mu-Chun Su, National Central University, Taiwan
Nary Subramanian, University of Texas at Dallas, USA

Giovanni Nardini, University of Pisa, Italy
Dmitry Ponomarev, SUNY Binghamton, USA
Martin Strohmeier, University of Oxford, UK

Invited Speakers

Bernard P. Zeigler, University of Arizona, USA
Paul Fishwick, University of Texas at Dallas, USA
Helena Szczerbicka, Universität Hannover, Germany

Contents

Front Velocity Modeling Approach to Column Chromatographic Characterization and Evaluation of Ketamine Enantiomers Separation with Simulated Moving Bed

Anderson Luis Jeske Bihain, Antônio José da Silva Neto
and Leôncio Diógenes Tavares Câmara

Abstract Simulated Moving Bed (SMB) is an efficient compounds separation process that operates in a continuous regime and works in a countercurrent flow of solid phase. Among several applications, this process has excelled in petrochemical resolution and especially nowadays in the separation of enantiomers from racemic mixtures, which are considered difficult to separate. In this work, the new Front Velocity approach to model an SMB process was proposed. To describe the mass transfer that occurs in the chromatographic process, the front velocity approach considers that convection is the dominant phase in the solute transport along the chromatographic column. "Front Velocity" is a discrete model (steps) where flow rate determines the advancing of liquid phase through the column. The steps are advancing liquid phase and mass transfer between the liquid and solid phases, the latter in the same time interval. Thus, the experimental volumetric flow is used for discretization of the control volumes moving along the porous column with the same velocity as the liquid phase. The mass transfer was represented by two different kinetic mechanisms without (linear type) and with maximum adsorption capacity (Langmuir type). The proposed approach was studied theoretically and evaluated by comparison with experimental data of ketamine anesthetic separation in SMB. The results were also compared with the simulation model using dispersive equilibrium. In the chromatographic column characterization step the new approach was associated with R2W inverse tool to determine the lumped mass

A.L.J. Bihain (✉)
Department of Mathematics, UNIPAMPA Federal University of Pampa,
Rio Grande do Sul, Brazil
e-mail: andersonbihain@unipampa.edu.br

A.J.d. Silva Neto · L.D.T. Câmara
Department of Mechanical Engineering and Energy, IPRJ-UERJ Polytechnic Institute of the
State University of Rio de Janeiro Nova Friburgo, Rio de Janeiro, Brazil
e-mail: ajsneto@iprj.uerj.br

L.D.T. Câmara
e-mail: dcamara@iprj.uerj.br

© Springer International Publishing Switzerland 2015
M.S. Obaidat et al. (eds.), *Simulation and Modeling Methodologies,
Technologies and Applications*, Advances in Intelligent Systems
and Computing 402, DOI 10.1007/978-3-319-26470-7_1

transfer parameters using only the experimental residence time of each enantiomer in the high performance liquid chromatography (HPLC) column. In the second step the mass transfer kinetic equations developed in the approach were applied in each column of the SMB process together with the values determined in the characterization of the chromatographic column, to perform the continuous separation process. The simulation results show good agreement between the modeling proposal and pulse experiments used to characterize the column in enantiomeric separation of ketamine over time. The simulations of the SMB separation show a discrepancy with the experimental data in the early cycles, but after these initial cycles, the model has good correlation with the experimental data. According to the study conducted the proposed approach proved to be a potential tool for prediction of the chromatographic behavior of a sample in a pulse experiment and to simulate the compounds separation in an SMB process despite small differences shown in the first SMB work cycles. Furthermore, the model equations can be easily implemented and applied in the analysis of the process, as it requires a small number of parameters and consists of ordinary differential equations.

Keywords Simulation moving bed · Inverse problems · Modeling and simulation · Chromatography

1 Introduction

Adsorption and reaction processes are extensively adopted by the food, textile, petrochemical, chemical, and pharmacological industries [1].

Therefore, many studies have been carried out with the aim to improve and create new separation techniques, and discover new substances to be used in the separation of a range of products [2].

Currently there are a variety of technical separations, among them is the simulated moving bed (SMB) process created in the 1960s by Universal Oil Products [3]. This process stands out for being a powerful tool to separate compounds that are difficult to separate, as when the difference of affinity between the molecules is very small. Another positive point is that the processes operate continuously requiring less solvent than batch chromatography [4]. The potential of this chromatography tool has been evidenced by means of a high number of studies and publications of the academic community.

The SMB technology has been applied in the petrochemical and food industry since its creation; however, recently it has been extensively employed by the pharmaceutical and fine chemical to separate large-scale enantiomers from racemic mixtures [5–8].

The need to separate enantiomers has increased, particularly when in some racemic mixtures one of these enantiomers has the desired therapeutic effect, the other is ineffective, or sometimes causes serious side effects.

Enantiomeric separation has become an important issue, especially in health [9, 10], as evidenced by high investments in the pharmaceutical market in research and development of chiral technologies stimulated by the need for separation of racemic compounds [10].

The development of SMB separation processes requires a thorough study, since it is necessary to determine some operating conditions such as the flow rate in each section, feed concentration, and the switch time of the position of the currents [11]. The determination of the operational conditions can become very costly to the operator of the equipment.

To resolve this issue many authors have formulated math models capable of predicting the SMB process of separation with statistically acceptable errors compared to experimental data. According to [12], two different approaches, discrete (mixed cells) and continuous (dispersion) models, can model chromatographic columns. Currently, the models used by researchers are robust and efficient, but they require a numerical treatment of partial differential equations, which carries high computational cost.

In order to get a new math model to predict the profiles of separating compounds of a mixture submitted to a chromatographic separation by SMB adsorption process, the new approach called Front Velocity was proposed [13]. This approach does not require equilibrium experiments, does not need application of adsorption isotherms for characterizing the components involved in the process, and is composed for ordinary differential equations.

To validate the proposed model, a theoretical analysis of kinetic equations in chromatographic column characterization was performed. And as a second step a study was conducted, where the developed approach was applied for separation of enantiomers of ketamine anesthetic [14, 15] and the results were compared with those obtained by conventional models.

2 Theory

The SMB process used by Santos [14, 15] for separation of enantiomers of racemic Ketamine consists of eight chromatographic columns connected in series, divided two by two per section. Each column is 0.77 cm in diameter and 20 cm in length as shown in Fig. 1. The greater retained enantiomer (R) is collected in the extract (Ex), while the least adsorbed enantiomer (S) is collected in the raffinate (R).

To separate the enantiomers from ketamine anesthetic [15] in the stationary and mobile phases were used, respectively, MCTA (microcrystalline cellulose acetate) and ethanol. Santos et al. [15] also used an HPLC (high performance liquid chromatography) column, where ketamine samples were injected for calibration and determination of purity. The HPLC column used was 0.46 cm in diameter and 20 cm in length.

To perform the simulation of separation process of ketamine carried by Santos [14, 15], a novel approach was developed. To represent the mass transfer that

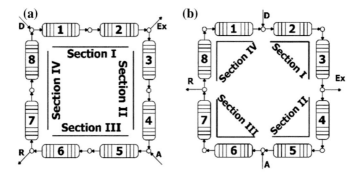

Fig. 1 SMB process with two chromatography columns per section

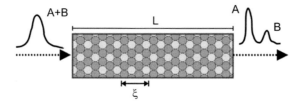

Fig. 2 Chromatographic column of length L, and volume control length ξ

occurs inside one chromatography column during a separation process, the *front velocity* approach establishes that convection is the dominant phase in the solute transport along the chromatographic column. *Front velocity* is a discrete model (mixed cells), where the flow rate determines the liquid phase advances along the column. The rate at which the liquid phase percolates the porous column (v) is the ratio between volumetric flow rates of the mobile phase through the porous medium as described by Eq. 1,

$$v = \frac{Q}{\varepsilon . A_T} \tag{1}$$

where Q, ε and A_T represent the volumetric flow rate, porosity, and the total area of the porous column, respectively (these data are obtained experimentally).

To perform calculation of the mass transfer, the chromatographic column was divided into control volumes (mixed cells) of length ξ, which move along the column with the same speed as the eluent, as can be seen in Fig. 2.

The necessary time to move the liquid phase along each control volume is obtained from Eq. 2, where V is the total volume of the column and n is the number of mixed cells.

$$\Delta t = \frac{\varepsilon . V}{n.Q} \tag{2}$$

To perform calculation of the mass transfer in the chromatography process, the equations with lumped mass transfer parameter kinetics (Eqs. 3, 4) assume that equilibrium is achieved everywhere at all times, so that the effects of axial dispersion and the mass transfer resistance were disregarded. Kinetic equations of mass transfer have been successfully employed in chromatographic processes [13, 16–18].

$$\left(\frac{\mathrm{d}C_i^p}{\mathrm{d}t}\right)_\zeta = -k_{\mathrm{ads}} \cdot C_i^p (q_m - q_i^p) + k_{\mathrm{des}} \cdot q_i^p \tag{3}$$

$$\left(\frac{\mathrm{d}q_i^p}{\mathrm{d}t}\right)_\zeta = k_{\mathrm{ads}} \cdot C_i^p (q_m - q_i^p) - k_{\mathrm{des}} \cdot q_i^p \tag{4}$$

$(q_m - q_i^p)$ is the Langmuir kinetic of mass transfer. C_i^p, q_i^p, q_m, k_{ads}, k_{des}, and t represent the concentration of compound i in the liquid phase at the p column, the concentration of compound i in the solid adsorbent phase at the p column, the maximum adsorption capacity, the mass transfer kinetic parameter of adsorption, the mass transfer kinetic parameter of desorption, and the time respectively. These equations (3 and 4) are applied in all mixture cells [12] and solved numerically utilizing a fourth-order Runge–Kutta method with a time step equal to 10–5 implemented in Fortran90.

The SMB process consists of four sections (Fig. 1), each with different volumetric flow rate, influenced by two input sms and two output streams (feed, desorbent, extract, and raffinate). To calculate the mass transfer in each column, first it is necessary determine those volumetric flow rate and after incorporate the mass balance at t entrance of each p column. After each change in the configuration of the streams (Fig. 1), the new mass balance of solutes at the nodes has to be recalculated.

AB has two streams of input and two outputs, the overall flow is necessarily written as Eq. 5.

$$Q^F + Q^D = Q^R + Q^E \tag{5}$$

To calculate the flow rates in each section and the mass balance for the first column of each section, Eqs. 6–9 are used.

Section I:

$$Q^I = Q^{IV} + Q^D, \qquad C_{iE}^I \cdot Q^I = C_{iS}^{IV} \cdot Q^{IV} \tag{6}$$

Section II:

$$Q^{II} = Q^{I} - Q^{Ex}, \qquad C_{iE}^{II} = C_{iS}^{I} = C_{i}^{Ex} \tag{7}$$

Section III:

$$Q^{III} = Q^{II} + Q^{F}, \qquad C_{iE}^{III} \cdot Q^{III} = C_{i}^{F} \cdot Q^{F} + C_{iS}^{II} \cdot Q^{II} \tag{8}$$

Section IV:

$$Q^{IV} = Q^{III} - Q^{R}, \qquad C_{iE}^{IV} = C_{iS}^{III} = C_{i}^{R} \tag{9}$$

D, Ex, F, and R represent the desorbent, extract, feed, and raffinate stream, respectively. E is the inlet concentration (concentration at the inlet of the first column of each section), S is the concentration at the last column of each section, and i is relative to the compound mixture (e.g., in the case of the racemic compounds, i is R or S).

To determine the lumped mass transfer parameters (k_{ads}, k_{des}) of kinetic equations (Eqs. 3, 4), the inverse toll Random Restrict Window (R2W) [19] was employed. R2W is a simple stochastic method that analyzes the cost function from random estimates, belonging to a defined area for the parameters you want to adjust, assuming new random searches, through a restricted area next window the best solution candidates the problem of interest.

The formulation of the inverse problem of mass transfer considered in this work seeks to minimize an objective function given by the sum of squared residuals between the experimental pulse experiment data and the calculated values (by front velocity) for the observable variable.

3   Results and Discussions

3.1   *Theoretical Analysis of Kinetic Model*

In the preparative chromatography step, samples are usually very small and dilute. Thereby, the chromatographic parameters of adsorption are generally independent of the injected mass, and can be easily represented by linear isotherms [20]. When the concentration of the compounds injected into the chromatographic columns increase, an overload can be observed in chromatogram, which is characterized by decrease in the retention coefficients of the solutes, decrease of theoretical plate numbers, and loss of chromatographic resolution [21].

It is necessary to know the chromatographic column overload condition in preparative chromatography to optimize the compounds separation [22]. A sample with increasing concentration is injected in the preparative chromatographic column to achieve overload condition. The results expected from concentration increase on

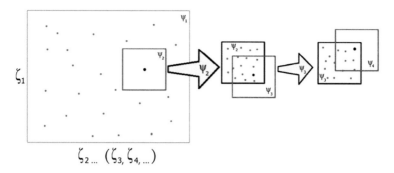

Fig. 3 Schematic representation of algorithm R2W

the chromatogram are distorted peaks and it may also occur as tail increase (peak tailing). The increased volume of injection and other factors such as the strength of the solvent used for dilution may result in deformation of the peak [23].

Initially, with the kinetic equations proposed in this paper, a theoretical analysis was conducted in order to check whether or not they are capable of reproducing the expected behavior in the chromatogram as the isotherms reproduce (Figs. 3, 4, 5, 6, 7, 8, 9).

In Figs. 4 and 5 the concentration over time simulated using stepwise kinetic equation[1] (Eqs. 3, 4) is compared with a linear isotherm (linear equilibrium condition). The time step size in the kinetic equation of mass transfer was assumed to be equal to 10^{-1}.

In Fig. 4 the immediate linear equilibrium[1] (linear isotherm) of mass transfer between liquid and solid phase corresponds to the curve with the highest peaks that show symmetric profiles. The maximum concentration of peaks almost reaches the feed concentration ($C_f = 15 \, g/L$) and indicates a peak broadening when the value of the equilibrium constant (k_{eq}) is increased. In Fig. 5 it may be noted that the residence times obtained are similar to those obtained from the linear isotherm. Furthermore, also by increasing the equilibrium constant, the peaks undergo enlargement, and

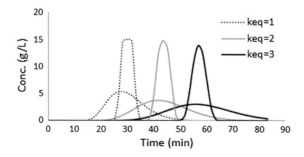

Fig. 4 Mass transfer model with linear kinetics versus immediate equilibrium: under variation of the equilibrium constant. $F = 30 \, ml/min$, $V_s = 300 \, ml$, $K_2 = 0,01 \, L/min$, $k_1 = k_2$, $CA = 15 \, g/L$

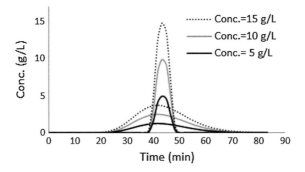

Fig. 5 Mass transfer model with linear kinetics versus immediate equilibrium: Under variation of the feed concentration. $F = 50$ ml/min, $V_s = 300$ ml, $k_{eq} = 2$, $k_2 = 0.01$ L/min, $k_1 = k_2 \cdot k_{eq}$

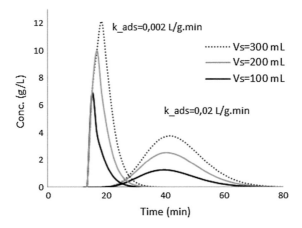

Fig. 6 Mass transfer model with linear kinetics. Effect of variation of the sample volume. $F = 50$ ml/min, $k_2 = 0.01$ L/g $\cdot$ min

when the feed concentration was increased the only effect was the changes in peak height, the same as the kinetic isotherm behavior.

In Fig. 6 are shown the results of increase of sample volume for two different mass transfer parameters (K_{ads}). The lowest value of k_{ads} leads to deformation (fronting peak) at the beginning of peaks.

The simulation results of mass transfer using the Langmuir kinetic model (Eqs. 3, 4) are shown in Figs. 7, 8 and 9. In Fig. 7 is observed the peak broadening when feed concentration was increased. In addition, the appearance of peak tailing was observed, which is Langmuir isotherm characteristic.

The increase in sample volume (Fig. 8) leads to a similar effect on the size and behavior of the peak, observed when a variation of feed concentration was applied (Fig. 7). Furthermore, the tailing peak is also clearly observed. The peak tailing

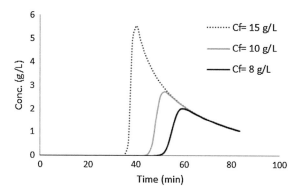

Fig. 7 Mass transfer model with Langmuir kinetics. Effect of variation of the sample concentration. $q_m = 10\,\mathrm{g/hv}$, $f = 30\,\mathrm{ml/min}$, $V_s = 300\,\mathrm{mL}$, $k_1 = 0.005\,\mathrm{L/g \cdot min}$, $K_2 - 0.01\,\mathrm{L/min}$

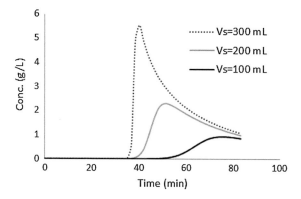

Fig. 8 Mass transfer model with Langmuir kinetics. Effect of variation of the sample volume. $q_m = 10\,\mathrm{g/hv}$, $f = 30\,\mathrm{mL/min}$, $C_F = 15\,\mathrm{g/L}$, $k_1 = 0.005\,\mathrm{L/g \cdot min}$, $K_2 - 0.01\,\mathrm{L/min}$

phenomenon is typically encountered when Langmuir isotherms are used in chromatographic models [24].

In a theoretical analysis performed to one chromatographic pulse as can be seen in Fig. 9, it is observed that the novel approach presented in this work has potential to represent the resistance to mass transfer as well as the saturation of the adsorbent phase (rectangular peak). The chromatogram of the simulation also shows a peak tailing, which is a behavior observed in the literature with the use of adsorption isotherms. Therefore, the characterization of a chromatography column cannot be linked exclusively with the isotherm application, but also with mass transfer kinetic equations.

The simulation performed in this section show that the use of models with lumped mass transfer parameters associated with linear kinetics equations and Langmuir leads to the same behavior observed in the literature, respectively, with

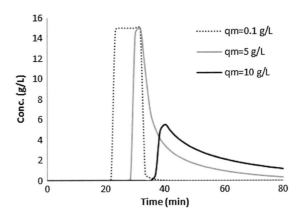

Fig. 9 Variation of the maximum adsorption capacity of the adsorbent phase. $F_c = 15$ g/L, $F = 30$ ml/min, $Vs = 300$ ml, $k1 = 0,005$ L/g.min, $K2 = 0,01$ L/min

the use of linear and Langmuir isotherms. Therefore, the characterization of a chromatographic column cannot be exclusively related to the implementation of the isotherms, but also to the kinetic models of mass transfer.

3.2 Characterization of the Chromatographic Column

A preview and important step in SMB chromatography process is the characterization of the columns through determination of the lumped mass transfer parameter kinetics (k_{ads}, and k_{des}). These parameters determine the rate of adsorption and desorption of molecules between the liquid and solid phases.

Santos et al. [15] via analysis in an HPLC system obtained the separation profiles of enantiomers of ketamine. To determine k_{ads}, and k_{des} parameters, the retention times (experimental) observed in the chromatogram [15] resulting from chromatographic pulse in an HPLC analysis system were used in this study. These data were combined with mass transfer equations (Eqs. 3 and 4) and with the inverse tool, Random Restricted Window (R2W) [19]. The results can be seen in Table 1. R2W is considered a simple stochastic inverse method that uses a search algorithm with a random distribution.

Table 1 Lumped mass transfer parameters obtained from the application of the inverse tool, R2W

	S enantiomer	R enantiomer
k_{ads}	0.00218857	0.00247352
k_{des}	0.05430671	0.02700426
k_{eq}	0.04030017	0.09159742
n	505	505
q_m	29,567	29,567
R^*	2.54E-05	1.87E-06

*Is the sum of squares of the residuals between the simulation and experimental data

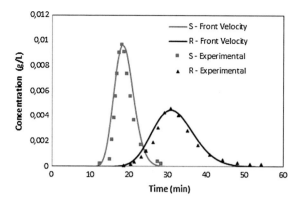

Fig. 10 Chromatogram of racemic ketamine injection in the analysis column packed with MCTA; anhydrous ethanol as mobile phase, flow rate 0.25 mL/min, loop 20 µl, the solution concentration 1.5 g/l. Results simulated with *front velocity* approach, compared to experimental data

In Fig. 10, the good correlation between the experimental data (chromatogram) and the simulation performed with front velocity approach is remarkable.

3.3 SMB Continuous Process

With the data obtained (Table 1) in the chromatographic column characterization stage, SMB was performed under the experimental conditions (Table 2) that were determined by Santos et al. [14] and Santos [15]. The kinetic equations (3–4) and mass balance equations (Eqs. 6–9) were applied to each column of the SMB to simulate separation of the enantiomers of the ketamine, allowing the results to be compared with those obtained by Santos et al. [14] and Santos [15].

The equations used in modeling of S in this work were solved numerically utilizing a fourth-order Runge–Kutta method and implemented in Fortran90. The total simulation time to SMB was approximately 1.5 min on a computer with Intel Core i5 processor (2.3 GHz) with a time step equal to 10^{-5}. The number of mixed cells (divisions of the columns) in each section is determined by volumetric flow rate (Fig. 1), and it is around 300–1000 equilibrium stages as can be seen in Table 3.

The differential partial equations of the dispersive equilibrium model used by Santos et al. [15] were solved by public subroutine PDECOL [25], which implements the finite element method for spatial discretization and the ordinary differential equations were solved by GEARIB time integrator [26]. The total simulation time was about 4 h, using an Intel Pentium IV (2.4 GHz) processor, with time step equal to 10^{-5}. Each column has been divided only into 30 elements.

Table 2 Operation condition evaluations according to solvent consumption, productivity, and purity

	Feed concentration	Switch time (s)	Solvent consumption (L/g rac.)	Productivity (q.rac./D.kg)	Volumetric flow at the streams (mL min^{-1})				Purity (%)	
					Ext.	Raff.	Feed	Q_{rec}	S	R
1	1.5	1500	2.67	10.65	0.47	0.43	0.18	1.10	99.83	99.71
2	2.5	1500	1.6	17.75	0.47	0.43	0.18	1.10	99.84	99.71
3	5	1500	1.46	19.79	0.44	0.39	0.1	1.10	99.99	99.90

Table 3 Number of mixed cells of the columns per section

SMB sections	Number of control volumes
Section I	322
Section II	562
Section III	437
Section IV	932

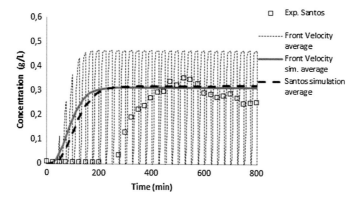

Fig. 11 SMB evolution of S enantiomer concentration in the extract stream over time (transient), under experimental condition 1

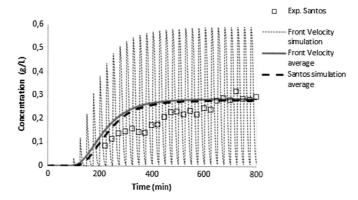

Fig. 12 SMB evolution of R enantiomer concentration in the raffinate stream over time (transient), under experimental condition 1

The good correlation between the stimulant using *front velocity* approach, and the classical model used by Santos et al. [15], over time in the separation process can be seen in Figs. 11 and 12. In addition, comparison with the experimental data in the extract and raffinate streams can be visualized, where there is a small deviation between the simulated and experimental data while the process is still in transience. When the process achieves pseudo steady state it reaches a better fit.

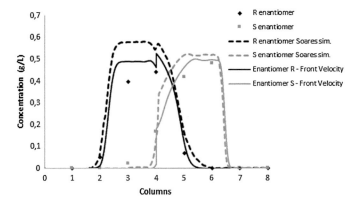

Fig. 13 SMB evolution of R enantiomer concentration in the raffinate stream over time (transient)

Table 4 The solvent consumption, productivity, and purity calculated with front velocity approach

Experimental condition	1	2	3
Raffinate purity (S) %	97.72	98.84	99.89
Extract purity (R) %	99.70	99.93	99.99
Productivity (q.rac./D.kg)	10.63	17.63	19.49
Solvent consumption (L/g rac.)	2.66	1.61	1.47

Figure 13 shows the concentration profiles of each enantiomer over the columns when the SMB reaches steady state. Good representation of the new approach is observed due to correlation with experimental data, and it may be noted that *front velocity* achieves a slightly better fit compared with Santos et al. [15] simulations.

The solvent consumption, productivity, and purity shown in Table 4 were calculated with the novel approach proposed in this work. These results are similar to experimental values (Table 2).

4 Conclusions

A set of programs for continuous simulation of SMB process, and also to characterize the chromatographic column was developed and used under the new proposed idea of modeling the pulse experiment to determine the kinetic constants of mass transfer with mass transfer kinetic equations, instead of performing equilibrium experiments and combining them with some kind of adsorption isotherm. This procedure satisfactorily performed the separation of enantiomers of anesthetic ketamine at SMB. The R2W algorithm was effective in determining the parameters of adsorption, desorption, and the maximum adsorption capacity of the adsorbent phase (k_{ads}, k_{des}, q_m).

The theoretical analysis proved that front velocity has potential to represent the resistance to mass transfer as well as saturation of the adsorbent phase. The simulations performed showed that the characterization of a chromatographic column cannot be exclusively related to adsorption isotherms, but with lumped mass transfer parameter models associated with different kinetics also represent the chromatographic phenomena in a preparative column.

The concentration profiles of the simulations proved to be consistent with the SMB process, and the simulated profiles of enantiomers were similar to the experimental data, showing a slightly more suitable behavior in relation to the experimental data than the classical model as adopted by Santos et al. [15].

The results showed the potential of front velocity in the prediction of SMB separations. In a shorter time than classical models, this approach performs a full simulation of the separation in an SMB process. The low computational cost is due to the use of ordinary differential equations in this approach that requires less parameters than classical models; furthermore, the ease of implementation and analysis, and the need to know just a few operational data of the real problem. Another relevant fact is that no performing equilibrium experiments are necessary to characterize the chromatographic column.

Acknowledgments The authors acknowledges the financial support from UERJ, UNIPAMPA, CNPq and CAPES.

References

1. Gomes, P.M., Figueirêdo, R.M., Queiroz, A.J.: Caracterização e Isotermas de Adsorção de Umidade da Polpa de Acerola em Pó. Revista Brasileira de Produtos Agroindustriais **4**(2), 157–165 (2002)
2. Zaijun, L., Xiulan, S., Junkang, L.: Ionic liquid as novel solvent for extraction and separation in analytical chemistry. Chem. Eng. Appl. Perspect. 154–180 (2011)
3. Broughton, D.B.: EUA Patent No. US002985589 (23 May 1961)
4. Gal, G., Hanak, L., Argyelan, J., Strbka, J., Szanya, T., Aranyi, A., Temesvari, K.: Simulated moving bed (SMB) separation of pharmaceutical enantiomers. Hung. J. Ind. Chem. **33**, 23–30 (2005)
5. Miller, L., Orihuela, C., Fronek, R., Honda, D.: Chromatographic Resolution of the Enantiomers of a Pharmaceutical Intermediate from the Milligram to the Kilogram Scale
6. Francote, E.R., Riechert, P.: Application of simulated moving bed to the separation of the enantiomers of chiral drugs. J. Chromatogr. A **769**, 101–107 (1997)
7. Francote, E.R., Richert, P., Mazzoti, M., Morbidelli, M.: Simulated moving bed chroma-tography resolution of a chiral antitussive. J. Chromatogr. A **769**, 239–248 (1998)
8. Biressi, G., Quattrini, F., Juza, M., Mazzotti, M., Schurig, V., Morbidelli, M.: Gas chromatography simulated moving bed separation of the enantiomers of the inhalation anesthetic eflurane. Chem. Eng. Sci. **55**, 4537–4547 (2000); J. Chromatogr. A. **849**, 309–317 (1999)
9. Park, T.H., Yoon, T.H., Kim, I.H.: Simplistic determination of operation parameters for simulated moving bed (SMB) chromatography for the separation of ketoprofen enantiomer. Biotechnol. Bioprocess Eng. **10**, 341–345 (2005)

10. Choi, Y.J., Han, S.K., Chung, S.T., Row, K.H.: Separation of racemic bupivacaine using simulated moving bed with mathematical model. Biotechnol. Bioprocess Eng. **12**, 625–633 (2007)
11. Gonçalves, C.V. Separação do Racemato N-Boc-Rolipram pelo Processo Cromatografico de Leito Movel Simulado Utilizando Fase Estacionaria Quiral Tris-(3,5-dimetilfenilcarbamato de celulose) Suportada em Silica. Campinas: UNICAMP: Doctoral Thesis (2008)
12. Antos, D., Seidel-Morgenstern, A.: Chem. Eng. Sci. **56**, 6667 (2001)
13. Bihain, A.L.J.: Desenvolvimento e avaliação de novas abordagens de modelagem de processos de separação em leito móvel simulado. Ph.D Dissertation, UERJ – Nova Friburgo (2013)
14. Santos, M.A.: Separação dos Enantiômeros do Anestésico Cetamina por Cromatografia Contínua em Leito Móvel Simulado. Campinas: UNICAMP: Doctoral Thesis (2004)
15. Santos, M.A., Veredas, V., Silva Jr, I.J., Correia, C.R., Furlan, L.T., Santana, C.C.: Simulated moving-bed adsorption for separation of racemic mixtures. Braz. J. Chem. Eng. **21**(01), 127–136 (2004)
16. Câmara, L.D.: Stepwise model evaluation in simulated moving-bed separation of ketamine. Chem. Eng. Technol. **37**(2), 301–309 (2014)
17. Bihain, A.L.J., Silva Neto, A.J., Santiago, O.L., Afonso, J.C., Câmara, L.D.T.: The front velocity modelling approach in the chromatographic column characterization of glucose and fructose separation in SMB. Trends Chromatogr. **7**, 33–41 (2012)
18. Bihain, A.L.J.: The front velocity approach in the modelling of simulated moving bed process (SMB). 4th international conference on simulation and modeling methodologies, technologies and applications. Wien Austrian (2014)
19. Câmara, L.D., Silva Neto, A.J.: Inverse stochastic characterization of adsorption systems by a random restricted window (R2W) method. International conference on engineering optimization (ENGOPT). Rio de Janeiro—RJ (2008)
20. Schmidt-Traub, H.: Preparative Chromatography: Of Fine Chemicals and Pharmaceutical Agents, 485p. Wiley-VCH®, Wheinheim (2005)
21. Perna, R.F.: Separação Cromatográfica dos Enantiômeros de Fármaco Verapamil em Processo Contínuo Multicolunas. Ph.D. Dissertation. UNICAMP (2013)
22. Rosa, P.C.P.: Estudo da separação cromatográfica dos enantiômeros do omeprazol em fase estacionária quiral Kromasil CHI-TBB (0,0-Bis[4-terc-butilbenzoil]-N, Ndialil-L-tartadiamida). Master Thesis, UNICAMP (2005)
23. Macherey-Nagel. Guia para colunas cromatográficas, pp. 62–66. (2000)
24. Sugata, S., Abe, Y.: An analogue column model for nonlinear isotherms: the double-glazed vessel model. J. Chem. Software **6**(4), 127–136 (2000)
25. Madsen, N.K., Sincovec, R.F.: PDCOL: general collection software for partial differential equations. ACM Trans. Math. Software **5**, 326–351 (1979)
26. Hindmarsh, A.: Preliminary documentation of GEARIB—solution of implicit systems of ordinary differential equations with banded jacobian. Report UCID—30130 (1976)

Modeling Hybrid Systems with Petri Nets

Debjyoti Bera, Kees van Hee and Henk Nijmeijer

Abstract The behavior of a hybrid system is a mixture of continuous behavior and discrete event behavior. The Simulink/Stateflow toolset is a widely used industrial tool to design and validate hybrid control systems using numerical simulation methods for the continuous parts and an executable Stateflow (combination of Statecharts and Flowcharts) for the discrete event parts. On the other hand, Colored Petri Nets (CPN) is a well-known formalism for modeling behavior of discrete event systems. In this paper, we show how the CPN formalism can be used to model a hybrid system. Then we consider the special case of Simulink/Stateflow models and show how they can be expressed in CPN.

Keywords Petri nets · Simulink · Colored Petri nets · CPN tools · Discrete event systems · Time-driven systems · Model checking · Performance analysis

1 Introduction

A hybrid system can be distinguished into two types of subsystems, a time-driven subsystem and a discrete event subsystem. The former defines one or more control algorithms together with their environment, while the latter defines discrete event logic in terms of an executable state machine.

A *time-driven subsystem* is usually defined as a set of mathematical equations, evaluated at discrete points in time. In general, such a subsystem models a set of controllers and its environment (also referred to as a plant). The goal is to define a

D. Bera (✉) · K. van Hee · H. Nijmeijer
Technische Universiteit Eindhoven, P.O. Box 513, 5600 MB Eindhoven,
The Netherlands
e-mail: d.bera@tue.nl

K. van Hee
e-mail: k.m.v.hee@tue.nl

H. Nijmeijer
e-mail: h.nijmeijer@tue.nl

© Springer International Publishing Switzerland 2015
M.S. Obaidat et al. (eds.), *Simulation and Modeling Methodologies,*
Technologies and Applications, Advances in Intelligent Systems
and Computing 402, DOI 10.1007/978-3-319-26470-7_2

17

mathematical model of a controller, given a model of its environment and a set of requirements that a controller must satisfy. Such models may contain both discrete (difference equations) and continuous parts (differential equations). An exact solution of such a model is in many cases not computable, therefore it is *approximated* by numerical simulation methods, i.e., by discretizing time into time steps, whose minimum is bounded by the resolution of the system. The results obtained from numerical simulation are used as a reference to validate the behavior of both the model and the implemented system (described in lower level languages like C). A *discrete event subsystem* is usually expressed as a finite state machine, whose state evolution depends entirely on the occurrence of discrete events (instantaneous) over time, like "button press," "threshold exceeded," etc. So the underlying time-driven dynamics (expressed as difference and/or differential equations) of a hybrid system are captured by the notion of time progression in a state (i.e., the time elapsed between consecutive event occurrences).

The Simulink/Stateflow toolset is an industrial tool that is widely used to design and validate hybrid control systems using numerical simulation methods for the time-driven parts and an executable Stateflow specification (combination of Statecharts [1] and Flowcharts) for simulating the discrete event parts. The Simulink toolset has only an informal semantics (c.f. [2, 3]) and of course an operational semantics in the form of an implementation in software. Since these models are eventually implemented as software and often integrated into larger systems, it becomes necessary to understand their logical structure and verify them in an exhaustive manner. The problem is even more serious for Stateflow because it has numerous complex semantics explained in a very verbose manner [3]. Furthermore, the semantics of Stateflow deviates from that of Statecharts, as a result, the many existing results on the formalization of Statecharts (c.f. [4–6]) are not directly applicable. As a result, runtime failures of Stateflow models are quite common in practice and these problems are usually hard to understand and analyze by designers. As a consequence several guidelines have been proposed by the industry restricting the language to safe subset and prescribing design patterns. However, these guidelines have no formal basis and rely solely on lessons learned from experience. So it becomes necessary to understand these models in a formal manner.

Many attempts have been made to address this shortcoming by proposing translations of Simulink models into existing formal frameworks such as automata and its various extensions [7–11]. It is only in [12] that a formal operational semantics of Simulink is defined. Unlike other approaches that focus on formalizing the solution method of equations encoded by a Simulink model, the semantics described here captures the behavior of the simulation engine itself, describing the outcome of a numerical simulation. Similar attempts have been made for Stateflow as well [7, 13–15].

In this paper, we present a method to model hybrid systems with the formalism and tools of Colored Petri Nets (CPN) [16]. Although there are several extensions of a Petri net for modeling hybrid systems such as hybrid Petri nets [17], differential Petri nets [18], etc., and their applications (for instance in bio-chemical networks [19]), there are several good reasons to apply the CPN formalism in this context such as (a) CPN has a simple and well-defined semantics, and (b) CPN as an extension of classical Petri nets has many analysis methods [20, 21] for verification of models. There are also specific analysis methods for Colored Petri Nets. Nonexperts often say that Petri nets are a very primitive modeling framework. This might be true for classical Petri nets although they generalize finite state machine models with concurrency. Classical Petri nets treat *states* and *events* in a symmetrical way, which is a big advantage. On the other hand, CPNs extend tokens with arbitrary complex data types and transitions have arbitrary complex functions to compute values of output tokens using input tokens. So they have a strong modeling power (of course Turing complete) and are supported by good tools for modeling and analysis (c.f. [22]). We also consider the special case of the Simulink/Stateflow toolset and express them as a CPN. The main difference between Simulink/Stateflow and CPN is that the former is developed as a tool and only afterwards scientists have tried to define formal semantics for it, while CPN were developed as a theory for concurrent systems in the 1960s and much later, in the 1990s, its supporting tools were developed. Although CPN was originally meant for modeling concurrent discrete event systems, they are very suitable for modeling continuous systems and hybrid systems as well. The advantage of Simulink/Stateflow is that they provide handy notations for often recurring modeling issues. Therefore we also introduce similar notations for CPN. Note that these notations are only syntactic sugar, i.e., their semantics are derived from the translation back to the basic syntax of CPN.

This paper is structured as follows: In Sect. 2, we introduce the Petri net and Simulink frameworks in an informal way. In Sect. 3, we give a compact notation for CPN and call it CPN*. In Sect 4, we show how models of hybrid systems can be expressed in CPN*. Here we also consider Simulink models as a special case. In Sect. 5, we study the relationship between Stateflow models and CPN* models and show how the latter can be derived from the former. In Sect. 6, we present our conclusions.

2 Overview of the Frameworks

In this section, we introduce the four modeling frameworks of interest in this paper, namely Petri nets, its colored and timed extension colored Petri net (CPN), the Simulink framework for modeling continuous and discrete timed behavior, and its Stateflow component for modeling discrete event systems.

2.1 Petri Nets

A *Petri net* (PN) is a bipartite graph consisting of two types of nodes, namely places (represented by a circle) and transitions (represented by a rectangle). We give an example of a Petri net in Fig. 1. The nodes labeled $P1$ and $P2$ are called places and

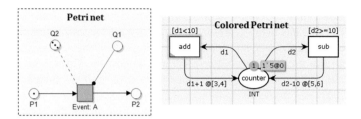

Fig. 1 An example of PN and CPN

the node labeled *A* is called a transition. A place can be connected to a transition and vice versa by means of directed edges called arcs. A *preset* (*postset*) of a node in a Petri net is the set of all nodes having an incoming (outgoing) arc to (from) it. The notion of a *token* gives a Petri net its behavior. Tokens reside in places and often represent either a status, an activity, a resource, or an object. The distribution of tokens in a Petri net is called its *marking* or *state*. We call the initial distribution of the tokens in a net as its *initial marking*. A Petri net having an initial marking is called a *net system*. Transitions represent *events* of a system. The occurrence of an event is defined by the notion of *transition enabling and firing*. A transition is enabled if all its input places have at least one token each. When an enabled transition fires it consumes one token from each input place and produces one token in each output place, i.e., changes the state. We lift the notion of transition enabling and firing to sequence of transitions. A given marking is said to be *reachable* by a net system if there exists an executable sequence of transitions leading to it. The set of all markings that are reachable by a net system is called the *set of reachable markings*. Apart from arcs between places and transitions there are two other types of arcs, namely *inhibitor* and *reset* arcs. An enabled transition having an inhibitor arc (represented as an edge having a rounded tip from transition *A* to place *Q*1) can fire only if the place associated with the inhibitor arc does not contain consumable tokens, i.e., place *Q*1. A transition connected by a reset arc (represented as a dotted edge from transition *A* to place *Q*2) to a place, removes all tokens residing in that place (i.e., *Q*2) when it fires.

A Petri net is called a *state machine net* if its transitions have a preset and postset of size equal to one. A *subnet* of a given Petri net is a subset of its places and transitions and the connections between them. A place of a Petri net is *safe* if in all reachable markings of its net system, the place contains at most one token. A net system with safe places is called a *safe net*.

2.2 Colored Petri Nets

A *colored Petri net* (CPN) is an extension of a Petri net with data and time. We give an example of a counter modeled as a CPN in Fig. 1. Each place has an inscription which determines the set of token colors (data values) that the tokens residing in that place are allowed to have. The set of possible token colors is specified by means of a type called the *color set* of the place (for e.g., the place *counter* has a color set of type integer denoted by *INT*). A token in a place is denoted by an inscription of the form $x`y$ (see token 1`5), interpreted as x tokens having token color y, i.e., as a multiset of colored tokens. Like in a standard Petri net, when a transition fires, it removes one token from each of its input places and adds one token to each of its output places. However, the colors of tokens that are removed from input places and added to output places are determined by *arc expressions* (inscriptions next to arcs).

The arc expressions of a CPN are written in the ML functional programming language and are built from typed variables, constants, operators, and functions. An arc expression evaluates to a token color, which means exactly one token is removed from an input place or added to an output place. The arc expressions on input arcs of a transition together with the tokens residing in the input places determine whether the transition is *color enabled*. For a transition to be color enabled it must be possible to find a binding of the variables appearing on each input arc of the transition such that it evaluates to at least one token color present in the corresponding place. When the transition fires with a given binding, it removes from each input place a colored token to which the corresponding input arc expression evaluates. Firing a transition adds to each output place a token whose color is obtained by evaluating the expression (defined over variables of input arcs) on the corresponding output arc. In our example, the variables $d1$ and $d2$ are bound to the value of the token (value 5) in place *counter*. When one of the transitions, say *add* fires, it produces a token with value $d1 + 1 = 6$ in the place *counter*. Furthermore, transitions are also allowed to have a *guard*, which is a Boolean expression. When a guard is present it serves as an additional constraint that must evaluate to true for the transition to be color enabled. In our example, the guard $d1 < 10$ ensures transition *add* can fire if the value of the token in place *counter* is less than 10.

In addition, tokens also have a timestamp that specifies the earliest possible consumption time, i.e., tokens in a place are available or unavailable. The CPN model has a *global clock* representing the current model time. The distribution of tokens over places together with their timestamps is called a *timed marking*. The execution of a timed CPN model is determined by the timed marking and controlled by the global clock, starting with an initial value of zero. In a timed CPN model, a transition is *enabled* if it is both color enabled and the tokens that determine this enabling are available for consumption. The time at which a transition becomes enabled is called its *enabling time*. The *firing time* of a timed marking is the earliest enabling time of all color enabled transitions. When the global clock is equal to the firing time of a timed marking, one of the transitions with an enabling time equal to the firing time of the timed marking is chosen nondeterministically for firing. The global clock is not advanced as long as transitions can fire (eager semantics). When there is no longer an enabled transition at the current model time, the clock is advanced to the earliest model time at which the next transition is enabled (if no such transition then it is a deadlock), i.e., the firing time of the current timed marking. The timestamp of tokens are written after the @ symbol in the following form $x`y@z$, interpreted as x tokens having token color y carry a timestamp z. If the place is safe, then we do not mention the number of tokens. When an enabled transition fires, the produced tokens are assigned a color and a timestamp by evaluating the time delay interval, inscribed on outgoing arcs of a transition (see time interval $@[a,b]$ or just $[a,b]$). The timestamp given to each newly produced token along an output arc is the sum of the value of the current global clock and a value chosen from the delay interval associated with that arc. Note that the firing of a transition is instantaneous, i.e., takes no time. Since the time domain is infinite and there is an infinite number of choices that can be made in a time interval, such a CPN model has an infinite state space. As a result, it is not possible to model check them directly. To be able to analyze such models, we need reduction methods as proposed in [23, 24].

2.3 Simulink

Signals and Blocks are the two basic concepts of a Simulink model. A *signal* is a step function in time, representing the behavior of a variable. So only at discrete points in time, the value of a variable may change. Furthermore, a signal in Simulink can have one of the following data types: integer, real, complex, or its multidimensional variants.

A *block* defines a functional relationship between a set of signals and possibly a variable representing its state (state variable). If a block has a state variable then it is a *stateful block*, otherwise it is a *stateless block*. The set of signals belonging to a block are either *input* or *output*. A block can have zero or more inputs and at most one output and one state variable.

A *stateless block* defines an *output function* $f : I_1 \times \ldots \times I_n \to O$, where $n \in \mathbb{N}$ is the number of inputs of the block, I_j is the type of input signal, for $j \in \{1, \ldots, n\}$, and O is the type of the output signal. A *stateful block* defines an *output function* $f : I_1 \times \ldots \times I_n \times S \to O$ and an *update function* $g : I_1 \times \ldots \times I_n \times S \to S$, where $n \in \mathbb{N}$ is the number of inputs of the block, I_j is the type of input signal, for $j \in \{1, \ldots, n\}$, S is the type of the state variable and O is the type of the output signal. The output function of a stateful block must be evaluated before its update function is evaluated. The result of evaluating the update function of a stateful block (see unit delay block as in Fig. 2) updates its state variable S which is expressed by S', so $S' = g(X, S)$. Furthermore, the *initial condition* of a stateful block is specified by the initial value of its state variable.

A block also has an associated *sample time* that states how often a block should repeat its evaluation procedure. The sample time of a block must be a multiple of the time resolution of a system called the *ground sample time*.

A Simulink model is a directed graph where a node corresponds to a block represented as either a triangle, rectangle, or circle and a directed edge between blocks corresponds to a signal. A signal models the data dependency between the output of one block and the input of another block. An output from a block can be connected to input of one or more other blocks by splitting it into two or more copies, which serves as an input to multiple blocks. However, the output signals of two or

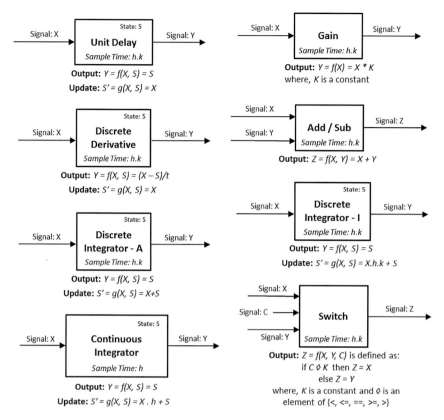

Fig. 2 Commonly used blocks in simulink

more blocks cannot join to become one input for another block. A network of blocks connected over signals in this manner is called a *system*. If all blocks of a system have the same sample time then the system is called an *atomic system*. A subset of blocks belonging to a system and having the same sample time is called an *atomic subsystem*.

A mathematical model of a dynamic system is a set of difference and differential equations. Such equations can be modeled as a system in Simulink. A simulation of such a system solves this set of equations using numerical methods. The *state* of a system is defined as the valuation of all its output variables (signals) and state variables. So from an initial state of the system the behavior in the state space is computed in an iterative manner over discrete time steps bounded by the ground sample time. A block having a sample time equal to the ground sample time is called a *continuous block*. All other blocks are *discrete blocks* and have a sample time that is equal to an integer multiple of the ground sample time. For continuous integration, the whole simulation is repeated several times within one ground sample time (depending on the numerical integration method, i.e., solver type) to get a better approximation and detect zero crossing events [12]. Note that discrete blocks change their outputs only at integer multiples of the ground sample time which implies that the input to continuous integration blocks remains a constant. However, without loss of generality, we neglect the numerical refinement of the ground sample time and consider only one integration step per ground sample time.

The output and update functions of a block can be programmed in Simulink using a low level programming language such as C. However, some commonly used constructs are available as predefined configurable blocks in Simulink. In Fig. 2, we give a few examples of commonly used blocks in Simulink along with their function definitions. We consider five stateful blocks (unit delay, discrete derivative, discrete integrator-A, discrete integrator-I, continuous integrator) and three stateless blocks (gain, add/sub, and switch). We will discuss two of them.

The *unit delay* block is a stateful block having one input signal X, a state variable S, and one output signal Y. A unit delay block serves as a unit buffer by delaying the current value of its input signal by one sample time, i.e., $h.k$ time units in the following way: The output function f assigns the current value of its state variable to its output Y. The update function g copies the current value of its input signal to its state variable. After every $h.k$ time units, the output function is evaluated and then its update function.

The *continuous integrator* block is a stateful block that receives as its input (signal X) the derivative of the state variable S (i.e., the rate of change of valuation of the state variable). The output function f assigns the current value of its state variable S to its output Y. The update function g, updates the value of its state variable by integrating the product of the derivative and the ground sample time. Note that Simulink blocks such as *transfer function* and *state space* are similar to continuous integrator blocks.

Simulink System. A *simulink system* is a set of Simulink blocks connected over shared input and output signals. In Fig. 3, we present an example of a cruise control system modeled in Simulink. The model is an adaptation of the example in the paper [12]. We consider this example because it is simple and covers all relevant modeling aspects of Simulink. For a detailed description of mathematical equations underlying this model, please refer to [12].

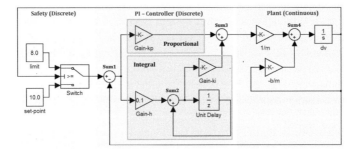

Fig. 3 Simulink model: Cruise controller

Simulation of Simulink System. The blocks of a Simulink system are executed in a sorted order. To determine this sorted order, we distinguish between two types of blocks, namely independent blocks and dependent blocks. If the output function of a block is defined over only its state variable (i.e., does not depend on its inputs, for instance the unit delay block), then we call it an *independent block*, otherwise we call it a *dependent block*. The order in which independent blocks are evaluated is not important. However, the order in which the output of dependent blocks are evaluated is important because the output of a dependent block must be computed only after all other blocks that produce an input to this block have been evaluated. This kind of dependency induces a natural ordering between blocks of a Simulink system which can be represented as a directed graph (blocks as nodes, dependency between blocks as directed edges) representing the order in which blocks of a Simulink system must be evaluated, i.e., a directed edge from block *A* to block *B* indicates that block *A* must be evaluated before block *B*. We call this graph the *dependency graph* of a Simulink system. For the example presented in Fig. 3, the dependency graph between blocks is shown in Fig. 4. The *block sorted order* is a sorted sequence of blocks whose ordering satisfies the dependency graph of a Simulink system. The sorted order of a Simulink system is determined by the simulation engine before the start of a simulation.

Each execution of the block sorted order is called a *simulation step*. In each simulation step, every block occurring in the model is visited according to the block sorted order and the following condition is checked for each block: if the time elapsed since this block's last execution equals its sample time, then the block produces its output, otherwise the block execution is skipped, i.e., the block's functions are not reevaluated. The time that is allowed to elapse between each simulation step must be a multiple of the ground sampling time called the *step size*. The value of the step size for a given Simulink model can be either specified explicitly or is determined by the simulation engine such that the bounds on approximation errors of integrator blocks are within some specified threshold. So there are two different simulation modes, namely *fixed step solver* and *variable step solver*. In a fixed step solver, a simulation step occurs every *step size* time units. In a variable step solver, a simulation step occurs at the earliest time instant when at least one block in the model exists such that the time elapsed since its last execution is equal to its sample time. For a description of the algorithm underlying the two solver modes, please refer to [25].

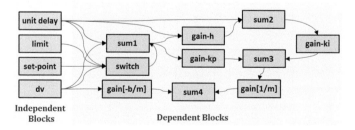

Fig. 4 Dependency graph: Cruise controller

2.4 Stateflow

Stateflow is almost a Simulink block except that it can have more than one output
signal. A Stateflow block models a combination of communicating hierarchial and
parallel state machines (resembling Statecharts) and flowcharts. The three basic concepts of Stateflow are states, transitions, and junctions.

A *state* in a Stateflow model has a hierarchial parent–child structure representing its decomposition, i.e., into substates. A state can be decomposed into either
exclusive OR substates connected with transitions (directed edges) or parallel AND
substates that are disconnected. The former is represented by substates with solid
borders and the latter by substates with dashed borders. The parent of these substates are referred to as a superstate. If a superstate has an OR decomposition then it
is active if exactly one of its substate is active. If a superstate has an AND decomposition then it is active if all its substates are active. In Fig. 5, we give an example of
AND/OR state decomposition in Stateflow and represent the parent–child relationship between a superstate and its substates as a tree, starting with a root node (not
explicitly represented in Statecharts). To distinguish between superstates that have
an AND/OR decomposition, we represent outgoing edges of an AND superstate by
dashed edges. All leaf nodes represent atomic substates, i.e., they have no children.
Note that an AND superstate having a child substate with an AND decomposition is a
redundant specification because this decomposition can be represented by the parent
AND superstate itself. Note that if a nonatomic substate is active then its superstate
is active as well.

Furthermore, a state in Stateflow has a set of associated actions. The action language consists of variable assignments or emission of events.

- Entry action: This action is executed when the state is entered.
- During action: This action is executed when no valid outgoing transitions exist in
  an active state.
- Exit action: This action is executed when a valid outgoing transition has been
  found.
- On event E action: This action is executed when an event E occurs in an active
  state.

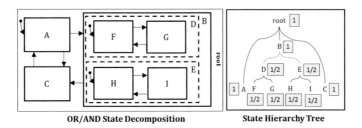

OR/AND State Decomposition   State Hierarchy Tree

Fig. 5 State hierarchy in stateflow

In addition, Stateflow defines a special type of state called a *junction*. In Stateflow models a junction is represented as a circle and is used to model logical decision patterns using a flowchart type notation. However, unlike a state, a junction does not have any hierarchy, nor any associated actions and must be exited instantaneously, when entered.

A *transition* is represented by a directed edge and models the passage of a system from one state to another. A transition may also have a junction in its source or(and) destination (also referred to as transition segment). The semantics of a transition is given in terms of an expression (specified as a label) having the form $E[C]\{Ac\}/At$, where E is the event, C is the guard condition, Ac is the condition action, At is the transition action, which means that a transition can occur only if an event E has occurred and the guard condition C is true. If the guard condition is true then the condition action Ac is executed. The transition action At is executed when a destination state has been found. So in case of a transition from one state to another state, transition action coincides with a condition action. However, in case of transitions connecting junctions, transition actions are executed (in the order they were visited) only when a valid path has been found to a state. Furthermore, all parts of a transition's expression are optional. Note that we will not consider the nested event broadcast feature in actions of Stateflow transitions because its preemption semantics (see [3]) is a bad practice and often leads to unpredictable results. So we will consider this feature without preemption.

Furthermore, transitions are allowed to cross superstate boundaries to a substate destination, referred to as *inter level transitions*. In Stateflow semantics, an initial state of a set of substates belonging to a parent superstate having an OR decomposition is specified using a special type of transition called a *default transition*. It is represented as a directed edge having no source to a single destination state. Furthermore, for a superstate, Stateflow has the notion of a *history junction* denoted as the symbol H with a circle around it. Such a junction remembers the last active substate of the superstate when it is exited and assigns this substate as the initial state. We give an example of a Stateflow in Fig. 6.

3   CPN*: A Compact Notation for CPN

Although the CPN formalism is expressive enough for most practical applications (as they are Turing complete), the modeling of large systems often lead to large models that are cumbersome and difficult to read. In this paper, we focus on modeling hybrid systems expressed partly as a set of difference and differential equations and partly by some kind of state transition diagram. In this context, we will identify few recurring CPN patterns and show how they can be graphically represented in a more compact and handy manner without changing the underlying semantics. We call the resulting CPN model as a CPN* model.

Given an arbitrary CPN model (see the example in Fig. 7.), we distinguish between two types of its places, namely Signal Places and State Places.

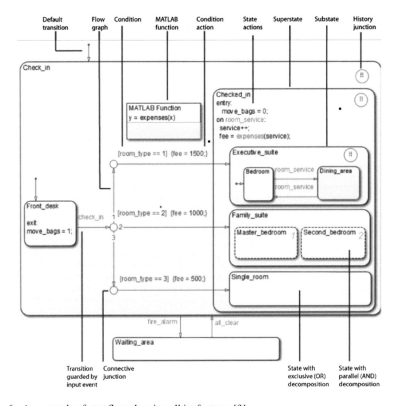

Fig. 6 An example of stateflow showing all its features [3]

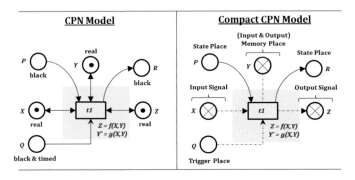

Fig. 7 A CPN and its CPN*

- **Signal Places**. A CPN place is called a *signal place* (a) if it is untimed and has a color type *real* having exactly one token in the initial marking of the CPN model, and (b) it has a preset that is equal to its postset, i.e., transitions that consume a token from it also produce back a token into it. As a consequence, a signal place has exactly one token in any reachable marking of the net system, i.e., it is always marked and safe. In models of hybrid systems, a marked signal place can be seen as modeling a *variable*, whose valuation is given by the color value of the token in that place. As a compact notation, we represent a marked signal place by a crossed circle.

- **Trigger Places**. A CPN place is called a *trigger place* if (a) it is timed and has a color type with only one value, which we call *black* (i.e., it can contain only *black* tokens like in a classical Petri net), and (b) it has a postset of size equal to one. In models of hybrid systems, a trigger place is used to model an event of a system. To make the distinction from other places of a CPN more graphically explicit, we will distinguish it by representing its incoming (outgoing) arcs to (from) this place by a dashed directed edge.

A transition of a CPN model is called a *function transition* if (a) it has associated with it signal places and exactly one trigger place, and (b) it defines one or more functions over its connected signal places (i.e., modeling variables). The function definition of such a transition induces a distinction between signal places as either an input, an output, or both. We call a signal place of a transition that is both an input and an output as a *memory signal place*.

In the CPN model of Fig. 7, we see the annotations of the transition $t1$ as $Z = f(X, Y)$ and $Y' = g(X, Y)$. The meaning of these annotations is that Z, X, Y are signal places and Y' is the same signal place as Y but Y' is the updated value after evaluation of function g. From this annotation, we can derive that X is an input and Z is an output for transition $t1$, which is indicated by the arrow on the arc between the signal place and the transition. For a memory signal place like Y, we have no direction on the arc. Note that these arcs and the use of directions are different from the meaning in the CPN model itself.

For the purpose of modeling Stateflow, we identify state places and super places of a given CPN model.

- **State Place**. A CPN place is called a *state place* if (a) it is untimed and has a color type *black*. We call a transition that has associated with it one or more state places as a *state transition*. So transition in Fig. 2 is also a state transition.
- **Super Place**. A *super place* is a subset of places of a CPN denoted by a solid box around its places. The set of all state places of a CPN is also a super place. We require that two super places of a CPN are either disjoint or one is contained in the other. A super place q is a child of super place p and p is a parent of super place q if and only if q is contained in p and no other super place r satisfies $q \subseteq r \subseteq p$. Due to this property, it is easy to see that the set of all super places forms a tree since if q is a child of p and q is a child of r then p and r are not disjoint. So either $p \subseteq r$ which means q is not a child of r, or $r \subseteq p$ which means q is not a child of p.
- **Super Place Net**. The subnet of a CPN model containing exactly the state places of a super place and all transitions that have their preset and postset contained in q is called the *super place net*. We require that super place nets are connected.

4 Modeling Hybrid Systems

In this section, we show how a hybrid system can be expressed as a CPN* model. We start with modeling arbitrary functions that do not have a notion of time and then show how these models can be extended to model functions having time as a parameter. Finally, we show how functions can be combined to model a system, both in a general way and in the Simulink framework.

4.1 Modeling Untimed Functions

From a modeling perspective, we distinguish between two types of functions, namely memoryless functions and memory functions.

The output of a *memoryless function* is a function of only its input variables. As an example, consider the memoryless function $Z = f(X, Y)$, with input variables X and Y and output variable Z. The memoryless function f is modeled in CPN* by assigning it to a function transition having two input signal places X and Y and an output signal place Z. We show the corresponding CPN* model in Fig. 8.

On the other hand, a *memory function* has also a variable modeling its memory (data store). The output of such a function depends not only on its input variables but also on its memory variable. Such a function may also define an additional *update function* to overwrite the current value of its memory variable. This update function depends on input variables or(and) the memory variable itself. As an example, consider the memory function $Z = f(X, Y)$, where X is an input variable, Z is the output variable, and Y is the memory variable with an associated update function $Y' = g(X, Y)$. We model a memory function f along with its update function g in CPN*, by assigning the two functions to a function transition having one input signal place X, one output signal place Z, and one memory place Y. We show the corresponding CPN* model in Fig. 8. Note that two or more transitions may share the same memory place.

So far we have seen how a function can be modeled using a function transition, signal places, and memory places. Using these concepts it is straightforward to model a set of functions related to each other by input/output variables and mem-

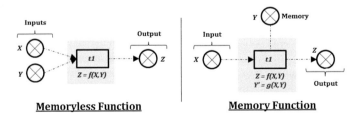

Fig. 8 Memory function and memoryless function

ory variables (by sharing common signal/memory places) as a network of function transitions. Such a network of function transitions defines a predecessor/successor relationship between them. Given a function transition, its predecessor (successor) transition is a transition whose output (input) signal place is the input (output) signal place of the given transition.

Although a network of function transitions is syntactically correct, it is semantically not well defined because all its transitions are always enabled. So it is not clear what the output of such a system really means, even if it is an acyclic network! For a cyclic net as shown in Fig. 9, the process (i.e., the successive values of the signals) is completely nondeterministic. Therefore, we use a trigger place as shown on the right-hand side of Fig. 9. Now the values of the signals X and Y can be computed as the sequence $(X_0, Y_0), \ldots, (X_n, Y_n)$, where for some $i \in \{1 \ldots n\}$, (X_i, Y_i) represents the value of the signals X and Y after firing transitions $t1$ and $t2$, i times, i.e., $(g \circ f)^i(X_0) = X_i$ and $f(g \circ f)^{i-1}(X_0) = Y_i$.

Next we show two ways of using a trigger place for a transition modeling an arbitrary function. A transition is *predecessor triggered* if every predecessor transition is connected only by an outgoing arc to the trigger place of this transition. A transition is *self-triggered* if it is connected to its own trigger place and there are no other transitions in either the postset or preset of this trigger place. Note that a transition with no predecessor transition must be self-triggered. Furthermore, the trigger place of a self-triggered transition has exactly one token in the initial marking.

In Fig. 10, we give an example of a CPN* with self-triggered and predecessor triggered transitions. The function h (modeled by transition $t2$) is a memory function having no predecessor. So it is self-triggered. This is also the case with function g (modeled by transition $t3$) except that it is memoryless. Furthermore, function g produces an output Y that is an input to the functions f (modeled by transition $t1$) and r (modeled by transition $t4$). As a result, transition $t3$ is connected to the trigger place of transitions $t1$ and $t4$. Since function f receives an input X from function h, its trigger place is connected to it.

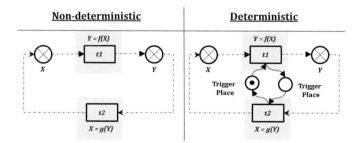

Fig. 9 Modeling a network of function transitions

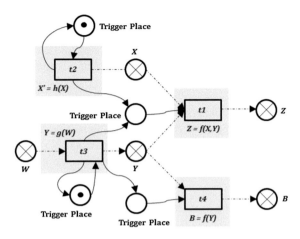

Fig. 10 CPN* with
self-triggered and
predecessor triggered
transitions

4.2 Modeling Timed Functions

In a hybrid system, the notion of state progression, i.e., the successive valuation of
its variables, are a consequence of the progression of time. So time is an important aspect of modeling such systems. Here functions have time as a parameter, for
instance in ordinary differential equations having time as a parameter. To model such
functions in CPN*, we must address the following two questions:

1. When Should the Functions of the System Be Evaluated? We capture this
aspect by making a distinction between functions, namely *timer functions* and *reactive functions*. Note that both timer functions and reactive functions can be either
memoryless or memory functions.

A *timer function* has an associated time attribute called the *sample time*, which
specifies how often its output must be computed. A *reactive function* has no oblig-
ation as to how often its output must be computed. So it only computes an output
whenever there is a change in the values of any of its input variables (indicated by a
token in its trigger place).

Modeling. A reactive function is modeled as a predecessor triggered transition. A
timer function is modeled as a self-triggered transition having a delay (corresponding
the sample time of the function) on the incoming arc to the trigger place. We give
an example of each type in Fig. 11.

2. How Can Time Be used as a Parameter in a Modeled Function? To model
a hybrid system, the function definitions must also allow the inclusion of time as a
parameter. However, the time parameter may either be relative to the model called
the *absolute model time* or relative to a function called the *relative function time*.
The former represents the time that has elapsed since the model began its execution,
while the latter represents the time that has elapsed since the last execution of a given
function.

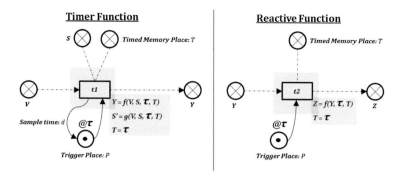

Fig. 11 Timed functions

Modeling. A transition modeling an arbitrary function may access the absolute model time by looking up the timestamp of the enabling token in its trigger place, denoted by τ. Note that by the eagerness property of CPN, this is the earliest token timestamp in the trigger place.

The relative function time of a given transition can be computed by first adding a memory place to it without any delays on any of its arcs. We call this memory place as a *timed memory place*. Its role is to store the absolute model time at which its connected function transition last fired. We model this by specifying an update function that produces as its output, the timestamp of the enabling token in the trigger place. Now the the relative function time can be computed by subtracting the color value of the token in the timed memory place from the timestamp of enabling token in the trigger place. See Fig. 11.

4.3 Modeling Simulink System and its Simulation Using CPN*

In this section, we give a formal semantics to Simulink using the CPN* formalism. First, we show how a Simulink block can be modeled in CPN* as a transition function having at most one output, which we will call a Petri net block (*P-block*). A few simple rules describe how a set of P-blocks can be connected to construct a system, which we will call a *P-system*. The blocks of a Simulink system are executed in a certain order called the *block execution order*. We show how the block execution order of a Simulink system can be modeled as a CPN*, on top of an existing P-system. The resulting system describes fully the behavior of a simulation in Simulink and we call it a *C-system*. Note that the execution of the block execution order is carried out by the underlying simulation engine, so it is an implementation detail and not a modeling concern for designers.

Modeling Simulink Blocks. It is straightforward to model the two basic types of Simulink blocks, namely stateless blocks and stateful blocks in the CPN* notation as memoryless functions and memory functions, respectively. Furthermore, since most of these blocks model functions having time as a parameter and have an associated sample time, we extend them with the notion of time as explained in Sect. 4.2. We call a CPN* model of a Simulink block a P-block. In Fig. 12, we give an example of modeling a discrete derivative block as a P-block having a sample time 3. Note that the initial condition of a stateful P-block is specified by the initial color value of its memory place(s).

Modeling Simulink System. Given a set of P-blocks, we model a P-system by fusing the shared input signal places of blocks with the output signal place of blocks. Note that two or more blocks are not allowed to have the same output signal place because in Simulink, blocks have only one output and this is modeled in P-blocks with a unique output signal place. To keep the figure readable, we consider only the *integral part* of the PI controller (with $h = 1$ and $k = 10$) and model it as a P-system in Fig. 13.

Modeling Simulation of Simulink System as Control Flow. We have seen how an arbitrary Simulink system can be expressed as a P-system. The enabling of P-blocks

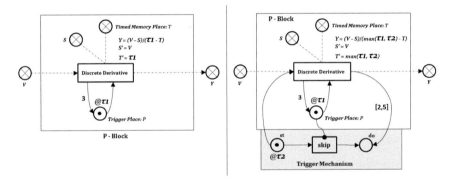

Fig. 12 P-block of the discrete derivative block

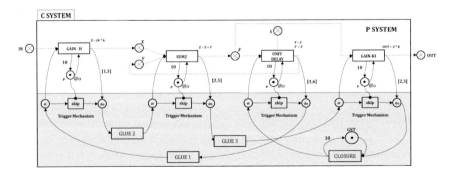

Fig. 13 C-system: integral part (PI controller)

in a P-system is determined by the timed availability of a token in the trigger place of function transitions. This means that in each simulation step of the model, multiple P-blocks with the same sample time can become enabled. In CPN semantics, the choice of executing a P-block is done in a nondeterministic manner. However, in a Simulink simulation, the blocks of a Simulink model that can produce their output at a simulation step (determined by their sample times), are evaluated according to their block sorted order.

Consider the P-system of the integral part of the PI controller in Fig. 13. This system has four blocks, namely one sum block (sum2), two gain blocks (gain-h and gain-ki), and one unit delay block. For this system, one of the block sorted order satisfying the dependency graph of Fig. 4, is the sequence: ⟨unit delay; gain-h; sum2; gain-ki⟩. The block execution order of P-blocks of a P-system is modeled by first adding a *trigger mechanism* to each P-block and then connecting the trigger mechanism of blocks according to the block sorted order. Furthermore, we also model that Simulink blocks have a nondeterministic *execution time* (i.e., the time it takes for a block to produce its output). In CPN*, such an execution time can be associated with a transition by specifying a time interval along an outgoing arc from it. We call the resulting system a *C-system*.

The *trigger mechanism* is modeled on top of a P-block by adding a timed input place *start* (labeled *st*), a timed output place *done* (labeled *do*), and a timed inhibitor transition (inhibitor arc that accounts for the availability of token in the trigger place of that block) called *skip*. If the token in place *start* is available and the token in the trigger place is unavailable, then the *skip* transition is enabled due to the timed inhibitor arc. Firing the skip transition does not progress time and the execution of the block is skipped in the current time step. If the *skip* transition is disabled (due to an available token in the trigger place), then the block transition (discrete derivative) fires at its enabling time and produces a token in the place *done*. Note that the functions requiring an absolute model time may compute it by taking the maximum of the timestamp of enabling tokens in the trigger place (called τ_1) and start place (τ_2). Furthermore, along the arc to the place *done* it is possible to specify the *execution time* of the block as a time interval. In both cases, one timed token is produced in the place *done*. In Fig. 12, we give an example of a P-block, extended with a trigger mechanism and modeling an execution time in the interval [2, 5].

Next, we model the block execution order of a system by introducing a new transition called the *glue transition* between the acknowledge and trigger places of successive blocks as they occur in the sorted order. The construction is carried out in the following way: For each block occurring in the block execution order, we add one glue transition that consumes a token from the place *done* of the preceding block and produces a token in the place *start* of this block. In this way, the C-System describes a simulation run of the system. In order to allow for more than one run of a simulation at a rate specified by the step size: (a) we add a transition labeled *closure* that consumes a token from the place *done* belonging to the last block in the sorted order and produces a token in the place *start* corresponding the first block in the sorted order, and (b) we initialize the place *done* of the last block occurring in the block sorted order with a token having a timestamp zero. Furthermore, to this closure transition, we connect a place called *GST* with bidirectional arcs and having one token. On the incoming arc to the place *GST*, we associate a delay corresponding the *step size* of the simulation (multiple of ground sample time). As a result, a simulation run can only occur once every *step size* time units. If the model has continuous blocks then the step size must equal the ground sample time. To simulate a variable step solver, the step size must be equal to the least sample time of all blocks in the system. In Fig. 13, we describe the C-system of the integral part of the PI controller having a step size of 10 time units. In [25, 26], we show some interesting properties of C-systems that can be verified by model checking.

5   Modeling Stateflow Using CPN*

Stateflow is a Simulink block except that (a) it can have more than one output signal, (b) its internal structure is defined as a hierarchial and concurrent state machine, and (c) its execution is either triggered by the simulation engine (in a simulation step), or triggered by events generated by Simulink/Stateflow blocks.

In this section, we will consider the relationship between Stateflow models and CPN*. For this we address the following three questions; First, we show how CPN* models can be constructed in *Stateflow style* by stepwise refinements using two refinement operations. Then we consider how an arbitrary CPN* model can be checked for conformance to the Stateflow style. Finally, we show how a CPN* model can be derived from a given Stateflow model.

5.1   Constructing CPN* Models in Stateflow Style

In this section, we define a Stateflow style construction method for CPN*. It is based on stepwise refinement with the two refinement techniques, namely place refinement and place duplication corresponding to an OR decomposition and an AND decomposition, respectively. For the construction, we will consider state machine nets that have only state places (We will add the signal and trigger places on top of the CPN* model, derived using the construction method).

Place Refinement. Given a place p in a CPN* model and a state machine net N, the *place refinement* operation replaces place p with the net N in the following way: Remove place p from the CPN* model and perform a net union of N (i.e., pairwise union of its places, transitions and relationships between them) with the given CPN* model. Next, connect all input arcs of p to places of net N and all output arcs of p to places of net N. Note that the modeler has the freedom to choose arbitrary places to connect these arcs.

Note that place p in the original CPN* model becomes a parent super place containing the set of state places belonging to the state machine net N in the resulting

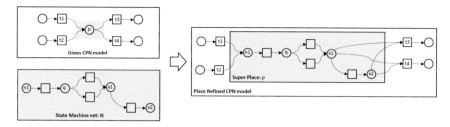

Fig. 14 Place refinement

CPN* model. We indicate this relationship by drawing a solid box around the nodes of the net N and labeling it with the name of the super place p. Furthermore, observe that the state machine net N is a super place net of the resulting CPN. See Fig. 14. Note that this operation corresponds to an OR operation of a superstate into its mutually exclusive substates in Stateflow.

Place Duplication. The place duplication operation provides a structured way to introduce concurrency into a CPN* model. The operation is defined in the following way:

Given a place p in a CPN* model and a set of places Q not belonging to this CPN* model, the *place duplication* operation adds the set Q to the set of places of this CPN* model in the following way: Remove place p from the CPN* model and add the set of places Q to the CPN* model such that for each place in the set Q, its preset (postset) is equal to the preset (postset) of p.

Note that place p in the original CPN* model becomes a parent super place containing the concurrent state places modeled by the set of places Q. We indicate this relationship by drawing a dashed box around the places of set Q and labeling it with the name of the super place p. We give an example in Fig. 15. Note that this operation corresponds an AND operation of a superstate into concurrent substates in Stateflow.

Construction Method. Start with a set of super places represented as a tree and associate each super place as having either an OR decomposition or an AND decomposition. Next we construct a CPN* model having no transitions and exactly one state place corresponding the root of the super place tree. Then set the *current node* to the root node of the tree and carry out the following steps in an iterative manner until all nodes have been *visited*. Note that after each iteration, we have a CPN* model.

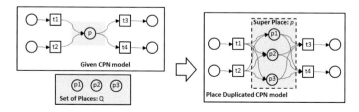

Fig. 15 Place duplication

- If the *current node* has an OR decomposition then

  - Construct a state machine net N containing a labeled place for each corresponding child of the current node.
  - For the place in the CPN* model corresponding the current node, carry out a place refinement with the state machine net N.

- If the *current node* has an AND decomposition then

  - Construct a set of places Q containing a labeled place for each corresponding child of the current node.
  - For the place in the CPN* model corresponding the current node, carry out a place duplication operation with the set of places Q.

- Mark the current node as *visited*
- If *there exits an unvisited child of a visited parent* then *set it as the current node*, otherwise *stop*.

We call the resulting CPN* model a *Stateflow Petri net* like in [4]. We also consider some interesting properties presented in [4]. To present them, we first give a few definitions.

Active Super Place and Correct Configuration. A super place is called an *active super place* if at least one of its state places contains a token. A marking of a Stateflow Petri net is said to be a *correct configuration* if and only if the following properties hold:

- The root super place is active.
- An active super place corresponding to an OR node has exactly one of its children active.
- An active super place corresponding to an AND node has all its children active.

The state places of a Stateflow Petri net satisfies an invariant property by construction. To see this, we first assign weights to each node of the state hierarchy tree according to the following rules:

- The root node of the tree has a weight equal to one.
- The children of an OR node in the tree inherit the weight of their parent.
- Each children of an AND node in the tree obtains a weight equal to *the weight of their parent* $\times$ *(number of children)*$^{-1}$.

Next to each node of the tree in Fig. 5, we indicate their node weights. For a Stateflow Petri net, it is easy to prove the following properties:

- A Stateflow Petri net is safe.
- In any reachable marking from an initial marking that is a correct configuration, the sum of the product of the number of tokens in a place (either zero or one) and its weight is a constant, i.e., the weight function is a place invariant (see [20, 21]).
- If a Stateflow Petri net is initially in a correct configuration then all its reachable markings are also correct configurations.

On top of a Stateflow Petri net, we may add signal places and trigger places.

5.2   Verifying Conformance of CPN* Models to Stateflow Style

Given a CPN* model, we verify that it conforms to the *Stateflow style* of construction by first deleting all its trigger places and signal places and then verifying if the resulting CPN* model satisfies the following property: *Is it possible to reduce the CPN* model to a single place by applying the refinement operations in their inverted form as reduction rules?* If this is true, then the CPN* model conforms to the Stateflow style, i.e., it is possible to derive the CPN* model by stepwise refinements with the two refinement operations. Note that in [4], a similar reduction is used without considering its inverted form as refinement rules.

5.3   Constructing CPN* Model from Stateflow Model

Now we present a *construction method* to derive a CPN* model from a given Stateflow model. However, before we use an arbitrary Stateflow model, we need to pre-process it due to unstructured and complicated semantics associated with flowchart notations (specified by junctions and transition segments modeling decisions and loop-based constructs) and redundant semantics associated with all kinds of actions in its states and transitions.

Flatten Stateflow. First, eliminate flowchart semantics in Stateflow by identifying a set of transition segments leading from a state to another state and defining their semantics (i.e., early return logic [3]) in terms of a recursive function. Then replace this set of transition segments by a single transition and associate it with the defined function. Note that in CPN, it is possible to express such a recursive function using the functional programming language ML. If a set of transition segments are such that they originate at a junction or terminate at a junction, then we treat the origin/termination junction as a state in Stateflow and refer to the transition replacing them as a junction transition. If both the origin and termination are a junction, then we do the same modification after adding a new transition (which we call closure transition) from the terminating junction to the origin and treating this connected pair of states as an AND decomposition of the root (see [3] for semantics of flowchart). Furthermore, to keep the distinction between transitions of Stateflow and CPN clear, we will refer to a transition in the former as an edge.

- Given a flattened Stateflow model, construct its state hierarchy tree.
- **Add Places and Super Place Notations**. Start with an empty CPN* model, (i.e., having no places and no transitions) and then add state places to it in the following way:

  – For each leaf node of the state hierarchy tree, add one corresponding state place to the CPN* model.

- For each parent superstate in the state hierarchy tree, draw a solid box around the state places in the CPN* model corresponding its children.

- **Super Place Net Construction**. Next add the transitions to the CPN* model in the following way:

  - For each edge of the Stateflow model add a transition to the CPN* model.
  - **Extended Stateflow**. Next, extend the Stateflow model in the following way: For each interlevel edge having a destination as a superstate, *recursively* extend this edge to the nested initial atomic substate (indicated by the default transition) of this superstate. If the destination superstate has an AND decomposition, then create an additional edge, one for each substate. Note that in an extended State-flow model, all interlevel edges have as their destination, an atomic substate (corresponding state places in our CPN* model).
  - For each transition in the CPN* model connect its postset to the state place corresponding the destination substate of the extended interlevel edge. If an extended edge has an atomic substate (leaf node) as its source, then add to the preset of the corresponding transition, the corresponding state place. If an extended edge has a super state as its source, then add reset arcs from the corresponding transition to every place in the corresponding super place.

In the resulting CPN* model, we associate signal places (modeling variables) and trigger places (modeling events) to transitions as they occur in the Stateflow model. A condition associated with an edge is expressed as a guard condition of corresponding transitions in the CPN* model. Since transition segments have been eliminated in the preprocessed Stateflow model, the definition of a condition action and transition action, coincide. These actions are specified as a function definition of a transition in CPN*. Furthermore, we eliminate the redundant definition of actions in states in the following way: An entry (exit) action of a state is modeled in CPN* as a function associated with all transitions in the preset (postset) of the corresponding state place. The during action of a state is modeled by adding a self-loop to a state place, i.e., a transition with preset and postset as the state place, and associating this transition to the function specified by the during action. An on event action of a state is modeled in a similar way but in addition, we associate a trigger place to the self-loop transition. It is a simple modeling exercise to express the semantics of a history junction in the CPN* model by duplicating each state places and using these copies to record the last active state place. Lastly, we model the triggering of a Stateflow block by the simulation engine in each simulation step. For this we add a memory place that is timed (i.e., can contain tokens with timestamps) and connect it to all transitions (except junction transitions) of the Stateflow model. Then by default, we have the *superstep semantics* (see [3]). For *step semantics*, we must add a delay equal to the ground sampling time of the system on each incoming arc to the newly added (timed) memory place.

6 Conclusions

We have shown how hybrid systems can be modeled in CPN*, a compact notation for CPN models. As a special case, we have considered the translation of
Simulink/Stateflow models into the CPN* formalism. The result is that the models
in both these formalism have the same size but the the CPN* formalism has many
advantages: First, it has a well-defined semantics and it is simple, as compared to verbose semantic descriptions of the Simulink/Stateflow toolset. Second, it is a strong
extension of the Simulink/Stateflow framework by not treating Stateflow elements as
a block, i.e., it is possible to use elements of a Stateflow block on the same footing as
Simulink blocks. Third, CPN* models exhibit true concurrency unlike the sequential semantics of Stateflow models. Lastly, there are numerous numerous Petri net
analysis techniques and tools available to verify CPN* models by model checking,

References

1. Harel, D.: Statecharts: a visual formalism for complex systems. Sci. Comput. Program. **8**, 231–
   274 (1987)
2. website, S.: http://nl.mathworks.com/help/simulink/
3. website, S.: http://nl.mathworks.com/help/stateflow/
4. Eshuis, R.: Translating safe petri nets to statecharts in a structure-preserving way. In: Caval-
   canti, A., Dams, D.R. (eds.) FM 2009. LNCS, vol. 5850, pp. 239–255. Springer, Heidelberg
   (2009)
5. Huszerl, G., Majzik, I., Pataricza, A., Kosmidis, K., Dal Cin, M.: Quantitative analysis of UML
   statechart models of dependable systems. Comput. J. **45**, 260–277 (2002)
6. Merseguer, J., Campos, J., Bernardi, S., Donatelli, S.: A compositional semantics for UML
   state machines aimed at performance evaluation. In: Proceedings. Sixth International Workshop on Discrete Event Systems, IEEE, pp. 295–302 (2002)
7. Agrawal, A., Simon, G., Karsai, G.: Semantic translation of simulink/stateflow models to
   hybrid automata using graph transformations. Electron. Notes Theoret. Comput. Sci. **109**, 43–
   56 (2004)
8. Tripakis, S., Sofronis, C., Caspi, P., Curic, A.: Translating discrete-time simulink to Lustre.
   ACM Trans. Embed. Comput. Syst. **4**, 779–818 (2005)
9. Denckla, B., Mosterman, P.J.:. Formalizing causal block diagrams for modeling a class of
   hybrid dynamic systems. In: IEEE CDC-ECC, pp. 4193–4198 (2005)
10. Tiwari, A.: Formal semantics and analysis methods for simulink stateflow models. Technical
    Report, SRI International (2002)
11. Zhou, C., Kumar, R.: Semantic translation of simulink diagrams to input/output extended finite
    automata. Disc. Event Dyn. Sys. **22**, 223–247 (2012)
12. Bouissou, O., Chapoutot, A.: An operational semantics for simulink's simulation engine. In:
    Proceedings of the 13th ACM SIGPLAN/SIGBED International Conference on Languages,
    Compilers, Tools and Theory for Embedded Systems. LCTES '12, ACM, pp. 129–138 (2012)
13. Hamon, G., Rushby, J.: An operational semantics for stateflow. Int. J. Softw. Tools Technol.
    Transf. **9**, 447–456 (2007)
14. Hamon, G.: A denotational semantics for stateflow. In: Proceedings of the 5th ACM Interna-
    tional Conference on Embedded Software, ACM, pp. 164–172 (2005)
15. Scaife, N., Sofronis, C., Caspi, P., Tripakis, S., Maraninchi, F.: Defining and translating a safe
    subset of simulink/stateflow into Lustre. In: Proceedings of the 4th ACM International Con-
    ference on Embedded Software, ACM, pp. 259–268 (2004)

16. Jensen, K., Kristensen, L.M., Wells, L.: Coloured petri nets and CPN tools for modelling and validation of concurrent systems. Int. J. Softw. Tools Technol. Transf. **9**, 213–254 (2007)
17. David, R., Alla, H.: Discrete, Continuous, and Hybrid Petri nets. Springer, Berlin (2010)
18. Demongodin, I., Koussoulas, N.T.: Differential Petri nets: Representing continuous systems in a discrete-event world. IEEE Transactions on Automatic Control **43**, 573–579 (1998)
19. Gilbert, D., Heiner, M.: From petri nets to differential equations—an integrative approach for biochemical network analysis. In: Donatelli, S., Thiagarajan, P.S. (eds.) ICATPN 2006. LNCS, vol. 4024, pp. 181–200. Springer, Heidelberg (2006)
20. Reisig, W.: Petri Nets: An Introduction. Springer, New York (1985)
21. Peterson, J.L.: Petri net theory and the modeling of systems. Prentice Hall PTR, Upper Saddle River (1981)
22. CPN Website: www.cpntools.org/
23. Bera, D., van Hee, K., Sidorova, N.: Discrete timed petri nets. Computer Science Report 13–03, Technische Universiteit Eindhoven (2013)
24. van Hee, K., Sidorova, N.: The right timing: reflections on the modeling and analysis of time. In: Colom, J.-M., Desel, J. (eds.) Petri Nets 2013. LNCS, vol. 7927, pp. 1–20. Springer, Heidelberg (2013)
25. Bera, D., van Hee, K., Nijmeijer, H.: Relationship between simulink and petri nets. Computer Science Report 14–06, TU Eindhoven (2014)
26. Bera, D., van Hee, K., Nijmeijer, H.: Relationship between simulink and petri nets. In: Obaidat, M., Kacprzyk, J., Oren, T. (eds.) Proceedings of SIMULTECH 2014: Fourth International Conference on Simulation and Modeling Methodologies. SCITEPRESS, Technologies and Applications (2014)

Automatic Tuning of Computational Models

Matteo Hessel, Fabio Ortalli, Francesco Borgatelli
and Pier Luca Lanzi

Abstract The aim of this paper is to present a methodology for automatic tuning of a computational model, in the context of the development of flight simulators. We will present alternative approaches to automatic parameter ranking and screening, developed in a collaboration between *Politecnico di Milano* and *TXT e-solutions* and designed to fit as much as possible the needs of the industry. We will show how the adoption of such techniques can make model tuning more efficient. Although our techniques have been validated against a helicopter simulator *case study*, they do not rely on any domain-specific feature or assumption, so they can, in principle, be applied in different domains.

Keywords Model tuning · Parameter screening · Machine learning · Feature ranking · Logistic regression · Multilayer perceptron

1 Introduction

1.1 Computational Models and Computer Simulations

The ever-increasing use of computer simulation in science and engineering has many theoretical and practical implications, changing our understanding of the dynamics

M. Hessel (✉)
Politecnico Di Milano, Milan, Italy
e-mail: matteo.hessel@mail.polimi.it

F. Ortalli · F. Borgatelli
TXT E-Solutions, Milan, Italy
e-mail: fabio.ortalli@txtgroup.com

F. Borgatelli
e-mail: francesco.borgatelli@txtgroup.com

P.L. Lanzi
DEIB, Politecnico Di Milano, Milan, Italy
e-mail: pierluca.lanzi@polimi.it

© Springer International Publishing Switzerland 2015
M.S. Obaidat et al. (eds.), *Simulation and Modeling Methodologies,
Technologies and Applications*, Advances in Intelligent Systems
and Computing 402, DOI 10.1007/978-3-319-26470-7_3

of complex systems and providing a new standard for testing and training whenever safety and costs are crucial elements (as it is in the aerospace, biomedical, pharmaceutical, and military industries). The development of computer simulations capable of accurately replicating the dynamics of complex systems in life-critical applications is a daunting problem. It requires first careful *mathematical modeling*, in order to take into account a reasonable number of key features of a system sufficient for predicting the evolution of all measures of interest. Second, it implies the implementation of algorithms for the *numerical resolution* of the model (and such algorithm must be robust enough to manage all the subtleties of error propagation: a constant problem when using an essentially discrete tool such as a computer for solving continuous problems). Finally, the development of a high-fidelity simulator requires also the *tuning* of the model's parameters: this is the process by which the impact of all elements, which have not been explicitly taken into account during modeling phase, is reduced to modifications of the parameters of the model. The first two steps of this process are far more developed than the last one, as decades (and in some cases centuries) of research in mathematics and physics have provided us with well-founded results and established methodologies, making it possible to model virtually any problem in any engineering domain. However, when an extreme accuracy is required (because, for example, the aim is not forecasting—a process for which a certain degree of imprecision is unavoidable—but, instead, the accurate replication in a simulated environment of some known phenomena) then these steps are not sufficient and *tuning* becomes the only way of filling the gap between model and reality. Relevant examples of this are the aerospace simulation industry (with its tight legal constraints in order to certify a simulator for pilot Training) and certain medical problems such as simulation-driven training [1] or patient-specific dose calculation in radiotherapy [2]. The main problem with this third step of *M&S* is that the tuning process is still dealt with manually, in spite of the complexity that arises when dealing with a huge number of parameter together with their nonlinear interactions. Proceeding by trial and error, guided only by the experience and intuition of the engineers involved in the process, makes obviously tuning *slow* and *human-intensive*; consequently, the identification of automatic techniques to support the process is a relevant research topic.

1.2 The Manual Tuning Process

Thanks to the collaboration with *TXT e-solutions*, we were also provided with reliable information regarding the effort required for tuning a simulator using traditional techniques, and could compare our results with meaningful benchmarks. The company also provided feedback on the quality of the resulting simulations allowing us to validate the new methodology. Traditional tuning takes a time that depends on the level of certification that the company want to achieve (each flight authority defines the standards in terms of accuracy that must be fulfilled in order to qualify for a specific level). In the case of a high-fidelity *Full Flight Simulator*

(*FFS*) tuning the model might require up to 500 days of work of an experienced engineer. It must be noted that the minimum level of experience to operate on a system of such complexity with a reasonably autonomy is about 1 year. Combined with the fact that the only way of gaining such experience is by trial and error (operating on a flight simulator of equivalent complexity), this means that training the engineers who work on tuning of flight simulators adds an additional burden on a process that is already slow and expensive. The overall year and a half of man work is further divided into a sequence of *static* and *dynamic* tests (the distinction is that in static tests stationary flights are simulated, while in dynamical tests real flight procedures such as takeoffs and landing are executed and evaluated), and many days are required for each test. Our methodology will be evaluated against a set of dynamic tests (the most complex ones), automatically tuning the parameters of the model in order to improve the accuracy of different takeoff procedures. The results, in terms of machine time, has been measured; and we show in the paper how automatic tuning can improve the time and effort required for developing a flight simulator, making the development of these complex systems cheaper.

1.3 The Automatic Tuning Problem

In order to overcome the limitations of the previously described methodology we need as first thing to define appropriately the problem we want solve. The terminology and the notations that we now introduce will then be used consistently in all the rest of the paper. Let P be a set of n parameters, and $\mathbf{A}$ be a generic assignment of parameters in P. Then assume for each $p_i \in P$ a range, and let us call any assignment $\mathbf{A}$ consistent if the parameters values are within the ranges. Let $\mathbf{S}$ be a set of domain-specific performance metrics, measuring different features of the simulation. Then, $\mathbf{S}(\mathbf{A})$ will represent the value of the performance metrics as computed if the simulation is executed with the parameters' assignment $\mathbf{A}$, and $\mathbf{Ref}$ will represent the set of expected values for the metrics (ground truth from observations). Finally, $E(\mathbf{A})$ will measure the *global simulation error*, depending upon the expected and computed values of the metrics in $\mathbf{S}$.

→ The *automatic tuning problem* is the problem of algorithmically identifying a *consistent* assignment $\mathbf{A}$, binding the values of all parameters in the set P, such that the *global simulation error* $E(\mathbf{A}) = f(\mathbf{S}(\mathbf{A}), \mathbf{Ref})$ is minimized.

Although in principle we would require this globally optimal solution, a more realistic setting slightly weakens the previous statement, requiring only to lower the global simulation error below some *threshold T* depending on the level of *certification* needed:

→ The *(weak) automatic tuning problem* is the problem of algorithmically identifying a *consistent* assignment $\mathbf{A}$, binding the values of all parameters in the set P, such that the *global simulation error* $E(\mathbf{A}) = f(\mathbf{S}(\mathbf{A}), \mathbf{Ref})$ is below threshold T.

Compliantly with this second specification of the tuning problem, it is understood that all our optimization procedures will not be run until some ideal global optimum is achieved but they will be early terminated as soon as the required accuracy is attained. Furthermore, we will even give up on ideal asymptotic convergence to the global optimum, as we will be satisfied with *local* optimization algorithms (as it must be the case given the highly *nonlinear* optimization problem we are facing).

1.4 Original Contributions and Innovations

This paper is an extended and revised version of the paper presented at the *4th International Conference on Simulation and Modeling Methodologies, Technologies and Applications (SIMULTECH)* and printed in the conference's proceedings. A more detailed analysis of one of the topics of this paper (screening) can be found in a previous paper, also authored by Hessel, Ortalli, Borgatelli, and dedicated to this exclusively: *Machine Learning for Parameter Screening in Computer Simulations* [3] (to appear in volume 8906 of the Springer's *Lecture Notes in Computer Science series*). In order to solve the general problem of automatic tuning, we present a novel decomposition of the *tuning* process in integrated but distinguished steps (*Screening* and *Optimization*). Our methodology is inspired by the manual process (which relies on knowledge of the impact of the different parameters to reduce the size of the parameter set). The algorithms employed for the first task—screening—are classic *statistical/machine learning* tools, but are used in an original way. Concerning optimization, the *hill climbing* algorithm with *adaptive variance* is a famous *nonlinear* optimization procedure, but the application domain is novel, and novel is also the *sequential masking* scheme using which we improved the results achieved using the standard algorithm. Overall, we believe that the work done in this area has the potential for deeply innovating the methodologies currently employed in industry for tuning. Although previous attempts of automatic model tuning can be found in literature, none, at the best of our knowledge, in the field of flight simulations (a sector with distinctive features—such as the strong regulatory action exerted by the Flight Authorities of the different countries). Furthermore, previous examples of automatic tuning only dealt with stationary processes, and not with dynamically varying inputs or to *man in the loop* schemes. Finally, previous attempts to use automatic tools for tuning relied mostly on the computational power of modern computers in order to solve the problem by brute force, limiting the range of application to problems of moderate complexity and making it very difficult to write effective *objective functions*. This last point, as we will see, is of extreme relevance in order to find set of parameters that offer high accuracy, and our decomposition of the problem into screening and optimization will allow the optimization procedure to *adapt* to the objective function, taking into account all and only the parameters that directly affect it.

2 Proposed Methodology

Our proposal in order to algorithmically solve the automatic tuning problems relies on the following decomposition in two subtasks: *screening* and *optimization*. The two subproblems must be solved in sequence, exploiting the results of the first phase in order to complete more efficiently and more effectively the second:

→ The *screening problem* is the problem of ranking the parameters of the model according to their relevance and using such ranking in order to define a subset $SP \subset P$ of *most* relevant parameters.

→ The *restricted tuning problem* requires to find the best consistent assignment $A|_{SP}$ of parameters in SP, exploiting the ranking of parameters in SP for optimization, while assuming fixed the values of all other parameters (those in P/SP).

In principle, if the screening problem is effectively solved, the restricted tuning problem should achieve the same accuracy of the full tuning problem at a lower cost. In our work we have observed something even more remarkable: The solution of the restricted tuning problem provides solutions that, for the same value of the objective function, exhibit a greater accuracy when observed and evaluated by domain experts. This is due to the fact that restricting the parameter set to those with direct impact on the objective function allows to partially overcome shortcomings in the objective function itself (an issue that is explained in more detail in a later section devoted to experimental results).

3 Screening

3.1 Introduction

Screening is a process with a long tradition in classical statistics, and multiple methods are available to pursue its goals. Most methods, though, make assumptions on the distribution of data, require prior knowledge, or have other disadvantages. *One-factor-at-the-time Designs* [4] require almost no interactions among factors. *Sequential bifurcation* [5] requires the sign of the contributes of all factors to be fixed and known. *Pooled ANOVA* [6] requires to know the fraction of relevant factors. *Design Of Experiments* [7] requires to evaluate an exponential number of configurations, introducing serious scalability issues. Although *fractional factorial designs* overcome such problem through controlled deterioration of the quality of results, the fraction evaluated must decay with exponential speed to keep constant the computational resources required. Furthermore, *sequential bifurcation* [8] and *Pooled Anova* assume further a low-order polynomial relations between input and output variables, and all previous approaches are usually based on a *two-level* scheme (implicitly assuming *linearity*—or at least *monotonicity*—for the functional form of the output, and making the choice of the two levels a sensitive decision). When all assumptions are satisfied and the required knowledge is available, all these methods can be very effective. However, in our research, we were looking for a fully general approach. Thus, we considered different Machine Learning (ML) approaches in order to accomplish this goal. *Feature selection*, for example, is the classic area of ML dealing with dimensionality reduction; yet, this is not what we are looking for. Indeed, feature selection algorithms do not focus on evaluating the impact of the different parameters on the simulation's output, as they only strive for the elimination of statistical redundancy. In order to exploit ML for solving the first task of our automatic tuning process, we have thus followed a completely different track, less direct but capable of discriminating among parameters in a way that is more useful for our aims. Build *feature ranking* from classifiers trained on a database of previously generated *<parameter-set, simulation error>* tuples, and only in a second moment deriving the subset of relevant parameters from such ranking. This second step can be done easily by sorting the parameters according to their weights and then drawing the cumulative function: the parameter set can be cut where the slope of the function slows down and at least a given percentage of the total weight is reached. This further decomposition of the general screening problem in *ranking* and *subsequent screening* has significant advantages. First, it actually produces feature rankings consistent with the domain experts' knowledge. Second, when defining the actual restricted parameter set SP, such feature rankings can easily be combined with prior knowledge (if desired) and the size of the set SP can be chosen by the analyst depending on the desired trade-off between computational effort and accuracy. Third, the availability of a ranking rather than just the binomial partition between *relevant* and *nonrelevant* parameters allows to improve the performances of the optimization procedures employed to solve the *restricted tuning problem*. In this paper we will be concerned mostly about ranking, referring to [3] for a more detailed analysis of the issues related to going from ranking to screening.

3.2   Database Generation

The *statistical/machine learning* techniques that we propose to use in this unusual application domain are traditionally applied in a very different context. In the usual environment, data is readily available in large amounts, and the main issue is how to learn meaningful relations from datasets that are massive in size but of low/poor quality (due to the presence of missing entries, errors, and other inconveniences). Instead, our novel application of the same techniques in the context of computer simulations is quite unique, because we have to generate ourselves the data that we are going to analyze, so the quality of data will be quite high, but other, potentially serious, issues appear. Indeed, if we generate ourselves the data that will be analyzed, special care must be devoted at reducing as much as possible the *bias* that the choice of a specific generation process introduces, otherwise we might end up learning just the bias itself. For this reason, the database generation step is absolutely a crucial step. In order to deal with this issue, we propose to generate randomly the database of parameter assignments (and corresponding simulation errors), using a normal distribution centered in the initial assignment which we denote as A_0 (such assignment might have been chosen at random, but, more likely, has been derived through physical considerations). The value of the standard deviation σ of the distribution employed in the generative process must depend on the range of reasonable values for the parameters (a data that can be easily estimated a priori from the domain experts). Such value has great importance as it is the parameter that defines the trade-off between the necessity of exploring as much as possible of the solution space and the need for avoiding physically meaningless assignments. An appropriate choice should guarantee us to learn relations, which are both significant in the neighborhood of the initial solution, where the optimal solution is likely to be, and not biased by the expectations of the analysts, in no way taken into account by the process.

$$\text{pnew}_i = p_i + \text{offset}_i, p_i = (A_0)_i \tag{1}$$

$$\text{offset}_i \sim N\left(0, \sigma_i^2\right) \tag{2}$$

$$\sigma_i^2 = \frac{|R_i|}{6} = \frac{(b-a)}{6} \tag{3}$$

There are however some problems in the previously defined scheme. Let us assume in the subsequent analysis, for sake of simplicity, that the ranges of the parameters span a segment which is symmetric with respect to the values the parameters in the initial assignment. The generative process that has been defined above is bound, eventually, to sample values outside the specified ranges. More precisely, by *Chebyshev's inequality* approximately one out of ten parameters will be sampled outside the range of feasible values, making the corresponding assignment *inconsistent*. All these *outlier* parameters must then be set on the boundary in order to make all assignments *consistent* with the given ranges. The database set up will then require only to execute the simulation with each of the assignments, simultaneously evaluating the performance of the simulator. The simulation error $E(\mathbf{A})$ can then be stored, and, possibly, discretized.

4   Ranking: Logistic Regression

Logistic Regression (LR) is a classical classification method, widely used to deal with discrete class variables. Therefore, in order to apply it to our database of simulations, the first step is to discretize the error class: this can be done in various ways, for example, according to some binary partition or following a (80-15-5) scheme. Here we present the most relevant features of the *logistic regression* model, focusing mostly on how to apply it in our specific context. We do not, however, go into great detail about the model itself as it is widely discussed in the literature [9] or in [10], and implementations of this algorithm are available in most data analysis packages (such as the *Matlab Statistical Toolbox* and *Weka*). Instead, we discuss how a trained LR model allows to extract the parameters' ranking we are looking for, and we discuss pros and cons of this choice of this specific predictive model.

4.1   The Logistic Regression Model

Logistic regression is a specific kind of *generalized linear model* [11], a class of statistical models in which linear regression is applied to an arbitrary function of the response variable instead of the response variable itself. In the case of logistic regression such *link function* is called *logit* function and has the expression:

$$\text{logit}(p) = \ln\left(\frac{p}{1-p}\right) \tag{4}$$

Conceptually, given a binomial class variable (*Low_Error High_Error*), the LR model approximates linearly the logit transformed conditional probability of the output *given one of the two values*. Exponentiation of such expression allows to isolate the probability, identifying a parametric distribution, that can then be fit to the data using *maximum likelihood* or *maximum* a posteriori *estimation*. In more formal terms this is expressed:

$$\text{logit}(P_{\text{Low\_error}}) = \beta_0 + \sum_i \beta_i * p_i \tag{5}$$

$$\text{logit}(P_{\text{Low\_error}}) = \beta \vec{X} \tag{6}$$

$$\ln\left(\frac{P_{\text{Low}_{\text{Error}}}}{1 - P_{\text{Low}_{\text{Error}}}}\right) = \beta \vec{X} \tag{7}$$

$$P_{\text{Low}_{\text{Error}}} = \frac{e^{\beta \vec{X}}}{\left(1 + e^{\beta \vec{X}}\right)} \tag{8}$$

If, however, a different *nonbinomial* discretization policy has been used (as we did in our experiments) this procedure must be appropriately modified. The extension is quite simple. Assuming K possible outcomes (and a technical condition known as *independence of irrelevant alternatives*), the *multinomial logit model* can be trained by running independently $K - 1$ binomial logistic regressions, leaving out just the last outcome Y_K, which can be then derived further exploiting the basic probability theory assumption that probabilities of all possible outcomes of an experiment must sum up to one.

4.2 Ranking Extraction

Choosing *logistic regression,* in the attempt of automatically learning some of the relation between the parameters' values and the simulation's output, makes it easy to extract a measure of relevance of each parameter of the model, providing us with a parameter ranking. Indeed, the coefficients β_i of the trained model are in a one-to-one correspondence with the parameters, and measures, according to a model, how likely it is for a variation of the value of a parameter to induce a variation in the class variable (which is our discretized simulation error). This is exactly what we are looking for, as, in order to tune efficiently a highly parametric computational model, we need to restrict the parameter set to those parameters with higher impact on the global simulation error. In a different setting, using the coefficients of the regression as ranking of the relevance of the parameters might be an unreliable and misleading approach, because, if the different factors are highly correlated, multicollinearity makes measuring the predictive power of each factor of a dataset much more complex. For this reason, more complex metrics have been developed in the literature to extract from the logistic regression model reliable rankings of the parameters, among the main approaches we recall *dominance analysis, likelihood ratio,* and *Wald statistics.* However, in the context of computer simulations, and given the specific generative process employed to set up our database multicollinearity is not an issue for us. Indeed, the parameters of the model are totally independent factors, making the ranking provided by the regression coefficients valid.

4.3 Pros and Cons

The logistic regression model has been widely used in most different settings and application domains, because it is capable of achieving good classification performance with a reasonable amount of training data. Together with the availability of efficient implementation of the training algorithm (both in the most basic form described in the previous section and in more advanced forms), this makes the logistic regression model a reasonable model to start to understand how the parameters impact on the behavior of the computer simulator that requires tuning. Most shortcomings of this approach, such as the unreliability of the choice of the coefficients as measure of relevance of the single factors, are due to multicollinearity, an issue that is not present in our peculiar context, making this model even more attractive. The main problem with the use of logistic regression for parameter ranking is that its expressive power is limited, therefore, in order to manage very complex objective functions, it might be required to resort to more expressive models. This could allow to learn more complex relations in the data, offering a better comprehension of the actual relevance of each parameter. This is the main reason that leads us to consider and evaluate an alternative approach, although the performance of the optimization algorithms that can be achieved using the LR model for screening are already considerably enhanced, with respect to the case in which the optimization algorithms are applied blindly. A detailed comparison between the performances achieved with each ranking model and without any of these is presented in the following sections of this paper.

5 Ranking: Multilayer Perceptron

The *Multilayer Perceptron (MP)* is a feedforward artificial neural network model. It was the first multilayer neural model to be fully understood and trained, and it is now widely employed in pattern recognition and machine learning. It is an extremely flexible model and, with little modifications, can be adapted both to discrete and continuous class variables. Given the importance of this model, also in this case we refer to the literature for more detailed analysis [12], and we point out that implementations of the ideas presented can be found in *Weka*, in the *RSNNS* [13], and in the *Matlab Neural Networks Toolbox*. As previously, we will now focus mainly on issues that more strictly concern our use of this model.

5.1 The Multilayer Perceptron Model

The computational unit of a multilayer perceptron is the *Perceptron*, a simple model of the behavior of biological neurons proposed by [14]. As it is shown in Fig. 1, a perceptron is a mapping of its inputs into a single output variable y. The output variable is computed by applying nonlinear transformation f (also known as *activation function*) to a linear combination of the inputs and a threshold b, with each term weighted by coefficients w_i. Within this general framework many different activation functions can be chosen, among the most common we recall the *step* function, the *sigmoidal* functions (such as the *logit* and *hyperbolic tangent*

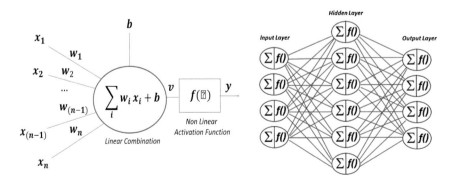

Fig. 1 A multilayer perceptron in the classical three-layered topology

functions), or the *rectifier/softplus* functions. The expressive power of a single perceptron is extremely limited as it can be used only to deal with linearly separable classification problems. However, these humble components, if combined in a more complex network, can exhibit extraordinary classification performances. When neurons are arranged in layers as in Fig. 1 (on the right), with each neuron transforming the outputs of the previous layer, we call the network a *multilayer perceptron*. For various reasons, first of all the easiness of training, this is the most common arrangement, and this is the scheme that we have employed in order to learn the relations in our database of simulations.

5.2 Neural Computation

To understand how a *multilayer perceptron* can be trained, and subsequently how we can extract a ranking of the parameters of the computational model from the trained network, we must first understand how *neural computation* works. Given $(L + 1)$ *node layers* (including the *input, output,* and *hidden* layers), and L *edge layers* (set of weighted connections between node layers), the computation of a multilayer perceptron can be described as a sequence of nonlinear transformations from vector x^0 to x^L:

$$x_0 \xrightarrow{w^1} x_1 \xrightarrow{w^2} \ldots \xrightarrow{w^L} x_L$$

Let us denote with N_j the number of nodes at layer j, with $x^j \in R^{N_j}$, for all $j = 0,$ ..., L the input of layer j, and with W^j the $N_j \times N_{j-1}$ matrix whose elements $W^j_{h,k}$ represent the weight of the edge connecting node h of layer j with node k of the previous layer $j - 1$.

The output is computed applying, for all edge layers from $j = 1$ to $j = L$, the expression:

$$x^j_h = f\left(v^j_h\right) = f\left(\sum_{k=1}^{N_{j-1}} W^j_{hk} x^{j-1}_k + b^j_h\right) \tag{9}$$

This is the paradigm of computation to be kept in mind to understand all the following.

5.3 Training a Multilayer Perceptron

The main problem in training a multilayer perceptron is how to determine the best values for the network's weights. However, this can be done efficiently using the *backpropagation* algorithm [15] on the dataset we are interested to analyze (in our case the set of previously evaluated simulations). If the database with continuous class variable is used, training proceeds according to a simple set of rules, applied at each iteration (also called *epoch*) to all instances in the dataset. First, the difference between the expected output ex and the actual output x^L must be computed. Then, the error is propagated from output to input layer, updating the values of the edge weights and the thresholds. During the process of training the network, the rate of convergence depends on the choice of a *hyperparameter* denoted as η (and called *learning rate*), which influences the entity of the updates done as consequence of prediction errors.

$$\delta_h^L = f'(v_h^L)(\mathrm{ex}_h - x_h^L) \tag{10}$$

$$\delta_h^{j-1} = f'\left(v_h^{j-1}\right) \sum_{h=1}^{n_j} \delta_h^j W_{hk}^j, \text{ for } j = L \text{ to } 1 \tag{11}$$

$$\Delta b_h^j = \eta \delta_h^j, \text{ for } j = 1 \text{ to } L \tag{12}$$

$$\Delta W_{hk}^j = \eta \delta_h^j x_k^{j-1}, \text{ for } j = 1 \text{ to } L \tag{13}$$

This algorithm is the most widely used for training these kinds of models: although convergence is quite slow, it can be made more efficient resorting to batching and multithreading. The algorithms being a form of gradient descent, do not guarantee convergence and the solution is likely a local optimum, it is however enough for our aims.

5.4   Ranking

Once we have a trained multilayer perceptron model we must be able to extract from it the ranking of parameters according to their impact on the output variable. In the case of a neural network, however, this is much more complicated than it was for the logistic regression model. In the literature different approaches are presented, each with its own pros and cons. We present a heuristic simple approach, easily applicable to MPs with any number of hidden layers. Maintaining the previously defined notations, and denoted as **R** the vector containing the parameters' ranks, we propose the following:

$$(R)_i: = \mathrm{rank}(p_i) \tag{14}$$

$$R = W^L W^{L-1} \ldots W^1 = \prod_{j=L \text{ to } 1} W^j \tag{15}$$

In the simplified case of a network made of linear perceptrons, using this policy would be equivalent to rank each parameter according to the contribution it gives a unitary value of that parameter to the value of the outcome variable. When applied to networks of nonlinear perceptrons, the metric has just a heuristic value, yet it has proved itself very effective in our experiments on flight simulations, yielding to even better results than logistic regression. The extension to N outcome variables or to a discrete outcome having N possible values is trivial (R becomes a matrix with obvious meaning).

5.5   *Comparison: Pros and Cons*

The multilayer perceptron's main strength is its expressive power: indeed, it has been proved that a multilayer perceptron with a single hidden layer is a *universal approximator* [16], thus any function can be approximated with arbitrary precision if the weights of edges are properly chosen. Furthermore, the MP can be trained on the original simulation error values, and does not require discretization as logistic regression does, although it is still possible to train the network on the discretized dataset. The greatest disadvantage of using a multilayer perceptron, when operating in our computer-simulated environment, is that training the model usually requires a larger amount of data, if compared to logistic regression. This can be a problem because in our application domain we are responsible of generating ourselves all data to be analyzed, and computer simulations can be computationally expensive. This constraint becomes even more critical because it is not always possible to speed up computation just using additional computational resources because most flight simulators operate in real time, and indeed, this was the case in our case study for validation. If a single function evaluation (i.e., a single computation of the simulation error for a given set of parameters) takes very long we advise to try logistic regression first, and resort to the multilayer perceptron only if needed, if not satisfied with the quality of the results. A simple check with a domain's expert can validate the resulting ranking by verifying that the parameters which are already known to have a significant impact are actually ranked high by the procedure.

6   Optimization

We have now reached the final step: the solution of the *(restricted) tuning problem* through automatic optimization procedures managing only the parameters in the previously defined subset SP. We present first a basic *stochastic hill climbing* procedure (searching for optimal values of all parameters in SP), then we describe two successive refinements of the algorithm which have proved themselves effective in a real tuning case study. The first refinement introduces *adaptive variance*, and corresponds to a *1 + 1 evolutionary strategy*. The third algorithms, which is by large the most efficient, is also capable of exploiting the *ranking* of parameters in SP in order to tune an increasing number of parameters at each iteration.

6.1 Local Stochastic Optimization

All proposed approaches are local and stochastic optimization strategies; this is no accident, and the choice is appropriate for different reasons as argued in the following:

Do We Have Any Choice? Often we do not; global optimization is more computationally demanding, therefore, in M&S, it might simply be impossible. The time for each function evaluation can be very long; executing different simulations in parallel can be unfeasible if all computational power available is required for executing a single computer simulation; finally, sometimes, e.g., in real-time simulations, no speedup of the single simulation is possible even if more computational power is available.

Sensible Initialization. If modeling is carried out in a sensible way, the initial assignment A_0 is not a completely random guess but a reasonable solution derived by physical considerations, and is thus (hopefully) near to the true optimal solution making local optimization less at risk of remaining stuck in non-global optima.

Prior Knowledge. When you recreate in a simulator some known observed reality (which is typical of flight or other life-critical simulations) you can recognize if you are stuck in a local optima because you know which the optimal value is (although you do not know *where* it is located within the very large solution space). Furthermore, we are not really interested in achieving the global optima, as much as we are in lowering the error below the given threshold, if we manage to do so the rest is irrelevant.

Black-box Optimization: Concerning the choice of stochastic optimization instead of deterministic procedures, there are two elements to be considered. First, stochastic procedures are less prone to getting stuck in local optima so that introducing some degree of randomness in an algorithm can be useful in general. Second, most stochastic algorithms need to know very little about the function (no need to compute derivatives or similar). Having devoted so much time to develop screening techniques capable of treating the model just as a black box, we do not want to start analyzing the equations now that efficient black-box optimization techniques are readily available.

6.2 Basic Stochastic Hill Climbing

The basic hill climbing procedure we now present in *pseudo-code* shall be thought of as a template to be refined in the following. Therefore, it is quite impressing that, as we will see, even in this basic form the procedure accomplishes a reasonably low error, showing the power of *screening* in making automatic tuning possible. The algorithm works by exploring the neighborhood of the best solution up to a given moment until a better one is found, then it moves to the new location and continues:

```
01 best = A₀
02 A = A₀
03 min-err = execute(A₀)
04 R = loadRange()
05 σ = computeStdDev(R)
06
07 while (error>threshold)
08
09    for (j=1 to |P|)
10        if(pⱼ in SP)
11            Aⱼ = bestⱼ,+ σⱼ × N(0,1)
12            Aⱼ = setWithinRange(Aⱼ,Rⱼ)
13        end-if
14    end-for
15
16    if(execute(A) < min-err)
17        min-err = execute(A)
18        best = A
19    end-if
20
21 end-while
```

best contains the best assignment found up to the current iteration, *min-err* stores the corresponding error. Exploration of the neighborhood of *best* is random: new assignments of each parameter in SP are generated at each iteration sampling from a normal distribution centered in. The function computeStdDev(R) defines once for all the value of the standard deviation used for each parameter, which is proportional to the range. The same precautions seen for the database generation, to guarantee the consistency of all assignments, are taken into account also in this circumstance when defining the relation between range and standard deviation (see Sect. 3.2). Function execute(A) runs the simulation with parameters' values specified by $\mathbf{A}$ and computes the simulation error $E(\mathbf{A})$. Note that, as required, all parameters in P/SP maintain their initial default, and that as the number of iterations grows to infinity the *best* parameter assignment eventually converges to the global optimum with probability 1.

6.3 Adaptive Variance

In the hill climbing procedure of the previous section, random mutations occur in every iteration in order to explore the neighborhood of the best solution found; the parameter controlling such mutations is the standard deviation of the normal distribution being used for sampling. Until this moment, the value of the standard deviation can be computed *statically*, depending only on the range of the different parameters considered. However, the optimal value of σ is not the same throughout the execution of the optimization procedure. Indeed, it is intuitively clear that the best value for σ should be greater when the error is high and we are exploring the solution space with a high success rate. Instead, it should become increasingly smaller when we are very near to the optimum and therefore very small corrections are needed in order to improve our solution (as any big variation will bring us far from it instead of helping us reaching it). This observation is extremely important and calls for a crucial important modification of the procedure described in Sect. 7.2, in order to allow parameter σ to *adapt online* during the execution of the optimization algorithm, maintaining an optimal range of exploration throughout the iterations. There are two main approaches to provide the previous procedure with the required capability. The most simple solution is to make the standard deviation decrease along with the simulation error; a more advanced solution will instead make it increase or decrease dynamically depending on the success rate (the fraction of mutations that are successful, i.e., that improve the best solution). The problem with the former approach still requires to define at *design time* the exact relation between error and standard deviation, while the latter is usually more powerful because it provides greater flexibility. A common policy when the second approach is chosen is to decrease σ when the *success rate* is below 0.2 and increase it otherwise (this strategy is known as the *1/5 success rule*; it can be proved optimal for several functional landscapes and it is widely recognized to give good results in practice in a wide range of circumstances).

```
18 if(execute(A) < min-err)
19      min-err = execute(A)
20      best = A
21      σ_J = σ_J × α
22 else
23      σ_J = σ_J
24 end-if
```

Among the many implementations of such rule, the simplest one [17] accumulates the knowledge about success and failure directly in the value of σ, substituting the *if-else* clause of lines 16–19 with the previous code. Reasonable values of parameter α fall between one and two. This implementation of the algorithm has the further advantage that it requires to set only one parameter, instead of the three hyperparameters (change rate, averaging time to measure success rate, update frequency) which are needed with classical implementations (following more narrowly the previous definition of the optimization procedure).

6.4 *Sequential Masking*

Although the hill climbing procedure, modified in order to adapt online the parameter controlling the entity of mutations, can already be used in practice to solve the restricted tuning problem, performance can be further improved by the third algorithm, sequential masking, which additionally exploits the ranking of parameters in SP during the optimization step. This third approach to automatic tuning works as follows:

1. A sequence of subsets, and thus of restricted tuning problems, SP1, SP2, …, SP is generated incrementally from the ranking of parameters in SP.
2. The hill climbing procedure with adaptive variance is executed for a fixed number of iterations for each subproblem, starting from the smallest (simplest) problem and using the result of each problem as initial guess of the subsequent problem.

The underlying idea is that challenges of increasing complexity are faced starting from increasingly good initial assignments. The increasing complexity is due to the fact that an increasing portion of the solution space is reachable. The overall behavior of the optimization procedure is shown in Fig. 2: starting from assignment A_0, we compute a sequence of solutions converging to the final solution of *the restricted tuning problem*.

The execution of the algorithm is controlled by two user-defined parameters: the number of restricted tuning problems that the algorithm must solve, which we can call the *granularity* of the algorithm, and the maximum number of iterations for each phase of the procedure (which must ensure to switch before starting to fit the subset SP_i at the expenses of the overall problem). The max number of iteration can be chosen to be different for each of the problems (higher for those considering more parameters).

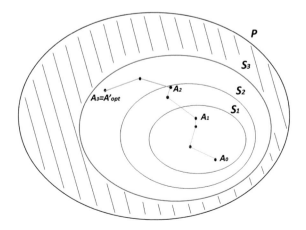

Fig. 2 Sequential masking with four subproblems

7 Case Study

Our case study for validation is the tuning of an industrial-level flight simulator of *TXT e-solutions* in order to accurately simulate the takeoff of a helicopter with one engine inoperative (breaking down during the execution of the procedure). This case study is a classic example of simulator with severe accuracy requirements, as flight simulators must be certificated by flight authorities of different countries, verifying that the execution of a certain set of flight procedures is adherent to the observed behavior of the aircraft. There are two main procedures for takeoff: *Clear area TO* and *Vertical TO*. We here analyze only the clear area procedure, showing the results of the screening and optimization algorithms (however, the techniques have been applied to both). In the following P is a set of 47 parameters, each associated to a range symmetric with respect to the initial assignment $\mathbf{A_0}$. The most relevant performance metrics in $\mathbf{S}$ are three: CTO distance (distance from starting point to takeoff-decision-point), GP1 (average climb in 100 feet of horizontal motion during the first phase), and GP2 (average climb in 100 feet of horizontal motion during the second phase). Expected values $\mathbf{Ref}$ of such metrics are specified on the helicopter's official manual. The global error $E(\mathbf{A})$ is the sum of the squared relative errors with respect to the different performance metrics.

7.1 Screening

A database of 1200 simulations has been generated with the rules established in Sect. 4.1, appropriately discretizing the values of the global simulation error. Then both the logistic regression and the neural network-based approach to parameter ranking are applied, using the implementations available in the open source machine-learning package Weka (and choosing the classic three-layered topology for the MP). Results were largely consistent: if the set of 10–20 most relevant parameters for LR is compared to the corresponding set within the MP ranking, about 85 % of the parameters figure in both sets. The main difference is a single small group of parameters, having similar physical meaning, such that all parameters are ranked very low by LR while MP seems to be able to discriminate more effectively between relevant and not relevant parameters. If the cumulative functions of the two approaches are compared this difference among the results is reflected in a steeper curve for the LR-based ranking, which concentrates most of the weight on fewer parameters (trend confirmed by other flight procedures). The resulting set SP of the most relevant parameters is thus smaller for logistic regression than it is for the multilayer perceptron.

7.2   Evaluating Screening

In the next section, we will go into details in the comparison of the performances of *adaptive hill climbing* and *sequential masking*, but, before, we use the basic optimization procedure in order to provide the reader with an intuitive proof of the impact of the previous screening techniques. In Fig. 3 three executions of the basic HC procedure are compared; the horizontal axis identifies the number of simulations executed, while the vertical axis measures the minimum error up to the given iteration. In all four cases the algorithm has gone through 100 iterations: the green and red lines represent the execution of the algorithm considering the results of screening (through LR and MP, respectively); the blue line, that instead converges to a relatively high simulation error and then stops improving, represents the execution of the procedure with SP = P (i.e., the procedure if applied directly to the full tuning problem with no screening).

An even more impressing result, in order to show that the screening procedures have been capable of effectively discriminating between relevant and irrelevant parameters, is that if we try to execute the same algorithm with the complement set of SP no improvement at all is achieved showing that no relevant parameter has been left behind by mistake. Overall these results show that LR and MP-based rankings truly identify the relevant parameters.

7.3   Evaluating Optimization

We now show the results of applying the two proposed algorithms to our case study. Both converge to an error of many orders of magnitude lower than the one

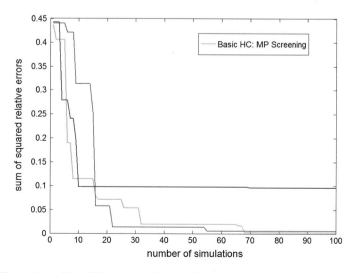

Fig. 3 Comparison of the different screening procedures

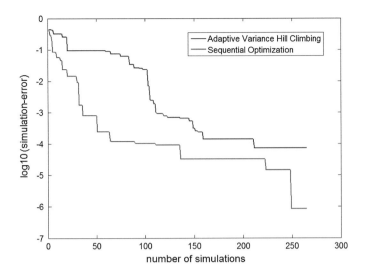

Fig. 4 Comparison of the optimization procedures

achieved by the basic procedure, and they also do so significantly faster. In order to make the results understandable the vertical axis is now the (base 10) logarithm of the error. The comparison of the two algorithms in Fig. 4 shows that they both are able to keep the error decreasing at a good, almost constant, average rate. Sequential masking achieves particularly impressive performances, however both algorithms proved themselves effective, significantly improving the basic hill climbing procedure, in Fig. 3, which instead, whether screening is used or not, hits sooner or later a *wall,* becoming unable of further improvements, or capable of doing so only at an unacceptable slow rate.

7.4 Benchmarks and Conclusions

In conclusion our approach has been proved effective in tuning a complex flight simulation model finding the optimal values of 50 parameters of the model. The entire process requires less than 2 days of machine time on a single desktop computer (with just a few hours actually dedicated to finding those values, and most of the time devoted to generating the database for parameter ranking). The main benchmark against which this result must be compared is manual tuning, which is still the state of the art in industrial applications. *TXT e-solutions*'s experienced engineers would require from 10 to 20 days to accomplish the same result. Concerning previous attempts to automatic tuning, there is little work done for the tuning of industrial-level computer simulators, and to the best of our knowledge, none in the area of flight simulations. The best related work is in medical context

[18]: this paper presents an evolutionary strategy for tuning, but the approach is used only for lower dimensional problem with just 15 parameters. Thanks to an integrated approach combining screening and optimization (tightly coupled especially in the *sequential masking* algorithm), our methodology allows to significantly expand the range of application of automatic techniques for parameter tuning. When comparing to other attempt of automatic tuning it is important to notice that combining screening and optimization is not only crucial in order to achieve fast convergence to a really low simulation error, but it is also crucial in order to avoid an issue we have anticipated in previous sections of the paper: the introduction of peculiar *side effects* invisible to the objective function but that can make simulations unrealistic to a human eye (such as odd small oscillations and vibrations difficult to control). The reason for such side effects is that parameters that have a low impact on the performance metrics and thus on the global simulation error are free to deviate randomly from their default values, because there is no selective *pressure* capable of limiting their erratic wandering. Our methodology solves the issue by restricting tuning to the set of parameters with direct impact on the performance metrics, so, the optimization step is *adapted* to the objective function, and therefore all nonfixed parameters are directed toward their optimal values instead of being free to roam around. The proposed methodology is therefore the first real alternative to manual tuning, allowing to speed up the process while preserving high-quality results.

References

1. Morgan, P.J., Cleave-Hogg, D., Desousa Lam, S., McCulloch, J.: Applying theory to practice in undergraduate education using high fidelity simulation. Med. Teach. **28**(1), e10–e15 (2006)
2. Lewis, J.H., Jiang, S.B.: A theoretical model for respiratory motion artifacts in free-breathing CT scans. Phys. Med. Biol. **54**(3), 745–755 (2009)
3. Kern, S., Muller, S.D., Hansen, N., Büche, D., Ocenasek, J., Koumoutsakos, P.: Learning probability distributions in continuous evolutionary strategies—a comparative review. J. Nat. Comput. **3**(1), 77–112 (2004)
4. Last, M., Luta, G., Orso, A., Porter, A., Young, S.: Pooled ANOVA. Comput. Stat. Data Anal. **52**(12), 5215–5228 (2008)
5. Fisher, R.A.: The Design of Experiments, xi 251 pp. Oliver & Boyd, Oxford, England. (1935)
6. Harrel, F.: Regression Modeling Strategies. Springer-Verlag (2001)
7. Bishop, C.: Pattern Recognition and Machine Learning, pp. 217–218. Springer Science +Business Media, LLC (2006)
8. Bettonvil, B., Kleijnen, J.P.C.: Searching for important factors in simulation models with many factors: sequential bifurcation. Eur. J. Oper. Res. **96**(1), 180–194 (1997)
9. Nelder, J., Wedderburn, R.: Generalized linear models. J Royal Statis Soc Series A (General) **135**(3), 370–384 (1972)
10. Haykin, S.: Neural Networks: A Comprehensive Foundation, 2 edn. Prentice Hall. ISBN 0-13273350-1 (1998)
11. Bergmeir, C., Benìtez, J.M.: Neural networks in R using the stuttgart neural network simulator: RSNNS. J. Statis. Softw. 46(7) (2012)

12. Rosenblatt, F.: The perceptron: a probabilistic model for information storage and organization in the brain. Psychol. Rev. **65**, 386–408 (1958)
13. Cybenko, G.: Approximations by superpositions of sigmoidal functions. Math. Control Signals Syst. **2**(4), 303–314 (1989)
14. Rumelhart, D.E., Hinton, G.E., Williams, R.J.: Learning representations by back-propagating
15. Hessel, M., Borgatelli, F., Ortalli, F.: Machine learning for parameter screening in computer simulations, p. 8906. Springer, Lecture Notes in Computer Science series
16. Errors. Nature 323 (6088): 533–536. doi:10.1038/323533a0, 1986
17. Vidal, F.P., Villard, P., Lutton, E.: Automatic tuning of respiratory model for patient-based simulation. MIBISOC'13—International Conference on Medical Imaging using Bio-inspired and Soft Computing (2013)
18. Zhang, A.: One-factor-at-a-time screening designs for computer experiments. SAE Technical Paper 2007-01-1660. doi:10.4271/2007-01-1660 (2007)

Enhanced Interior Gateway Routing Protocol with IPv4 and IPv6 Support for OMNeT++

Vladimír Veselý, Vít Rek and Ondřej Ryšavý

Abstract Enhanced Interior Gateway Routing Protocol seized to be Cisco proprietary since its release in form of IETF's informational draft. EIGRP has history of hybrid routing protocol widely deployed by Cisco's customers due to its performance and advanced features. This paper introduces free-available simulation module that complies with RFC specification and offers platform for EIGRP's further testing and evaluation within OMNeT++ simulator.

Keywords EIGRP · Routing protocol · Hybrid distance vector · Simulation module · OMNeT++

1 Introduction

The Automated Network Simulation and Analysis project (ANSA) running at our university is dedicated to develop the variety of software tools that can create simulation models based on real networks and subsequently allow for formal analysis and verification of target network configurations. It might be used by public as the routing/switching baseline for further research initiatives using simulator for verification. This paper extends our previous work involving EIGRP [1] by updated version of our simulation module, which is a part of the ANSA project and which extends functionality of the INET framework [2] in OMNeT++ [3].

Network layer serves the purpose of end-to-end data delivery. Routers employ routing tables to make correct routing decisions in order to pass packet closer to

V. Veselý (✉) · V. Rek · O. Ryšavý
Brno University of Technology, Brno, Czech Republic
e-mail: ivesely@fit.vutbr.cz

V. Rek
e-mail: xrekvi00@stud.fit.vutbr.cz

O. Ryšavý
e-mail: rysavy@fit.vutbr.cz

© Springer International Publishing Switzerland 2015
M.S. Obaidat et al. (eds.), *Simulation and Modeling Methodologies,
Technologies and Applications*, Advances in Intelligent Systems
and Computing 402, DOI 10.1007/978-3-319-26470-7_4

65

receiver. Dynamic routing protocols maintain up-to-date content of routing tables by exchanging updates about known networks.

Routing protocols for traditional wired networks could be divided into three categories: (a) **distance-vector** where routing is based on information provided by neighbors and each route has one attribute representing distance of network from a given router; (b) **path-vector** which is the same as distance vector but routes have more than one attribute; and (c) **link-state** where every router maintains independent view on topology and computes the shortest path tree toward all other nodes. Additional typology of routing protocol is according to type of deployment: (a) **interior gateway protocols** (**IGP**) for routing within one administrative domain; (b) **exterior gateway protocols** (**EGP**) for routing between autonomous systems (AS). Among typical represents belong:

- Routing Information Protocol (RIPv2 for IPv4 [4], RIPng for IPv6 [5])— Distance-vector routing protocol that works with hop-count as the metric. Routes with metric 16 or more are considered unreachable;
- Babel [6]—Babel is distance-vector protocol specialized (but not exclusively) for wireless networks that has different metric criteria than wired networks. Metric may represent cost, number of host, or any other implementation dependent route attribute. Nevertheless, routes with infinity metric 0xFFFF are considered unreachable. Babel currently supports both IPv4 and IPv6;
- Intermediate System to Intermediate System (IS-IS) [7]—The first link-state protocol ever, which is also capable of working with different metrics simultaneously. IS-IS was originally intended to be used with connectionless mode network service protocol (concurrent of IP) for ISO/OSI networks, however, later was developed implementation for both IPv4 and IPv6 protocols. IS-IS is by design agnostic to used address-family and single instance can carry routing updates for various network protocols. Formerly used IS-IS metrics were delay and link erroneous, current revision employs only speed of the link;
- Open Shortest Path First (OSPFv2 for IPv4 [8], OSPFv3 for IPv4/6 [9])—OSPF started as IP alternative to IS-IS and later become industrial standard link-state routing protocol that has wide-spread deployment. OSPF uses cost as the metric, where cost is derived from the interface bandwidth. OSPF supports only IP routing updates;
- Border Gateway Protocol (BGPv4) [10]—Extends distance-vector idea by having multiple attributes accompanying the single prefix update. BGPv4 is currently the only one EGP that is being used and it is often referred as policy-control routing protocol.

Enhanced Interior Gateway Protocol (EIGRP) is the backward compatible successor of previous Cisco proprietary Interior Gateway Protocol (IGRP). It is categorized as a hybrid routing protocol which means that it is a crossover between distance-vector (topology is known based on announcement from neighbors) and link-state protocols (instead of periodic updates, topology changes are propagated immediately). Down below follows the list of main beneficial features of EIGRP:

- EIGRP employs **Diffusing Update Algorithm (DUAL)** [11] that effectively propagates any topology change and minimizes path recomputational time;
- Currently, EIGRP is the only routing protocol that guarantees loop-free topology even during the time when topology is actively converging toward a new routing state;
- EIGRP leverages its own reliable transport protocol (even for multicast data transfer);
- In the contrary to other distance-vector protocols, EIGRP is capable of sending event-driven partial bounded updates;
- It has neighbor discovery and recovery mechanism to determine route reachability via particular adjacent node;
- EIGRP contains protocol-dependent modules that allow operation over different network protocols (including IPv4 and IPv6).

The EIGRP was introduced in 1993 as a result of joint effort of Cisco and SRI International [12]. Initial and later measurements revealed that it outperforms other routing protocols (i.e., speed of convergence, network bandwidth utilization, queuing delay) [13]. Despite its beneficial aspects (or maybe because of them) it had been protected as one of the major Cisco intellectual properties by a bunch of patents for nearly 20 years. In the beginning of 2013, basic EIGRP design and functionality were submitted as a publicly available IETF informational draft [14].

This paper has the following structure. The next section covers a quick overview of existing EIGRP implementations (either real or simulation ones). Section 3 deals with our contribution, mainly necessary theory, proposed design, and subsequent implementation. Section 4 presents validation scenarios proving correctness of the implementation. The paper is summarized in Sect. 5 together with unveiling our future plans.

2 State of the Art

Currently none of vendors other than Cisco supports EIGRP in its active network devices. Despite positive campaign targeting wider EIGRP acceptance, many manufacturers and customers remain skeptical and rely on a long-time proven open solutions like OSPF or IS-IS. One of the first publicly available open-source EIGRP routing demon is being developed at the University of Žilina [15] within the scope of Quagga project [16].

A freely available demonstration tool called Easy-EIGRP [17] exists rather for educational purposes.

OPNET simulator has contained EIGRP simulation modules even before its public IETF release. However, its functionality is limited and it lacks IPv6 support for EIGRP. Nevertheless, OPNET and its simulation models were used to conduct several measurement studies comparing different routing protocols including EIGRP [18].

Previously described state of EIGRP deployment affirmed our decision to offer academic and enterprise community with a full-fledged EIGRP implementation with all usually employed features.

The current status of unicast routing support in OMNeT 4.5 and INET 2.3 framework is according to our best knowledge as follows: (a) the IPv4 (named networkLayer) and IPv6 (pragmatically called networkLayer6) layers are already parts of INET framework; (b) the framework contains OSPFv2 as the only available IGP routing protocol.

During ANSA project development we have extended original simple router module to be dual-stack capable and enhanced it with a variety of dynamic routing protocols (RIP, RIPng, IS-IS, OSPFv3, PIM), thus creating **ANSARouter** as the compound simulation module based on the standard behavior of Cisco routers.

The basic goal behind our effort is to support EIGRP dynamic routing protocol. Hence, we have decided to add missing functionality in form of simulation module directly connected to networkLayer and networkLayer6 as depicted in Fig. 1.

OMNeT++ state-of-the-art prior to this paper is the result of ongoing research covered in our other articles.

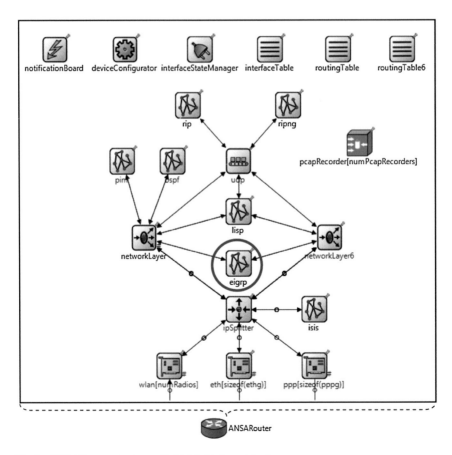

Fig. 1 ANSARouter structure with highlighted contribution

3 Contribution

We have implemented OMNeT++ compound simulation module supporting EIGRP behavior and functionality. This section provides a short theoretical background, overview of design, and some implementation notes.

3.1 Theory of Operation

An EIGRP process computes a successor for every destination. A **successor** represents the next-hop router where the route to the destination via successor is loop-less and with the shortest distance. **Feasible successor** (FS) or so-called backup next-hop is the router that provides loop-less route but with higher distance. To determine whether particular router is a feasible successor, the router is working with two parameters—a feasible and a reported distance. **Feasible distance** (FD) is the best known distance from a destination network to a given EIGRP router (historical minimum). **Reported distance** (RD) is distance from destination network advertised by a given EIGRP router neighbor. The router is using FD and RD to decide whether the feasible condition is satisfied or not. **Feasible condition** (FC) assumes that any route with $RD < FD$ is without any doubts loop-less. The **passive state** is the state of the destination network when the successor is known and the route is converged and usable. **Active state** is in contrast to the previous definition when the destination network does neither have a successor nor FS and the router is actively searching and computing a new successor.

The EIGRP employs composite metric which takes into account multiple route attributes. The basic composite metric consists of following four parameters: (a) bandwidth (abbr. Bw is minimal bandwidth enroute); (b) delay (abbr. Dl is accumulative sum of route delays), (c) load (abbr. Lo is maximal traffic load in range from 1 to 255 on the links toward destination where lower is considered better), (d) reliability (abbr. Re is minimal reliability in range from 1 to 255 on the links towards destination where higher is considered better). Parameters (a) and (b) are static, parameters (c) and (d) are dynamically recomputed every 5 min on certain EIGRP versions. Parameters are accompanied with K-values called weights which are unsigned byte long values, where $K_4 = K_5$. Usually, Cisco routers are using default composite formula (1) for metric computation without dynamic parameters:

$$K_1 \cdot \text{Bw} + K_3 \cdot \text{Dl} \qquad (1)$$

Complete composite formula (2) including all parameters looks like this:

$$\left(K_1 \cdot \text{Bw} + \frac{K_2 \cdot \text{Bw}}{256 - \text{Lo}} + K_3 \cdot \text{Dl} \right) \cdot \frac{K_5}{\text{Re} + K_4} \qquad (2)$$

The new revision of EIGRP establishes two new parameters: (a) jitter (abbr. Ji is accumulative delay variation enroute measured in microseconds where lower is preferred); (b) energy (abbr. En is accumulative energy consumption in watts per transferred kilobit where lower is preferred). Both parameters are accompanied with K_6 weight. A new wide metric is 64 bit long in opposite to older 32 bit long standard metric and it also solves problem of standard metric when taking into account delay on links faster than 1 Gbps. Wide metric composite formula (3) is then:

$$\left(K_1 \cdot \mathrm{Bw} + \frac{K_2 \cdot \mathrm{Bw}}{256 - Lo} + K_3 \cdot \mathrm{Dl} + K_6 \cdot (\mathrm{En} + \mathrm{Ji})\right) \cdot \frac{K_5}{\mathrm{Re} + K_4} \qquad (3)$$

When employing multicast for communication on local segment, EIGRP has either reserved address 224.0.0.10 for IPv4 or FF02::A for IPv6. EIGRP routers exchange following messages during operation:

- *EIGRP Hello*—Detects EIGRP neighbors with their settings (*K*-values, autonomous system number, timers, and authentication) and checks their aliveness. Sent periodically every 5 s by default. Hold timer (period after which neighbor is considered dead) is 3× longer, and by default it is 15 s. Neighbor announces its own hello and hold intervals which will obey during its operation;
- *EIGRP Update*—Carries routing information that might cause receivers to start DUAL. Sent either as unicast or multicast;
- *EIGRP Ack*—Used for acknowledging *EIGRP Update*, *Query* and *Reply* messages. It is reused *EIGRP Hello* message with empty structure;
- *EIGRP Query*—If network transits to active state and router starts to search for a new successor then router starts DUAL and sends *EIGRP Queries* to neighbors usually as multicast;
- *EIGRP Reply*—This message contains the routing answer to previous *EIGRP Query*.

DUAL functionality could be described in form of finite-state machine that reacts on events and messages and transits from one state to another with accompanied response action (change of distance D). Basic version (shown in Fig. 2) consists of 5 states and 18 transitions depending on the fact whether router is successor (S) or not (¬ S) and whether feasible condition is satisfied or not (more details in [14]).

3.2 Design and Implementation

The EIGRP implementation works with three tables:

- **Neighbor Table (NT)**—Stores information (e.g., IP address, router-id, uptime, hold-time, query count, etc.) relevant to all adjacent EIGRP routers;

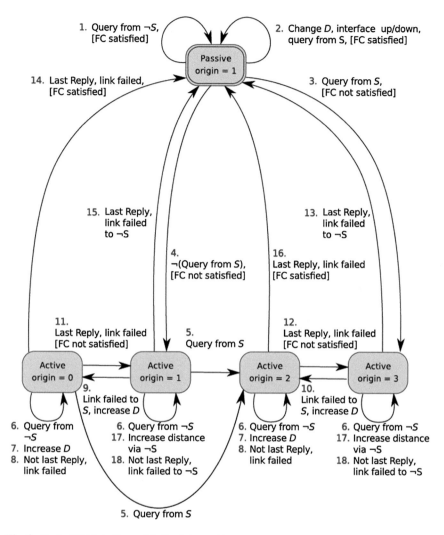

Fig. 2 Basic DUAL in form of finite-state machine

- **Topology Table (TT)**—The main routing information base from point of view of a given router. It contains each known network and relevant routes, their states, and next-hop addresses together with their FD and RD;
- **Routing Table (RT)**—A routing table is the gathering place of best routes from different routing sources, thus, the best EIGRP routes are installed here from TT.

The compound EIGRP simulation module is divided into components depicted in Fig. 3 and their brief description is in Table 1.

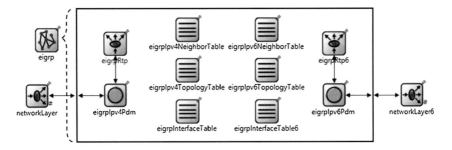

Fig. 3 EIGRP simulation module structure

Table 1 Description of EIGRP submodules

Name	Description
eigrpIpv*Pdm	The protocol-dependent module (PDMs) sends and receives EIGRP messages that contain routing information. It mediates control exchange between routing table and topology table. It leverages different network protocols as carriers for EIGRP messages. Currently IPv4 and IPv6 is supported
eigrpRtp*	EIGRP uses Cisco Reliable Transport Protocol (RTP) to ensure reliable transfer of EIGRP messages. It uses sequence number and positive acknowledgement scheme to detect any gaps in transfers. There is separate RTP module for each PDM
eigrpIpv* NeighborTable	This module is model of neighbor table. It maintains state of all EIGRP adjacencies (i.e., neighbor address, state, hold timer, RTP sequence number)
eigrpIpv* TopologyTable	EIGRP routing information base, which includes all learned routes, their state (either active or passive), FDs and computed successors (in form of their addresses)
eigrp InterfaceTable*	Simulation module keeps settings relevant to any interface on which EIGRP is enabled (i.e., separate hello and hold timers, query count)

4 Testing

In this section, we provide information on testing and validation of our implementation.

Only two scenarios are described here really thoroughly because of limited space. Nevertheless a rich set of test scenarios (including the one verifying our newly implemented stub feature) is accompanied with the published source codes.

We compared results with the behavior of the referential EIGRP implementation running at Cisco routers. For this reason, we built exactly the same topology and observed (using Switched Port Analyzer and Wireshark) relevant message exchanges between real devices (Cisco 7204 as routers with c7200-adventerprisek9-mz.152-4. M5 IOS implementing EIGRP rev. 10 with EIGRP TLV 2.0 and host stations with Windows 7 OS).

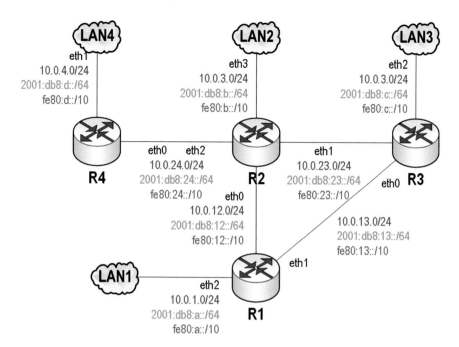

Fig. 4 EIGRP testing topology

Testing topology (see above Fig. 4) consists of four EIGRPRouters (marked R1, R2, R3, and R4) and four ANSA_DualStackHosts (LAN1, LAN2, LAN3, and LAN4) which substitutes whole separate LAN segment with dedicated IP networks. EIGRPRouter is ANSARouter equipped with only above described eigrp routing module. Router interfaces are marked "eth*." Each one of them has single IPv4 address, one IPv6 global unicast address, and IPv6 link-local address.

In the first scenario, we would like to show how metric changes are being propagated. In the second scenario, we focus on topology changes. IPv4 and IPv6 routing events were recorded and evaluated for both scenarios. Following thorough description contains EIGRP events for both IPv4 and IPv6 routes.

4.1 Scenario 1: Metric Change

A typical EIGRP message exchange of freshly booted router consist of following phases (hereafter numbered with #X):

(#1) Routers establish neighborship by sending and receiving *EIGRP Hello* messages. Whenever a new neighbor is discovered, all relevant information is recorded and stored in NT. EigrpInterfaceTable with its settings is in Fig. 6. We can observe eigrpIpv*NeighborTable content on router R2 prior to Scenario 1 events (few seconds before the metric change) in

Fig. 5 R2's IPv4 and IPv6 neighbor tables prior to scenario 1 events

Fig. 5. Please notice that neighborship is bound to link-local addresses in case of IPv6 adjacencies;

(#2) Whenever neighborship is established, routers exchange *EIGRP Updates* containing routing information to build their TTs and determine best routes toward known destinations. Reception and processing of any update is confirmed by *EIGRP Ack*. Figure 7 shows converged state of the topology from the router R2's `eigrpIpv*TopologyTable` point of view. Routes have known FD, successors, and are in passive states. Please notice that addresses of successors for IPv6 are link-local ones, same applies also for next-hops in IPv6 RT.

We scheduled bandwidth alternation R3's eth2 interface facing LAN3 changes its Dl attribute from 1 to 100 ms in order to show how the change of metric influences topology (for instance content or R2's RT is depicted in Fig. 8). In simulator, we uses `scenarioManager` to accomplish this goal, in case of real network, we change interface configuration.

(#3) R3 initiates DUAL, which discovers that LAN3 is only reachable via eth2 and propagates metric change to its neighbors R2 and R1 by sending *EIGRP Update* for LAN3's network (either 10.0.3.0/24 or 2001:db8:c::/64);

(#4) R2 acknowledges update with *EIGRP Ack*. R2's DUAL is unable to find FS, hence route transits to active state and router sends ordinary *EIGRP Query* to R1 and R4 and poison reverse *EIGRP Query* with maximal metric toward R3. Same previous steps apply also for R1 where situation is similar—acknowledgment toward R3, DUAL marks network as active, query to R2 and poison reverse query to R3;

(#5) R1 receives *EIGRP Query* from R2 and it acknowledges it with *EIGRP Ack*. Following next, R1 responds with *EIGRP Reply* with a new metric via successor R3. Same situation repeats on R2 when replying to R1 query;

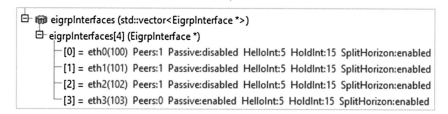

Fig. 6 R2's interface table settings

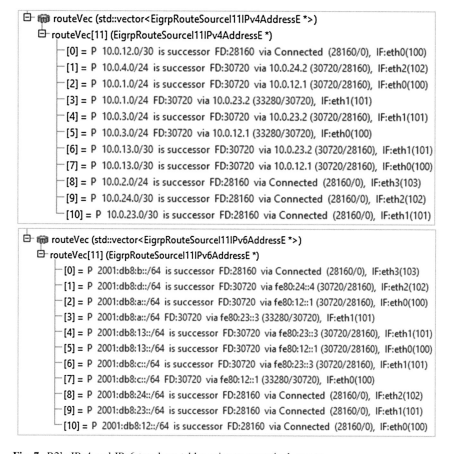

Fig. 7 R2's IPv4 and IPv6 topology tables prior to scenario 1 events

(#6) R3 receives queries from R1 and R2 and it acknowledges them. Following
      next, R3 finds out FS (itself) and responds with *EIGRP Replies* to R2 and R1;
(#7) R4 receives *EIGRP Query* from R2 and confirms it with *EIGRP Ack*. DUAL
      is unable to determine FS, thus route transits to active state. Because of

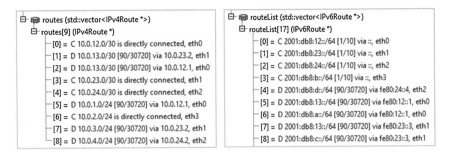

Fig. 8 R2's IPv4 and IPv6 routing tables prior to scenario 1 events

split-horizon rule, there is no neighbor to query. Hence, R2 is marked as a successor due to infinity FD. The network transits back to passive state with a changed metric via new and old successor R2. R4 sends poison reverse *EIGRP Reply* back to R2;

(#8) R1 and R2 receive and acknowledge *EIGRP Replies* which they exchanged and store a new metric in TT;

(#9) R1 and R2 receive *EIGRP Reply* from R3 and store a new metric in TT. Because all neighbors of R1 and R2 responded to their queries, DUAL stops. Next, they both R1 and R2 update records in RTs to reflect changed metric situation of LAN3's network (for IPv4 10.0.3.0/24 or for IPv6 2001:db8:c::/64). Topology is converged and state of R2's routing table is depicted in Fig. 9.

4.2 Scenario 2: Topology Change

This scenario begins exactly same as the previous one with phase #1, when neighbors are discovered, and phase #2, when topology converges by initial routing information exchange (same content of R2's NT, TT, and RT as on Figs. 5, 6, 7 and 8).

```
⊟ 📠 routes (std::vector<IPv4Route *>)              ⊟ 📠 routeList (std::vector<IPv6Route *>)
  ⊟ routes[9] (IPv4Route *)                           ⊟ routeList[17] (IPv6Route *)
     [0] = C 10.0.12.0/30 is directly connected, eth0    [0] = C 2001:db8:12::/64 [1/10] via ::, eth0
     [1] = D 10.0.13.0/30 [90/30720] via 10.0.23.2, eth1   [1] = C 2001:db8:23::/64 [1/10] via ::, eth1
     [2] = D 10.0.13.0/30 [90/30720] via 10.0.12.1, eth0   [2] = C 2001:db8:24::/64 [1/10] via ::, eth2
     [3] = C 10.0.23.0/30 is directly connected, eth1    [3] = C 2001:db8:b::/64 [1/10] via ::, eth3
     [4] = C 10.0.24.0/30 is directly connected, eth2    [4] = D 2001:db8:d::/64 [90/30720] via fe80:24::4, eth2
     [5] = D 10.0.1.0/24 [90/30720] via 10.0.12.1, eth0   [5] = D 2001:db8:13::/64 [90/30720] via fe80:12::1, eth0
     [6] = C 10.0.2.0/24 is directly connected, eth3    [6] = D 2001:db8:a::/64 [90/30720] via fe80:12::1, eth0
     [7] = D 10.0.3.0/24 [90/284160] via 10.0.23.2, eth1   [7] = D 2001:db8:13::/64 [90/30720] via fe80:23::3, eth1
     [8] = D 10.0.4.0/24 [90/30720] via 10.0.24.2, eth2   [8] = D 2001:db8:c::/64 [90/284160] via fe80:23::3, eth1
```

Fig. 9 R2's routing table after scenario 1 events

We scheduled link failure (R2's eth1) of interconnection between routers R2 and R3 for this scenario. The goal is to show how topology change is propagated from the source to other routers. Once again, we accomplish this with the help of scenarioManager in simulator. In case of real network, we just shut down the interface. In both cases, R3's eth1 remains operational.

We have decided to omit all acknowledgments in subsequent text in order to make it clearer and easier to read. Nevertheless, all routers correctly confirm reception of *EIGRP Update, Query,* and *Reply* messages by sending *EIGRP Ack.* Scenario continues in following manner:

(#3)   Eth1 comes down on R2. EIGRP process goes through TT and transits all networks reachable via successor, i.e., R2-R3 interconnection and LAN3 (10.0.23.0/30 or 2001:db8:23::/64 and 10.0.3.0/24 or 2001:db8:c::/64) on eth1 interface to active state. R2 sends *EIGRP Queries* to neighbors R1 and R4. Load balancing is enabled, thus R1-R3 interconnection (i.e., 10.0.13.0/30 or 2001: db8:13::/64) is reachable via two routes in the RT—the one that leads through R3 is removed and neighbors are notified by *EIGRP Update* messages;

(#4)   R4 receives *EIGRP Query* from R2. DUAL cannot find FS for routes and because of split-horizon rule there is no other neighbor to ask. Hence, R4 sends *EIGRP Reply* stating that R2-R3 interconnection and LAN3 are unreachable from its perspective;

(#5)   R1 receives *EIGRP Query.* Dual finds out FS and responds back with *EIGRP Reply.* Moreover, the route to 10.0.23.0/30 or 2001:db8:23::/64 via R2 is removed from RT and *EIGRP Update* about this is sent to neighbors R3 and R2. Routes on this router remain in passive state;

(#6)   Integrated optimization prevents information from particular updates to be passed to DUAL. Namely previously sent *EIGRP Update* from R1 to R3, from R2 to R1, from R1 to R2, and from R2 to R4;

(#7)   R2 receives *EIGRP Reply* from R4 and from R1. All replies has been received, thus routes to R2-R3 interconnection and LAN3 have a new successor in R2's TT and that is R1. Those routes are propagated to R2's RT and information about change is sent to neighbors as *EIGRP Update*;

(#8)   R4 receives *EIGRP Update* from R2 and inserts R2 as a new successor to its RT. Because of RT change, poison reverse *EIGRP Update* is sent back to R2;

(#9)   Same optimization as in case of phase #6. EIRGP Updates from R2 to R1 and from R4 to R2 are omitted from DUAL processing. Content of R2's NT, TT, and RT does not change for the rest of scenario and it shown in Figs. 10, 11 and 12);

(#10) Hold timer expires on R3, thus neighborship is terminated and R2 is removed from R3's NT. Also R3 sends goodbye *EIGRP Hello* as a preventive notification. All affected networks reachable via R2 (i.e., 10.0.24.0/30, 10.0.2.0/24, 10.0.4.0/24, or 2001:db8:24::/64, 2001:db8:b::/64 and 2001:db8: d::/64) transit to active state and *EIGRP Query* is sent to remaining neighbor R1. Only exception is R1-R2 interconnection (i.e., 10.0.12.0/30 or 2001:db8:12::/ 64) that has another FS due to load balancing. However, its second route is removed from R3's RT and *EIGRP Update* is sent to R1;

⊟ 🔟 neighborVec (std::vector<EigrpNeighborI11IPv4AddressE *>)
  ⊟ neighborVec[2] (EigrpNeighborI11IPv4AddressE *)
    ├ [0] = ID:2 Address:10.0.12.1 IF:eth0(100) HoldInt:15 SeqNum:12
    └ [1] = ID:3 Address:10.0.24.2 IF:eth2(102) HoldInt:15 SeqNum:5

Notice that neighbor R3 on eth1 is missing.

⊟ 🔟 neighborVec (std::vector<EigrpNeighborI11IPv6AddressE *>)
  ⊟ neighborVec[2] (EigrpNeighborI11IPv6AddressE *)
    ├ [0] = ID:2 Address:fe80:12::1 IF:eth0(100) HoldInt:15 SeqNum:12
    └ [1] = ID:3 Address:fe80:24::4 IF:eth2(102) HoldInt:15 SeqNum:5

Fig. 10 R2's IPv4 and IPv6 neighbor tables after scenario 2 events

⊟ 🔟 routeVec (std::vector<EigrpRouteSourceI11IPv4AddressE *>)
  ⊟ routeVec[8] (EigrpRouteSourceI11IPv4AddressE *)
    ├ [0] = P 10.0.12.0/30 is successor FD:28160 via Connected (28160/0), IF:eth0(100)
    ├ [1] = P 10.0.4.0/24 is successor FD:30720 via 10.0.24.2 (30720/28160), IF:eth2(102)
    ├ [2] = P 10.0.1.0/24 is successor FD:30720 via 10.0.12.1 (30720/28160), IF:eth0(100)
    ├ [3] = P 10.0.3.0/24 is successor FD:33280 via 10.0.12.1 (33280/30720), IF:eth0(100)
    ├ [4] = P 10.0.13.0/30 is successor FD:30720 via 10.0.12.1 (30720/28160), IF:eth0(100)
    ├ [5] = P 10.0.2.0/24 is successor FD:28160 via Connected (28160/0), IF:eth3(103)
    ├ [6] = P 10.0.24.0/30 is successor FD:28160 via Connected (28160/0), IF:eth2(102)
    └ [7] = P 10.0.23.0/30 is successor FD:33280 via 10.0.12.1 (33280/30720), IF:eth0(100)

⊟ 🔟 routeVec (std::vector<EigrpRouteSourceI11IPv6AddressE *>)
  ⊟ routeVec[8] (EigrpRouteSourceI11IPv6AddressE *)
    ├ [0] = P 2001:db8:b::/64 is successor FD:28160 via Connected (28160/0), IF:eth3(103)
    ├ [1] = P 2001:db8:d::/64 is successor FD:30720 via fe80:24::4 (30720/28160), IF:eth2(102)
    ├ [2] = P 2001:db8:a::/64 is successor FD:30720 via fe80:12::1 (30720/28160), IF:eth0(100)
    ├ [3] = P 2001:db8:13::/64 is successor FD:30720 via fe80:12::1 (30720/28160), IF:eth0(100)
    ├ [4] = P 2001:db8:c::/64 is successor FD:33280 via fe80:12::1 (33280/30720), IF:eth0(100)
    ├ [5] = P 2001:db8:24::/64 is successor FD:28160 via Connected (28160/0), IF:eth2(102)
    ├ [6] = P 2001:db8:23::/64 is successor FD:33280 via fe80:12::1 (33280/30720), IF:eth0(100)
    └ [7] = P 2001:db8:12::/64 is successor FD:28160 via Connected (28160/0), IF:eth0(100)

Fig. 11 R2's IPv4 and IPv6 topology tables after scenario 2 events

(#11) R1 receives *EIGRP Query* and *Update* from R3. DUAL finds FS for all queried routes in R1's TT and thus no network transits to active state. *EIGRP Reply* is sent to R3 as response;

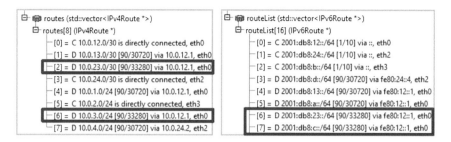

Fig. 12 R2's IPv4 and IPv6 routing tables after scenario 2 events

(#12) R3's DUAL collects all (single) *EIGRP Replies* (from R1). R3's TT is updated with a new successor and affected networks transit back to passive state. The best routes are introduced to R3's RT and *EIGRP Update* is sent to R1;

(#13) Processing of update is optimized just as in case of phase #6 and #9 on R1. Topology is converged.

4.3   Test Summary

Comparison for Scenario 1 can be observed in Tables 2 and 3 for IPv4 and IPv6. Similarly description for Scenario 2 is in Tables 4 and 5 for IPv4 and IPv6. In completely revisited comparisons, we have focused on messages processed mostly by router R2. Nevertheless, messages that are not shown and were processed by other routers are also in correct order and without any significant deviations between simulation and real time.

The correlation of messages between simulation and real network suggests correctness of our EIGRP implementation.

Validation testing against the real-life topology shows just reasonable time variations. Slight difference could be observed in case of Scenario 2 for IPv6. Phase #5 precedes phase #4. The vindication is that phases #4 and #5 run parallel and are independent. Hence, in real-topology message for phase #5 may be dispatched earlier by IOS.

Simulation results are influenced by the fact that EIGRPRouter has simpler control-plane and it is not delayed by any other traffic. Hence, some timestamps have sub 0 ms differences (same tables with milliseconds accuracy are available at [19]).

Time variation observable on real Cisco devices is caused by three factors: (a) control-plane processing delay and internal EIGRP optimizations; (b) packet pacing that guarantees constant bandwidth consumption by EIGRP process and avoids potential race conditions between EIGRP instances; and (c) inaccuracy in timing of certain event in real-life network.

Table 2 Timestamp comparison for IPv4 routing in scenario 1

Phase	Message	Sender → Receiver	Simulation (s)	Real (s)
#3	Update	R3 → R2	0.000	0.000
#4	Query	R2 → R1	0.000	0.030
#5	Reply	R1 → R2	0.000	0.057
#7	Reply	R4 → R2	0.001	0.074

Table 3 Timestamp comparison for IPv6 routing in scenario 1

Phase	Message	Sender → Receiver	Simulation (s)	Real (s)
#3	Update	R3 → R2	0.000	0.000
#4	Query	R2 → R1	0.000	0.062
#5	Reply	R1 → R2	0.000	0.097
#7	Reply	R4 → R2	0.001	0.132

Table 4 Timestamp comparison for IPv4 routing in scenario 2

Phase	Message	Sender → Receiver	Simulation (s)	Real (s)
#3	Query	R2 → R1	0.000	0.000
#4	Reply	R4 → R2	0.000	0.021
#5	Reply	R1 → R2	0.000	0.046
#7	Update	R2 → R1	0.000	0.074
#8	Update	R4 → R2	0.001	0.124
#10	Hello	R3 → R2	10.924	10.277
	Query	R3 → R1	10.924	10.299
#11	Reply	R1 → R3	10.924	10.349

Table 5 Timestamp comparison for IPv6 routing in scenario 2

Phase	Message	Sender → Receiver	Simulation (s)	Real (s)
#3	Query	R2 → R1	0.000	0.000
#4	Reply	R4 → R2	0.000	0.059
#5	Reply	R1 → R2	0.000	0.033
#7	Update	R2 → R1	0.000	0.121
#8	Update	R4 → R2	0.001	0.179
#10	Hello	R3 → R2	14.587	14.564
	Query	R3 → R1	14.587	14.575
#11	Reply	R1 → R3	14.587	14.617

Nevertheless, the routing outcomes of simulated and real network are exactly same when taking into account accuracy in order of seconds and EIGRP messages are in confluence on phases that depends on each other.

5   Conclusion

We presented an overview of the theory behind EIGRP routing protocol. The main contribution of this work is a new OMNeT++ compound module routing both IPv4 and IPv6. Module mimics Cisco's EIGRP protocol implementation based on the available specification and from reverse-engineering observations. We introduce a simulation scenario and relevant results to demonstrate its compliance with the reference Cisco IOS implementation. EIGRP is beneficial namely for large enterprise networks because it generally consumes less resources than link-state IGPs. It is the one of the best distance-vector IGPs available and with its public release we can expect that more companies will tend to use it. For such entities, we offer polished simulation models for a reliable comparison on their network functionality which now includes also EIGRP.

We plan to carry on work toward extending simulation module with stuck-in-active support and further tune EIGRP. Additional plan is to conduct comparative evaluation of our models against those in OPNET simulator.

More information about the ANSA project is available on homepage [20]. All source codes including EIGRP implementation could be downloaded from GitHub [21]. Real packet captures, which serve as a baseline for results reproduction, could be downloaded from Wiki of above-mentioned GitHub repository.

Acknowledgments This work was supported by following research grants and institutions:

- FIT-S-14-2299 supported by Brno University of Technology;
- VG20102015022 supported by Ministry of the Interior of the Czech Republic;
- IT4Innovation ED1.1.00/02.0070 supported by Czech Ministry of Education Youth and Sports.

References

1. Veselý, V., Jan, B., Ryšavý, O.: Enhanced interior gateway routing protocol for OMNeT++. In: SciTePress (ed.) Proceedings of the 4th International Conference on Simulation and Modeling Methodologies, Technologies and Applications (SIMULTECH 2014), Wien, pp. 50–58 (2014)
2. OMNeTpp/INET: INET Framework|Main/Welcome to the INET Framework. http://inet.omnetpp.org/. Accessed 2014
3. OMNeTpp: OMNeT++Network Simulation Framework. http://www.omnetpp.org/. Accessed 2014
4. Malkin, G.: RFC 2453: RIP Version 2. https://tools.ietf.org/html/rfc2453. Accessed Nov 1998
5. Malkin, G., Minnear, R.: RFC 2080: RIPng for IPv6. https://tools.ietf.org/html/rfc2080. Accessed Jan 1997
6. Chroboczek, J.: RFC 6126: The babel routing protocol. http://tools.ietf.org/html/rfc6126. Accessed April 2011
7. Oran, D.: RFC 1142: OSI IS-IS Intra-domain routing protocol. http://tools.ietf.org/html/rfc1142. Accessed Feb 1990
8. Moy, J.: RFC 2328: OSPF Version 2. http://tools.ietf.org/html/rfc2328. Accessed April 1998

9. Coltun, R., Ferguson, D., Moy, J., Lindem, A.: RFC 5340: OSPF for IPv6. http://tools.ietf.org/html/rfc5340. Accessed July 2008
10. Rekhter, Y., Hares, S.: RFC 4271: A border gateway protocol 4 (BGP-4). http://tools.ietf.org/html/rfc4271. Accessed Jan 2006
11. Garcia-Lunes-Aceves, J.J.: Loop-free routing using diffusing computations. IEEE/ACM Trans Netw **I**(1), 130–141 (1993)
12. Albrightson, R., Garcia-Luna-Aceves, J., Boyle, J.: EIGRP a fast routing protocol based on distance vectors. In: Proceedings Networld/Interop, vol. XCIV, pp. 136–147 (1994)
13. Xu, D., Trajkovic, T.: Performance analysis of RIP, EIGRP, and OSPF using OPNET. http://summit.sfu.ca/item/10841. Accessed August 2011
14. Savage, D., Slice, D., Ng, J., Moore, S., White, R.: Enhanced interior gateway routing protocol. https://tools.ietf.org/html/draft-savage-eigrp-01. Accessed 7 Oct 2013
15. GitHub/janovic: janovic/Quagga-EIGRP. https://github.com/janovic/Quagga-EIGRP. Accessed 2013
16. nonGNU: Quagga Software Routing Suite. http://www.nongnu.org/quagga/. Accessed 2013
17. SourceForge: Easy-EIGRP|Free software downloads at SourceForge.net. http://sourceforge.net/projects/easy-eigrp/. Accessed 2013
18. Wu, B.: Simulation based performance analyses on RIPv2, EIGRP, and OSPF Using OPNET. In: Math and Computer Science. http://digitalcommons.uncfsu.edu/macsc_wp/11. Accessed Aug 2011
19. Veselý, V., Vít, R.: EIGRP comparison tables between simulation and real topology with milliseconds accuracy. http://nes.fit.vutbr.cz/ivesely/PreciseEigrpTables.pdf. Accessed 2014
20. Brno University of Technology: In: ANSAWiki|Main/HomePage. http://nes.fit.vutbr.cz/ansa/pmwiki.php. Accessed Jan 2014
21. GitHub/kvetak: In: kvetak/ANSA. https://github.com/kvetak/ANSA. Accessed Dec 2013

Simulating LTE/LTE-Advanced Networks with SimuLTE

Antonio Virdis, Giovanni Stea and Giovanni Nardini

Abstract In this work we present SimuLTE, an OMNeT++-based simulator for LTE and LTE-Advanced networks. Following well-established OMNeT++ programming practices, SimuLTE exhibits a fully modular structure, which makes it easy to be extended, verified, and integrated. Moreover, it inherits all the benefits of such a widely used and versatile simulation framework as OMNeT++, i.e., experiment support and seamless integration with the OMNeT++ network modules, such as INET. This allows SimuLTE users to build up mixed scenarios where LTE is only a part of a wider network. This paper describes the architecture of SimuLTE, with particular emphasis on the modeling choices at the MAC layer, where resource scheduling is located. Furthermore, we describe some of the verification and validation efforts and present an example of the performance analysis that can be carried out with SimuLTE.

Keywords LTE · LTE-advanced · System-level simulator · OMNeT++

1 Introduction

The Long-Term Evolution (LTE) of the UMTS [1] is the de facto standard for next-generation cellular access networks. In an LTE cell, a base station or eNodeB (eNB) allocates radio resources to a number of User Equipments (UEs), i.e., handheld devices, laptops, or home gateways, using Orthogonal Frequency

A. Virdis (✉) · G. Stea · G. Nardini
Dipartimento di Ingegneria dell'Informazione, University of Pisa, Pisa, Italy
e-mail: a.virdis@iet.unipi.it

G. Stea
e-mail: giovanni.stea@unipi.it

G. Nardini
e-mail: g.nardini@ing.unipi.it

© Springer International Publishing Switzerland 2015
M.S. Obaidat et al. (eds.), *Simulation and Modeling Methodologies,
Technologies and Applications*, Advances in Intelligent Systems
and Computing 402, DOI 10.1007/978-3-319-26470-7_5

Division Multiplexing Access (OFDMA) in the downlink, and Single-Carrier Frequency Division Multiplexing (SC-FDMA) in the uplink. On each Transmission Time Interval (TTI, 1 ms), a time/frequency frame of resource blocks (RBs) is allocated to the UEs, in both directions. Each RB carries a variable amount of bytes to/from an UE, depending on the selected modulation and coding scheme. The latter, in turn, is chosen by the eNB based on the Channel Quality Indicator (CQI) reported by the UE, which measures how the UEs perceive the channel. The new Release 10, called LTE-Advanced (LTE-A) [2], besides promising larger bandwidths due to spectrum aggregation, envisages different transmission paradigms, such as Coordinated Multi-Point (CoMP) Scheduling or device-to-device (D2D) communications.

Resource allocation and management in LTE/LTE-A is a key performance enabler: carefully selecting which UEs to target, using which modulation scheme, etc., clearly affects the efficiency of the system. Moreover, intercell interference coordination, cooperative transmission, antenna selection, resource partitioning among macro and micro-/picocells, energy efficiency and battery saving schemes, etc., enable a wealth of declinations of the classical point-to-multipoint LTE scheduling problem. The performance evaluation of resource management schemes for LTE/LTE-A is normally carried out via simulation. This is because—on one hand—testbeds are hardly available, especially to the academia, and—on the other —the entire set of protocols, features, and functions of this standard, and their complex interactions, defy any attempt at unified analytical modeling, whereas neglecting or abstracting away some aspects incurs the risk of oversimplification.

Most of the research on LTE/LTE-A is evaluated via home-brewed "disposable" simulators, which are never made available to the community. This makes the claims backed up by such results unverifiable, hence unauthoritative. In the recent past, more structured efforts have been made to provide the research community with reference, open-source simulation frameworks for LTE/LTE-A. Leaving aside physical layer simulators—e.g., [3], or [4]—which are outside the scope of this paper, there are three main contributions. The first one is [5], written in MATLAB, which is limited to the downlink and does not consider some salient aspects of LTE simulation, e.g., realistic traffic generators.

The second one is *LTE-Sim* [6], written in object-oriented C++. This includes many more functionalities, such as the uplink, handover, mobility, UDP, and IP protocols, several schedulers and some traffic generators (e.g., VoIP, CBR, trace-based, full buffer). This simulator, however, comes with some limitations. Hybrid ARQ (H-ARQ) seems not to be supported. Moreover, scenarios are written as static C++ functions. Thus, they are hard to verify for correctness, and they are compiled together with the simulator. Mixing models, scenarios, and experiment definition in a simulator is hardly desirable from a software engineering standpoint. The main drawback of *standalone* simulators such as the above two is that they are costly to extend and not interoperable. Even granting a state-of-the-art code structure, they force users to invest an often unreasonable time into modeling, coding, and verifying any extra functionality, an effort which is seldom reusable. For instance, the current release of LTE-Sim lacks TCP. The effort required to develop the code to run a web-browsing simulation on LTE-Sim (e.g., to back up a claim in a scientific paper) entails modeling TCP, HTTP, server, content, and user behavior, and is probably large enough to discourage most scientists. The same can be said for the setup of a Wi-Fi-core-LTE simulation scenario. Moreover, these simulators offer no support to *simulation workflow automation*. This includes, ideally, the ability to define parametric scenarios, to manage multiple repetitions with independent initial conditions, to define and collect measures, to efficiently store, retrieve, parse, analyze, and plot simulation results, etc. The lack of such instruments (and the ensuing need to make up for it with home-brewed, error-prone, or however unstructured solutions) is a major cause of delay and errors in simulation studies, especially large-scale ones, as shown in [7, 8].

A third contribution has been realized within the framework of *ns-3* [9]. ns-3 is being developed by a large community of scientists with heterogeneous research interests, and it already includes a considerable number of network modules (e.g., Wi-Fi, WiMAX, 802.11 s mesh networks, etc.). The ns-3 LTE simulator [10], henceforth *ns-3-LTE* for want of a better name, is an ongoing effort, which has been designed to allow end-to-end simulation with realistic traffic models. It includes both the Radio Access Network and the Evolved Packet Core. Recently, the SAFE (Simulation Automation Framework for Experiments) framework has been released, providing functionalities for planning, control, and result storage of experiments in ns-3 [11]. However, the SAFE project—still ongoing at the time of writing—appears to require a nontrivial effort on the user side to be integrated with ns-3.

Recently, some LTE open-source *emulators* have appeared. *LibLTE* [12] is a library for Software-defined Radio UEs and eNodeBs, still in pre-alpha development status. *OpenLTE* [13] includes a simple implementation of an eNB relying on dedicated hardware boards, which however lacks the whole user plane at the time of writing. The *OpenAirInterface* project [14] offers a complete physical layer implementation, working both with open-source hardware equipment and in a simulated environment (single machine or multimachine via LAN), a mostly complete layer 2 implementation for both eNBs and UEs, and a testing suite with customizable scenarios. The project appears to be in rapid development, although still lacking complete documentation. It is well known that emulation packages often allow a greater degree of realism than simulators, often at the price of a more limited scalability and flexibility.

Our contribution is a new LTE simulator, called SimuLTE, which has been developed for the OMNeT++ simulation framework [15] and has been released under LGPL at http://www.simulte.com. We chose OMNeT++ for several reasons: first, it is a stable, mature, and feature-rich framework, much more than ns-3. It was released in 2001, with the explicit goal of automating as many steps of the simulation workflow as possible, so as to enable large-scale simulation studies and bridge the gap between writing down models and getting credible results. Second, it is perfectly modular, which made it easy to write all the required code from scratch, and makes it easy to extend it. Third, it already includes a large amount of simulation models, e.g., INET (http://inet.omnetpp.org/), which boasts an impressive protocol matrix, all the TCP/IP stack, mobility, wireless technologies, etc. Thanks again to the modular structure, this allows users to simulate *mixed* scenarios, of which LTE/LTE-A constitutes only one part, and get a feeling of the true end-to-end performance of simulated applications. Fourth, OMNeT++ has a smooth learning curve: novice users can exploit graphic interfaces to setup and run fairly complex scenarios in no time, while advanced users may exploit a structured topology-definition language NEtwork Description (NED). Last, but not least, it is supported by a large and active community of users, from both academia and networking industries, which is a guarantee that any setup time invested in learning the ropes can be amortized over a long future.

SimuLTE simulates the data plane of the LTE Radio Access Network and Evolved Packet Core. It consists of over 40,000 lines of code (i.e., roughly double as many as LTE-Sim, which however also factors in many extra functionalities, e.g., mobility, applications, IP/UDP, event queues, etc., which instead we took from the OMNeT++ kernel and INET). In this paper, we describe its modeling approach and show some nontrivial results that can be obtained with it regarding resource allocation. As for the modeling, besides giving a high-level description, we focus on the modeling approaches that differ from those of the other simulators. Specifically, the H-ARQ functionalities—which are not described in other simulators—are introduced in detail. Furthermore, we detail the scheduling model, which is based on *virtual schedulers*, which can be run concurrently and then be chosen among by an allocator. As for performance evaluation, we profile the simulator execution time in several scenarios, we show how scheduling heuristics for multiband resource allocation problems fare against the optimum, and we highlight results related to multicell environments which defy analytical modeling.

The rest of the paper is organized as follows. Section 2 describes the OMNeT++ framework. Section 3 provides an overview of SimuLTE. We describe some SimuLTE validation and verification in Sect. 4 and performance results in Sect. 5. Finally, Sect. 6 concludes the paper and outlines future work.

2 The OMNeT++ Framework

SimuLTE is based on the OMNeT++ simulation framework, which can be used to model several things—i.e., not only networks. We focus only on the aspects of OMNeT++ which are more relevant to understanding SimuLTE. The interested reader is referred to http://www.omnetpp.org for more details.

The basic OMNeT++ building blocks are *modules*, which can be either *simple* or *compound*. Modules communicate through *messages*, which are usually sent and received through a *connection* linking the modules' *gates*, which act as interfaces. A *network* is a particular compound module, with no gates to the outside world, which sits at the top of the hierarchy. Connections are characterized by a bit rate, delay and loss rate, and cannot bypass module hierarchy: with reference to Fig. 1, simple module 3 cannot connect to 2, but must instead pass through the compound

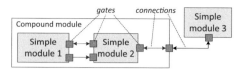

Fig. 1 OMNeT++ module connection

module gate. Simple modules implement model behavior. Although each simple module can run as an independent coroutine, the most common approach is to provide them with event handlers, which are called by the simulation kernel when modules receive a message. Besides handlers, simple modules have an initialization function and a *finalization* function, often used to write results to a file at the end of the simulation. The kernel includes an event queue, whose events correspond to messages (including those a module sends to itself).

OMNeT++ allows one to keep a model's *implementation, description, and parameter values* separate. The implementation (or *behavior*) is coded in C++. The description (i.e., gates, connections, and parameter definition) is expressed in files written in Network Description (NED) language. Moreover, a modular development is favored by allowing inheritance of both module description and behavior. The parameter values are written in initialization (INI) files. NED is a declarative language, which exploits inheritance and interfaces, and it is fully convertible (back and forth) into XML. NED allows one to write *parametric* topologies, e.g., a ring or tree of variable size. NED files can be edited both textually and graphically through a Graphic User Interface (GUI), switching between the two views at any time. INI files contain the values of the parameters that will be used to initialize the model. For the same parameter, multiple values or intervals can be specified.

As far as coding is concerned, OMNeT++ is integrated into eclipse, thus it comes with an IDE which facilitates development and debugging. In fact, modules can be inspected, textual output of single modules can be turned on/off during execution, and the flow of messages can be visualized in an animation, all of which occurs automatically. Events can be logged and displayed on a time chart, which makes it easy to pinpoint the causes or consequences of any given event.

As for simulation workflow automation, OMNeT++ clearly separates *models* (i.e., C++ and NED files) from *studies*. Studies are generated from INI files, by automatically making the Cartesian product of all the parameter values and, possibly, generating replicas of the same instance with different seeds for the random number generators. Note that the IDE allows one to launch and manage multiple runs in parallel, exploiting multiple CPUs or cores, so as to reduce the execution time of a simulation study. Finally, data analysis is rule-based: the user only needs to specify the files and folders she wants to extract data from, and the *recipe* to filter and/or aggregate data. The IDE automatically updates and redraws graphs when simulations are rerun or new data become available. The performance of OMNeT++ has been compared to that of other simulators, notably ns-3 [16]. The result is that the two are comparable as for run time and memory usage, with ns-3 faring marginally better within the limits of the analyzed scenarios.

3   Overview of SimuLTE

SimuLTE simulates the data plane of the LTE/LTE-A Radio Access Network and
Evolved Packet Core. Due to lack of space, we assume that the reader is familiar
with the LTE architecture (a table of acronyms is reported in the appendix for ease
of reference), and we only describe the RAN modeling. SimuLTE allows simulation of LTE/LTE-A in Frequency Division Duplexing (FDD) mode, with heterogeneous eNBs (macro, micro, pico, etc.), using omnidirectional and anisotropic
antennas, possibly communicating via the X2 interface. Realistic channel models,
MAC, and resource scheduling in both directions are supported. In the current
release, the Radio Resource Control (RRC) is not modeled.

The general structure of the three main nodes is shown in Fig. 2. SimuLTE
implements eNBs and UEs as compound modules. These can be connected with
each other and with other nodes (e.g., routers, applications, etc.) in order to com-
pose networks. The binder module is instead visible by every other node in the
system and stores information about them, such as references to nodes. It is used,
for instance, to locate the interfering eNBs in order to compute the intercell
interference perceived by a UE in its serving cell. UE and eNB are further composed of modules. Every module has an associated description file (.ned) defining
its structure, and may have a class definition file (.cpp,.h) which implements the
module functionalities.

The UDP and TCP modules, taken from the INET package, implement the
respective transport layer protocols, and connect the LTE stack to TCP/UDP
applications. As Fig. 2 shows, TCP and UDP applications (*TCP App* and *UDP
App*) are implemented as vectors of *N* modules, thus enabling multiple applications
per UE. Each TCP/UDP App represents one end of a connection, the other end of
which may be located within another UE or anywhere else in the topology.
SimuLTE comes with models of real-life applications (e.g., VoIP and Video on

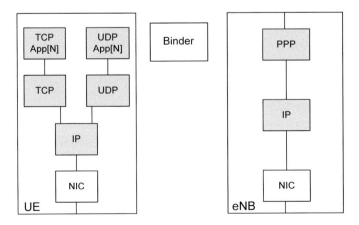

Fig. 2 UE and eNB module structure. Shaded modules are taken from the INET framework

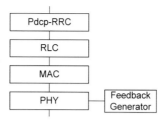

Fig. 3 NIC module architecture on the UE

Demand), but any other TCP/UDP-based OMNeT++ application can also be used. The IP module is taken from the INET package as well. In the UE it connects the *Network Interface Card* (NIC) to applications that use TCP or UDP. In the eNB it connects the eNB itself to other IP peers (e.g., a web server), via PPP (Point-To-Point Protocol).

The NIC module, whose structure is shown in Fig. 3, implements the LTE stack. It has two connections: one between the UE and the eNB and one with the LTE IP module. It is built as an extension of the *IWirelessNic* interface defined in the INET library, so as to be easily plugged into standard scenarios. This allows one—among other things—to build hybrid connectivity scenarios, e.g., with nodes equipped with both Wi-Fi and LTE interfaces. With reference to Fig. 3, each of the NIC sub-modules on the left side represents one or more parts of the LTE protocol stack, which is common to the eNB as well. The only module in the UE that has no counterpart in the eNB is the *Feeback Generator*, which creates channel feedbacks that are then managed by the PHY module. The communication between modules takes place only via message exchange, thus each action starts from a message handler. Cross-module calls are used only in the form of *getter/setter* functions. This allows us to maintain a strong control over the interactions between modules, thus limiting possible buggy behaviors. In the following we describe each sub-module with their related functionalities. We will use *downstream* and *upstream*, respectively, to identify the flow of traffic from an upper to a lower layer within the same module and vice versa, and *downlink* and *uplink* for the traffic flow from the eNB to the UE and vice versa.

3.1 PDCP-RRC Module

The PDCP-RRC module is the connection point between the NIC and LTE IP modules. The PDCP-RRC module receives data from upper layers in the down-stream direction and from the RLC layer in the upstream. In the first case it performs RObust Header Compression (ROHC) and assigns/creates the Connection Identifier (CID) that uniquely identifies, together with the UE ID, a connection in the whole network. A Logical Connection Identifier (LCID) is kept for each 4-tuple

in the form <sourceAddr, destAddr, sourcePort, destPort>. When a
packet arrives at the PDCP-RRC module, the correct LCID is attached to it (if there
exists one), otherwise a new one is created storing the new 4-tuple. Then, the packet
is encapsulated in a PDCP PDU and forwarded to the proper RLC port, depending
on the selected RLC mode, as described in Sect. 3.2. In the upstream, a packet
coming from the RLC is decapsulated, its header is decompressed and the resulting
PDCP PDU is sent to the upper layer.

3.2 RLC Module

The RLC module performs multiplexing and demultiplexing of MAC SDUs
to/from the MAC layer, implements the three RLC modes: *Transparent Mode*
(TM), *Unacknowledged Mode* (UM), and *Acknowledged Mode* (AM) as defined in
[17], and forwards packets from/to the PDCP-RRC to/from the proper RLC mode
entity. RLC operation is the same on both the eNB and the UE. The structure of the
RLC module is shown in Fig. 4. There are three different gates connected with the
PDCP-RRC module, one for each RLC mode. The TM submodule has no buffer, as
it forwards packets transparently, while AM and UM have their own set of
transmission/reception buffers, one for each CID associated to that RLC mode.

3.3 MAC Module

The MAC module is where most of the intelligence of each node resides. Its main
tasks are buffering packets from upper (RLC) and lower layers (PHY), encapsu-
lating MAC SDUs into MAC PDUs and vice versa, managing channel feedback,
adaptive modulation and coding (AMC), and scheduling. Those operations are the
same at the UE and the eNB, with scheduling and channel-feedback management
being the only exceptions. The high-level structure of the MAC is shown in Fig. 5.
In the downstream, MAC SDUs coming from the RLC layer are stored in *MAC*

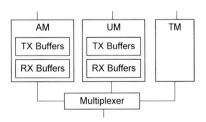

Fig. 4 RLC module representation

Fig. 5 MAC packet flows

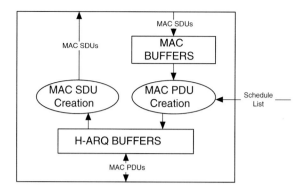

Buffers, one for each CID. On each TTI some connections are scheduled for transmission, according to the schedule list composed by the scheduler. MAC SDUs from the scheduled connections are then encapsulated into MAC PDUs, which are then stored in the H-ARQ buffers and forwarded to the physical layer. The detailed structure of H-ARQ buffers will be explained later on. In the upstream, MAC PDUs coming from the physical layer are stored into H-ARQ buffers. They are then checked for correctness, decapsulated and forwarded to the RLC.

The entities involved in eNB scheduling are shown in Fig. 6. Resource scheduling is performed at the eNB for both the uplink and the downlink. For the uplink, decisions are notified to the UEs via *grant* messages, i.e., in the current release the Physical Downlink Control Channel (PDCCH) is not directly simulated. Each UE reads the grants and decides which local connection will be able to use the granted resources, if any. UEs in turn request uplink resources via the *Random Access* (RAC) Procedure, which is again implemented through messages generated by the MAC module. The Physical Random Access Channel (PRACH) is not directly simulated. In order to properly make scheduling decisions, both downlink and uplink schedulers need the status of user connections and the channel quality

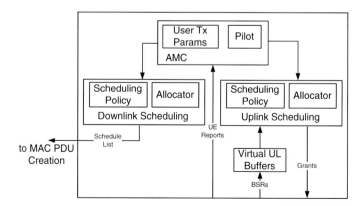

Fig. 6 MAC scheduling operation

perceived by each UE. The downlink connection status is inferred directly from MAC buffers, while in the uplink the status is inferred from *virtual uplink buffers*. The latter are kept synchronized with the real MAC buffers on the UE side via Buffer Status Reports (BSR), as defined in [18], which are control elements attached to uplink transmissions.

The Adaptive Modulation and Coding (AMC) entity stores channel status information from the UE reports, deciding (*Pilot* entity) and storing (*User Tx Params* entity) transmission parameters such as modulation and coderate, and computing the amount of bytes per resource block for each user based on those parameters, as defined in [19].

As far as H-ARQ is concerned, *Transmission* and *Reception H-ARQ Buffers* (from now on *TxHbuff* and *RxHbuff*) store MAC PDUs that are being sent and received, respectively. This means that a MAC PDU is stored therein until it is received correctly or the maximum number of retransmissions is reached. Reception is notified via H-ARQ feedback messages. H-ARQ buffers on the eNB maintain MAC PDU information for each connected UE in both downlink and uplink, for each H-ARQ process and for each codeword (hence, these modules require no modification to support Single-User MIMO). The general structure shown in Fig. 7 is valid for both Tx- and Rx-Hbuffs, hence we will use the neutral *Hbuff* to denote both. A Hbuff contains K buffers, one for each active connection to another node. Hence an eNB has as many buffers as connected UEs in each direction, whereas a UE has up to one per direction, as it communicates directly only with its serving eNB. Each buffer contains N *processes* ($N = 8$, usually), one per H-ARQ process. Finally, each *process* contains two *units*, to support SU-MIMO. A unit contains the information related to a transmitting/receiving MAC PDU. The status of the unit is stored in a different manner on TX- and RX-buffs as it depends on the feedback management procedure. The finite state automata for transmission is represented in

Fig. 7 Common *Hbuff* structure

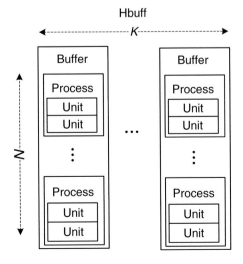

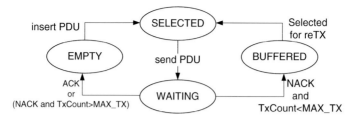

Fig. 8 Finite state automata for unit status in TX

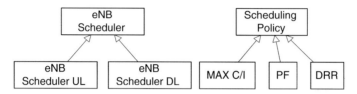

Fig. 9 eNB schedulers and Sch. Policy hierarchy

Fig. 8 (the one for reception is similar, *mutatis mutandis*). ACK and NACK are the reception of the corresponding H-ARQ feedback. *TxCount* is the transmission counter, increased in the SELECTED state and reset in the EMPTY one. *MAX_TX* is the maximum number of transmissions before discarding a PDU.

The hierarchy of eNB schedulers is shown in the left part of Fig. 9. The *eNB Scheduler* base class implements operations that are common to the DL and UL, such as data structures initialization, allocation management via the allocator, and statistics collection. The two classes *eNB Scheduler UL* and *DL* extend the base class by implementing the `rtxSchedule()` method, which manages retransmissions. In addition the *eNB Scheduler UL* manages RAC requests.

At the beginning of each TTI the eNB prepares a *schedule list* in each direction, which shares available resources among the active connections, according to a given *policy*. This is performed by a member of the eNB Scheduler called *Scheduling Policy*. Its main function, `schedule()`, consists of two steps: first, the scheduling policy is applied on a set of *virtual* connections, without modifying MAC buffers; then the decisions are stored and passed to the MAC, which enforces them by fetching the data from its buffers and constructing the PDUs (see again Fig. 6). The scheduling policy builds the schedule lists by repeatedly polling the allocator for given amounts of bytes transmitted at the CQI of the connection being examined. This allows one to envisage general schedulers, that decide *both* the priority order of connections *and* the amount of data to be scheduled from each: for instance, schedulers that only serve *urgent* data from each connection (instead of attempting to empty each queue sequentially), or set fixed quanta.

The fact that scheduling occurs in two steps allows a user to run multiple schedulers in parallel, choosing which schedule list to commit among a set of alternatives. The scheduling policy hierarchy is shown in the right part of Fig. 9. To implement a new scheduling policy, a developer only needs to implement the two steps of the `schedule()` function. Three well-known scheduling policies are included in the current release, namely *Maximum Carrier over Interference* (Max C/I), *Proportional Fair* (PF), *Deficit Round Robin* (DRR). Finally, we observe that scheduling policies may redefine the `rtxSchedule()` function. Retransmissions can be scheduled either at a higher/lower priority than first transmissions, or jointly with them.

3.4 PHY Module

The PHY module implements functions related to the physical layer, such as channel-feedback computation and reporting, data transmission and reception, air channel emulation and control messages handling. It stores the physical parameters of the node, such as the transmission power and antenna profile (i.e., omnidirectional or anisotropic). This allows one to define *macro-, micro-, pico-eNBs*, with different radiation profiles. Each physical module on the eNB and the UE has an associated channel model C++ class that represents the physical channel as perceived by the node itself. The channel model is an interface that defines two main functions: `getSINR()`, which returns the *Signal to Interference plus Noise Ratio* (SINR), and `error()`, which checks if a packet has been corrupted. We describe the current implementation of the channel model later on, but one can easily implement its own model by implementing the channel model interface.

On the UE side, some tasks related to physical layer procedures are performed by an independent *Feedback Generator* module. This module generates channel feedback (i.e., CQIs), which can be configured to be periodic or aperiodic. Furthermore, CQIs can be either wideband or per-sub-band: in the latter case, the user can configure the number of sub-bands and of reported CQIs. The feedback generation process exploits the function provided by the channel model interface. Note that the physical LTE channels, such as the *Physical Downlink Control Channel* (PDCCH), *Physical Uplink Control Channel* (PUCCH), and *Physical Random Access Channel* (PRACH) are not modeled down to the level of OFDM symbols, to keep both memory and CPU usage limited. Their functionalities are instead implemented via control messages sent between the eNB and UE nodes. Any limitation of those channels, (e.g., on the maximum number of UEs that can be scheduled simultaneously on the PDCCH), can nevertheless be emulated by imposing constraints on the messages themselves.

In the downstream, MAC PDUs are received from the MAC layer and encapsulated in an *Air Frame* packet. Packets are marked with a *Type*, i.e., *data* or *control*, then they are directly sent to their destination module, selected according to the control information attached to the packet. In the upstream, a received Air Frame packet is selectively processed depending on its type: control packets are directly forwarded to the upper layer (assuming correct reception), whereas data packets are tested against the `error()` function of the channel model before being marked and sent to the upper layer.

Our implementation of the channel model interface is called *Realistic Channel Model*. The SINR is computed as $\mathrm{SINR} = P_{RX}^{eNB} / \left(\sum_i P_{RX}^i + N \right)$, where P_{RX}^{eNB} is the power received from the serving eNB, P_{RX}^i is the power received from the interfering eNB i, N is the Gaussian noise. Furthermore, P_{RX} is computed as $P_{RX} = P_{TX} - P_{\mathrm{loss}} - F - S$, where P_{TX} is the transmission power, P_{loss} is the path loss due to the eNB-UE distance (which depends on the frequency), and F and S are the attenuation due to *fast* and *slow* fading, respectively [20]. The above operations are performed within the `getSINR()`, on a per-RB basis. This allows one to simulate the effects of interference on every single RB, e.g., taking into account the transmissions from neighboring eNBs, so as to evaluate LTE-A interference coordination mechanism such as the Enhanced intercell interference coordination (eICIC) or Coordinated Multi-Point (CoMP). The `error()` function compares the CQI used for transmitting a packet with the one computed at the moment of reception. An error probability P_{err} is then obtained using realistic BLER curves and the two previously computed CQIs. A uniform random variable $X \in [0, 1]$ is sampled and the packet is assumed to be corrupted if $X < P_{\mathrm{err}}$, and correct otherwise.

The choice of the channel model usually arises strong feelings among scientists and practitioners in wireless networks. For this reason, besides proposing one, we took care of implementing it in an easily modifiable way. Modifications can be done in either of the following ways:

- extend the base *LteChannelModel* class, and specifically implement the two functions `getSINR()` and `error()`. This requires some work, but allows maximum flexibility.
- Modify our *Realistic Channel Model*, and specifically the functions for computing pathloss, fading, and shadowing. This requires minimal modifications, at the price of less flexibility.

Note that, whichever the choice, the rest of the code is obviously unaffected.

4   Validation and Verification

Validating the simulation model of a complex system and checking its implementation for correctness are fundamental tasks. In this section we will first show some validation of the SimuLTE model, and then describe the verification techniques that were used during/after the development of the simulator.

4.1   Model Validation

Validating the tons of data produced by a simulator is a very common problem. One possible approach is to compare these values against measurement obtained from real systems. Unfortunately, to the best of our knowledge, none of these measurements are publicly available. Another possible solution, as suggested in [21], is to compare simulation results with theoretical ones obtained in simple reference scenarios. We thus choose a single cell in two scenarios: a single-user one (*Scenario I*) and a multiuser one (*Scenario II*). In both cases UEs are static and measurements are made at distances to the eNB ranging from 500 to 2000 m. In *Scenario II* UEs are placed within a 50 m-radius circle centered at the specified distance. The main simulation parameters are described in Table 1. We compare the throughput of the simulated scenarios against a reference throughput obtained as $\text{Tpt} = S(\text{MCS}, N_{\text{PRB}})/\tau$ [21] where τ is the duration of a TTI (i.e., 1 ms), $S(\text{MCS}, N_{\text{PRB}})$ is the transport block size for the measured CQI and for N_{PRB} physical resource blocks.

As we show in Fig. 10, simulation results match the reference ones, in both scenarios. Note that the results obtained with distances greater than 500 m are not directly comparable with the ones of the above work, as the two pathloss models are different.

Table 1 Simulation parameters

Parameter	Value
N_{PRB}	25
Pathloss model	ITU-R, rural macro
Mobility model	Stationary (OMNeT++ model)
eNB Tx power	30 dBm
Noise figure	5 dB
Cable loss	2 dB
Distance eNB—UE	500, 1000, 1500, 2000 m
Simulation time	2 s

Fig. 10 Cell throughput for Scenario I and Scenario II compared against theoretical reference throughput

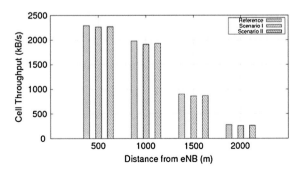

4.2 System Verification

We describe here the verification techniques used to ensure correctness of the implementation. The latter take advantage of both well-known verification techniques and instruments made available by the OMNeT++ framework.

First of all, during the whole development process, antibugging has been used to enforce the consistency of the variables that define the crucial elements of the system. For example, for each scheduling operation, we repeatedly check the number of allocated RBs, verifying that none of them is in use at the beginning of a TTI and that their number never exceeds the system capacity. Whenever a consistency check fails, the simulation is stopped and an error is reported. This avoids obtaining results from a system that traversed an inconsistent state.

Second, we monitored the values of many statistics *during* some critical scenarios, in order to check for odd behaviors throughout the transient phase. In this case we take advantage of OMNeT++ *vectors,* which allow the user to store all the values of a metric, together with the time at which they were recorded. This way we can easily trace key metrics over time.

Finally, we checked module interactions for correctness using structured testing. For each module we defined a *stub* counterpart, with the same interfaces to other modules, but no operations. Stubs are helpful to test other modules in isolation. For instance, we can monitor the interactions between MAC and PHY layers alone, without the added burden of the other layers, for which stubs are used. Moreover, each simulation can be configured to produce a hash value, the latter depending on the way messages are exchanged between modules. We can thus define a set of reference scenarios and obtain their associated hash values. After each modification to the system (e.g., an enhancement or a new version), we can then run the same reference scenarios again and check for any changes by just comparing the obtained hash values. Note that a mismatch in a hash value, does not necessarily imply an error. Rather, it highlights a change in the communication pattern that may possibly be the subject of further investigation.

5 Performance Evaluation

We present three contributions. First, we show how the simulation times vary with the number of UEs and cells and with the traffic profile. Second, we show that intercell interference increases the number of allocated RBs in a cell in a way which analytical formulas fail to predict. Finally, we formulate the MaxC/I scheduling problem in a multiband setting as an optimization problem, we show how to solve it optimally by interfacing SimuLTE and CPLEX, and we use it to benchmark a simple heuristic.

Fig. 11 Simulation scenario

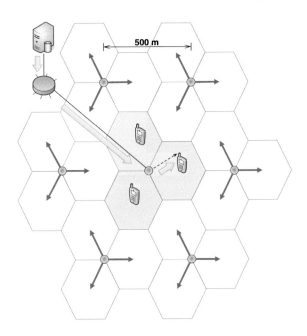

5.1 Profiling of Execution Time

We setup a scenario with a varying number of UEs and cells. eNBs use anisotropic antennas, radiating at 120° angles and transmitting at 46 dBm, and employ MaxC/I scheduling. The simulation scenario is shown in Fig. 11, where each circle represents a cluster of three colocated cells. CQIs are wideband and reported periodically. The UM RLC is employed, and the MAC fragment is set to 20 bytes. The overall number of available RBs is 50. UEs are uniformly distributed among cells and receive VoIP traffic (40-byte application packets on each 20 ms, with alternated talkspurts and silences), which is generated by a server and forwarded to the eNBs through a router. 20 s of the above system are simulated on an Intel(R) Core(TM) i7 CPU at 2.80 GHz, with 8 Gb of RAM, a Linux Ubuntu 12.04 operating system, and OMNeT++ version 4.2.2. Each run is repeated five times in independent conditions. 95 % intervals are displayed unless negligible.

Figure 12 reports the running time for 3, 9, and 21 cells. The latter depends almost linearly on the number of UEs in all cases. Increasing the number of cells influences the overall running time, as computing interference requires more operations.

In the scenario with nine cells, Fig. 13 shows how the simulation time depends on the traffic profile: we vary the packet size and the period, while keeping the same application layer bitrate. However, Fig. 14 shows that the MAC-level per-UE bitrate is quite different, due to the protocol overheads. In particular, the MAC PDU

Fig. 12 Execution time with varying number of cells

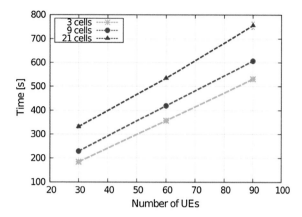

Fig. 13 Execution time with varying offered load

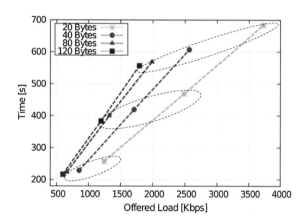

Fig. 14 Bitrate per UE at MAC level

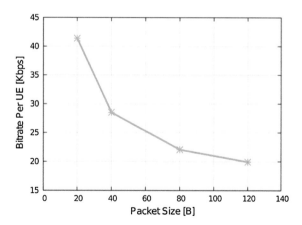

size is 33 bytes larger than the application packet (header compression is not used). In Fig. 13, bottom, mid, and top markers on each curve are for 30, 60, and 90 UEs, respectively. For the same MAC-layer offered load (i.e., the same abscissa), smaller packets sizes require shorter simulation times. However, for the same number of UEs (i.e., the same group of markers), the simulation time depends on the *number* of transmitted packets, hence a smaller packet size (i.e., more packets per unit of time) implies longer simulation times.

5.2   Intercell Interference

SimuLTE allows in-depth analysis of the effects of the intercell interference. We consider three colocated cells as in Fig. 13. UEs receive CBR traffic at a rate of 0.2 Mbps. Figure 15 reports the average number of RBs allocated by a cell. We compare the results with those obtained using the formula in [22], using the same channel model in both cases. The formula evaluates the interference based on the probability that two cells use the same RB, given the amount of RBs allocated by the two cells and assuming that each cell picks its RBs at random in the frame. Since the interference depends on the allocated RBs and vice versa, the formula is updated until it reaches a steady state. We note that there is a point (at 65-70 UEs) where the simulative curve has a step. This happens because the more a cell fills its own frame, the more it generates interference on UEs of neighboring cells, which in turn report lower CQIs. Accordingly, neighboring cells need to allocate more RBs to serve their UEs, thus producing additional interference that further decreases CQIs and causes retransmissions. Moreover, transmissions may occur simultaneously on RBs with low and high interference. Since wideband CQIs are used, high-interference RBs may be corrupted, causing retransmission, hence even more allocated RBs. At low loads, the system is still able to satisfy the additional requests, whereas at high loads this results in a positive feedback that rapidly drives

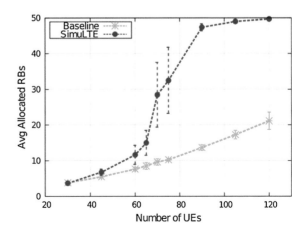

Fig. 15 Average allocated RBs per cell

the system toward saturation. This effect cannot be observed using the analytical formula because it does not account for the impact of errors and retransmissions.

5.3 Multiband Scheduling

Many resource allocation problems in LTE/LTE-A can be formulated as optimization problems. Hereafter, we describe a typical one, and we show how to interface SimuLTE with optimization solver CPLEX to solve it, believing this to be useful to researchers working in this area, for both the results and the way we obtained them.

MaxC/I scheduling is a relatively simple problem if we assume wideband CQIs and a number of transmitted bytes which is linear with the number of allocated RBs, with the CQI representing the line slope. In fact, in this case, all it takes to obtain the maximum throughput is to allocate enough RBs to empty the queue of each UE, in decreasing order of CQI (taking some extra care when padding is involved). Therefore, MaxC/I scheduling can be obtained at a complexity of $O(N \log N)$, N being the number of UEs. When per-sub-band CQI are used, however, obtaining the maximum throughput becomes an NP-hard problem, as shown in [23], hence some heuristics are used to obtain suboptimal solutions. Recall that a UE can use *one* CQI in a TTI (typically, the minimum of all the CQIs in the sub-bands where it is allocated some RBs). The maximum throughput problem can be formulated as:

$$\max \sum_{i=1}^{N} \left(r_i \cdot \sum_{j=1}^{M} x_{i,j} - p_i \right)$$

$$s.t. \quad \sum_{j=1}^{N} \left(r_i \cdot \sum_{j=1}^{M} x_{i,j} - p_i \right) \leq Q_i \qquad \forall i \quad (i)$$

$$p_i \leq r_i - 1 \qquad \forall i \quad (ii)$$

$$r_i \leq CQI_{i,j} + (1 - b_{i,j}) \cdot CQI_{max} \quad \forall i \quad (iii) \qquad (1)$$

$$x_{i,j} \geq b_{i,j} \qquad \forall i,j \quad (iv) \,,$$

$$\sum_{i=1}^{N} x_{i,j} \leq B_j \qquad \forall j \quad (v)$$

$$r_i, p_i \in \mathbb{Z}^+ \qquad \forall i \quad (vi)$$

$$b_{i,j} \in \{0,1\}, x_{i,j} \in \mathbb{Z}^+ \qquad \forall i,j \quad (vii)$$

where M is the number of sub-bands, each one having B_j RBs. r_i is the rate that will be used by UE i (assumed to be a number of bytes per RB), $CQI_{i,j}$ is the CQI of UE i in sub-band j (translated to a number of bytes per RB as well), p_i and Q_i are the padding and queue length for UE i. $b_{i,j}$ is a binary variable that is equal to 1 if UE *i* is allocated RBs in sub-band *j*, and $x_{i,j}$ is the number of RBs that it has allocated in that sub-band.

The above one is a mixed-integer nonlinear problem, which can be linearized using standard enumeration techniques and solved through CPLEX (which only solves *linear* problems). Therefore, we can use the optimal solutions to benchmark heuristics, provided that we interface SimuLTE and CPLEX. There are two ways to do so: one is to use its C++ API, which requires integrating it into eclipse and learning to use it, both being nontrivial. The other, which we chose, is to manage files via `tmpfs`, i.e., save files in volatile memory: while this is less efficient, it only requires one to be able to write down a scheduling problem in the LP file format used by CPLEX (which is very intuitive) *and* to write a parser for SOL CPLEX output solution files, which are XML-based.

At each TTI, we do the following: first we obtain the constants, i.e., queue size, per-sub-band CQIs, etc. Then, we generate an LP file describing the linearized problem. The problem is then solved by invoking CPLEX with the LP file as an input. The resulting SOL file can be easily parsed via xml-related functions given by the `omnetpp` environment. Figure 16 shows a snippet of the output file that defines the final allocation: each line specifies the *name* of a variable (x_{i_j}, in this case) and its *value*, i.e., the RBs allocated to user *i* within sub-band *j*. Finally, the allocation is enforced in the simulated system and the rest of the operations for the TTI are executed.

The multiband MaxC/I heuristic that we choose to benchmark is the following: we sort all UEs by decreasing *sum of per-sub-band CQIs*, and we allocate enough RBs to empty their queue (as if multiple CQIs could be used simultaneously). Then we use the minimum CQI among the selected sub-bands. We compare the above solutions in a scenario with six RBs, a number of UEs ranging from 10 to 75, and VoIP traffic. Both solutions achieve the same throughput, since the system is still far from saturation conditions, but the heuristics uses up to 30 % more RBs, as shown in Fig. 17.

Fig. 16 Snippet of CPLEX XML output file

```
[...]
<variable name="x4_1" value="1"/>
<variable name="x4_2" value="1"/>
<variable name="x4_3" value="1"/>
<variable name="x4_4" value="1"/>
<variable name="x4_5" value="1"/>
<variable name="x5_0" value="1"/>
[...]
```

Fig. 17 RB occupancy
comparison between the
heuristic and optimum
solutions

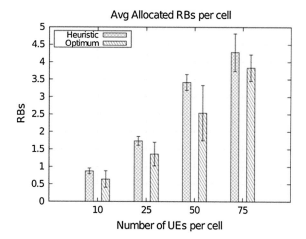

Fig. 17 RB occupancy comparison between the heuristic and optimum solutions

6   Conclusions

This paper described SimuLTE, an open-source LTE/LTE-A simulator based on the
OMNeT++ framework. The current release supports the data plane of the radio
access network, with a full protocol stack, a realistic physical layer, and scheduling
capabilities. We have described the modeling approach in detail, especially
focusing on the MAC and resource allocation models. Then we explained some
validation of the model and the main verification techniques that have been used.
Moreover, we have shown some examples of performance evaluation that can be
obtained through SimuLTE. There are at least two directions in which this work can
be extended, most of which are being pursued at the time of writing:

- As explained in the introduction, there are other LTE simulators and emulators
  available for the public domain. A comparison of SimuLTE against these is on
  the agenda. We expect this comparison to reinforce validation, to provide
  insight regarding the different features and capabilities of the simulators, and to
  benchmark their efficiency.
- The LTE-A technology is evolving, and SimuLTE evolves accordingly. We are
  in fact adding several models to SimuLTE. Among the planned future releases
  we mention support to device-to-device (D2D) communications and Coordinated Multi-Point transmission (CoMP).

Acknowledgments We would like to thank all those who have contributed to the SimuLTE code
as former students of the University of Pisa, namely Matteo Maria Andreozzi of NVIDIA, UK,
Daniele Migliorini of Aruba, Italy, Giovanni Accongiagioco of CNR, Italy, Generoso Pagano of
INRIA Grenoble Rhône-Alpes, France, Vincenzo Maria Pii of ZHAW, Switzerland. Many thanks
also to Andras Varga, Rudolf Hornig, Levente Mészáros and Gábor Tabi of SimulCraft for support
and discussion on the code.

Appendix

See Table 2.

Table 2 LTE-related acronyms used in the paper

Acronym	Definition
AM	Acknowledged RLC mode
BLER	Block error rate
BSR	Buffer status report
CID	Connection identifier
CoMP	Coordinated multi-point
CQI	Channel quality indicator
DL	Downlink
DRR	Deficit round robin
eNB	Evolved node-B
EPC	Evolved packet core
e-ICIC	Enhanced intercell interference cancelation
FDD	Frequency division duplexing
H-ARQ	Hybrid automatic repeat reQuest
LCID	Logical connection identifier
LTE	Long-term evolution
LTE-A	Long-term evolution advanced
MaxC/I	Maximum carrier over interference
PDCCH	Physical downlink control CHannel
PDCP	Packet data convergence protocol
PDU	Protocol data unit
PF	Proportional fair
PRACH	Physical random access CHannel
RAC	Random access procedure
RB	Resource block
RLC	Radio link control
RRC	Radio resource control
RU	Remote unit
SDU	Service data unit
SU-MIMO	Single-User MIMO
TM	Transparent RLC mode
TTI	Transmission time interval
UE	User equipment
UL	Uplink
UM	Unacknowledged RLC mode

References

1. 3GPP—TS 36.300. Evolved universal terrestrial radio access (E-UTRA) and evolved universal terrestrial radio access network (E-UTRAN). Overall description; Stage 2
2. 3GPP—TS 36.913. Requirements for further advancements for evolved universal terrestrial radio access (E-UTRA) (LTE-Advanced)
3. Mehlfuerer, C., Wrulich, M., Ikuno, J.C., Bosanska D., Rupp, M.: Simulating the long term evolution physical layer. In: Proceedings of the 17th EUSIPCO, Glasgow, UK (2009)
4. Bouras, C., Diles, G., Kokkinos, V., Kontodimas, K., Papazois, A.: A Simulation Framework for Evaluating Interference Mitigation Techniques in Heterogeneous Cellurar Environments. Wireless Personal Communications, Springer (2013)
5. Ikuno, J.C., Wrulich, M., Rupp, M.: System level simulation of LTE networks. In: Proceedings of IEEE VTC-Spring, Taipei, Taiwan, pp. 1–5. May 2010
6. Piro, G., Grieco, L.A., Boggia, G., Capozzi, F., Camarda, P.: Simulating LTE cellular systems: an open-source framework. IEEE Trans. Veh. Technol. **60**(2), (2011)
7. Perrone, L.F., Cicconetti, C., Stea, G., Ward, B.: On the automation of computer network simulators. In: Proceedings of SIMUTools'09, Rome, Italy, 3–5 March 2009
8. Kurkowski, S., Camp, T., Colagrosso, M.: MANET simulation studies: the incredibles. ACM SIGMOBILE MCCR **9**(4), 50–61 (2005)
9. Ns-3 homepage. http://www.nsnam.org/. Accessed April 2014
10. Baldo, N., Miozzo, M., Requena-Esteso, M., Nin-Guerrero, J.: An open source product-oriented LTE network simulator based on ns-3. In: Proceedings of ACM MSWiM'11, Miami, US, Nov 2011
11. Perrone, L.F., Main, C., Ward, B.C.: SAFE: simulation automation framework for experiments. In: Proceedings of IEEE WSC 2012, Berlin, DE, 9–12 Dec 2012
12. Gomez, I.: libLTE: Open source 3GPP LTE library. http://sourceforge.net/projects/liblte/ (Oct 2013)
13. Wojtowicz, B.: openLTE: an open source 3GPP LTE implementation. http://sourceforge.net/projects/openlte/ (Sept 2014)
14. Institut Eurecom. OpenAirInterface. https://twiki.eurecom.fr/twiki/bin/view/OpenAirInterface/WebHome
15. Varga, A., Hornig, R.: An overview of the OMNeT++ simulation environment. In: Proceedings of the SIMUTools '08, Marseille, France, March 2008
16. Weingartner, E., Vom Lehn, H., Wehrle, K.: A performance comparison of recent network simulators. In: Proceedings of IEEE ICC'09, Dresden, DE, 14–18 June 2009
17. 3GPP—TS 36.322. Radio link control (RLC) protocol specification
18. 3GPP—TS 36.321. Medium access control (MAC) protocol specification
19. 3GPP—TS 36.213. Physical layer procedures
20. Jakes, W.C.: Microwave Mobile Communications. Wiley (1975)
21. Zhou, D., Baldo, N., Miozzo, M.: Implementation and validation of LTE downlink schedulers for ns-3. In: Proceedings of SIMUTools '13, Cannes, France, March 3013
22. Fantini, R., Sabella, D., Caretti, M.: An E3F based assessment of energy efficiency of relay nodes in LTE-advanced networks. In: Proceedings of IEEE PIMRC 2011, pp. 182–186. 11–14 Sept 2011
23. Accongiagioco, G., Andreozzi, M.M., Migliorini, D., Stea, G.: Throughput-optimal resource allocation in LTE-advanced with distributed antennas. Comput. Netw. **57**, 3997–4009 (2013)

Sensitivity Estimation Using Likelihood Ratio Method with Fixed-Sample-Path Principle

Koji Fukuda and Yasuyuki Kudo

Abstract The likelihood ratio method (LRM) is an efficient indirect method for estimating the sensitivity of given expectations with respect to parameters by Monte Carlo simulation. The restriction on application of LRM to real-world problems is that it requires explicit knowledge of the probability density function (pdf) to calculate the score function. In this study, a fixed-sample-path method is proposed, which derives the score function required for LRM not via the pdf but directly from a constructive algorithm that computes the sample path from parameters and random numbers. The boundary residual, which represents the correction associated with the change of the distribution range of the random variables in LRM, is also derived. Some examples including the estimation of risk measures (Greeks) of option and financial flow of funds networks showed the effectiveness of the fixed-sample-path method.

Keywords Monte Carlo simulation · Sensitivity · Likelihood ratio · Score function · Fixed-sample-path principle

1 Introduction

Given a system of interest, it is a major concern for engineers and designers to understand how to make the system behavior "desirable" by changing the parameters. To this end, knowledge about the relationship (or the sensitivity) between the system parameters and the system behaviors is required. However, for complicated and probabilistic systems, the relationship between parameters and system behavior is often unclear, so Monte Carlo simulation is needed to estimate the relation.

K. Fukuda (✉) · Y. Kudo
Central Research Laboratory, Hitachi, Ltd, Kokubunji-shi, Tokyo, Japan
e-mail: koji.fukuda.jf@hitachi.com

Y. Kudo
e-mail: yasuyuki.kudo.ec@hitachi.com

© Springer International Publishing Switzerland 2015
M.S. Obaidat et al. (eds.), *Simulation and Modeling Methodologies,
Technologies and Applications*, Advances in Intelligent Systems
and Computing 402, DOI 10.1007/978-3-319-26470-7_6

Let X be a random variable describing the system behaviors under consideration and x be its sample value (sample path). Here, X can be a multidimensional vector and/or a family of random variables indexed by "time" t (random process). Therefore, X should be denoted as $\mathbb{X}(t)$ in nature, but we henceforth use X to avoid cumbersome notation. Assume X is dependent on the system parameters z, where $z = (z_i)_{i=1...N}$ is an N-dimensional vector, and let $f(x, z)$ be the probability density function (pdf) of X. There exists a behavior evaluation function $a(x)$ that maps the system behavior x to a real value $a(x)$. We call the expectation value $A(z)$ of $a(x)$, i.e.,

$$A(z) = \mathbb{E}[a(X)] = \int a(x) f(x, z) \, dx, \tag{1}$$

a "system evaluation function."

To calculate $A(z)$ of real interesting systems, the fact that density function $f(x, z)$ is often unknown becomes an obstacle. However, even if $f(x, z)$ itself is unknown, in many cases, the system behavior, which makes the distribution of X, is known and modeled, and Monte Carlo simulation is applicable.

Let $\mathbb{W} = (W_j)_{j=1...M}$ be an M-dimensional vector of random numbers and $w = (w_j)_{j=1...M}$ denote its sample value. We suppose that the simultaneous probability density function $g(w)$ of $\mathbb{W}$ is well-known, and we can easily generate random numbers of this distribution on computers. Examples of $\mathbb{W}$ include M number of independent random numbers from a uniform distribution on $(0, 1)$ or the standard normal distribution $N(0, 1)$. Assuming that there exists a function $x = x(w, z)$, i.e., a constructive algorithm to compute x from the parameters z and random numbers $w =$

$$(w_j)_{j=1...M'}$$
$$A(z) = \mathbb{E}[a(X)] = \int_\Omega a(x(w, z))g(w)dw \tag{2}$$

holds, where $\Omega \subset \mathbb{R}^M$ denotes the support of $g(w)$. Using the L set of random numbers $(w^{(k)})_{k=1...L} = (w_j^{(k)})_{j=1...M, k=1...L}$, we can estimate $A(z)$ by

$$A(z) \cong \frac{1}{L} \sum_{k=1}^{L} a\left(x\left(w^{(k)}, z\right)\right). \tag{3}$$

Now, our goal is to estimate the sensitivity

$$\frac{\partial A(z)}{\partial z} = \left(\frac{\partial A(z)}{\partial z_i}\right)_{i=1...N} \tag{4}$$

of the behavior evaluation function with respect to all of the N number of parameters z by Monte Carlo simulation. A direct method to this end is the finite differential method (FDM), which reruns a set of Monte Carlo simulations under a small variation Δz_i of the parameter z_i and approximates

$$
\begin{aligned}
\frac{\partial A(z)}{\partial z_i} &\cong \frac{A(z + \Delta z_i) - A(z)}{\Delta z_i} \\
&\cong \frac{1}{\Delta z_i} \left[\frac{1}{L} \sum_{k=1}^{L} a\left(x\left(w'^{(k)}, z + \Delta z_i\right)\right) - \frac{1}{L} \sum_{k=1}^{L} a\left(x\left(w^{(k)}, z\right)\right) \right],
\end{aligned}
\tag{5}
$$

where $z + \Delta z_i = (z_1, z_2, \dots, z_i + \Delta z_i, \dots, z_N)$. The problem of FDM is that the convergence speed of Eq. (5) is slow. The variance of estimation value $A(z)$ by Eq. (5) is proportional to L^{-1}, whereas that of estimation value $\frac{\partial A(z)}{\partial z_i}$ by Eq. (3) is proportional to $L^{-1/4}$ (for independent sampling) or $L^{-1/3}$ (for common sampling) [1]. Moreover, since the rerun of a set of Monte Carlo simulations is needed for each parameter z_i, the computational time might be impractical if the number of parameters N is large.

There exist two known indirect methods for estimating the sensitivity more efficiently than FDM: the pathwise derivative method (PDM) [2–5] and the likelihood ratio method (LRM) [2, 3, 6–8]. PDM (also called the "infinitesimal perturbation method") is based on the idea of differentiating Eq. (2) with respect to the parameters z,

$$
\begin{aligned}
\frac{\partial A(z)}{\partial z} &= \frac{\partial}{\partial z} \int_{\Omega} a(x(w, z)) g(w) \, dw \\
&= \int_{\Omega} \frac{\partial a(x(w, z))}{\partial z} g(w) \, dw.
\end{aligned}
\tag{6}
$$

Assuming this holds, we estimate the sensitivity by

$$
\frac{\partial A(z)}{\partial z} \cong \frac{1}{L} \sum_{k=1}^{L} \frac{\partial a(x(w, z))}{\partial z}.
\tag{7}
$$

PDM is quite effective in terms of its small variance and fast convergence speed. However, it has a limited range of application because interchanging the order of integration and differentiation in Eq. (6) requires that $a(x(w, z))$ is (almost surely) continuous with respect to z, which is often not the case.

In comparison, LRM (also called the "score function method") is based on the idea of differentiating Eq. (1) with respect to the parameters z,

$$
\begin{aligned}
\frac{\partial A(z)}{\partial z} &= \frac{\partial}{\partial z} \int a(x) f(x, z) \, dx \\
&= \int a(x) \frac{\partial f(x, z)}{\partial z} \, dx \\
&= \int a(x) h(x, z) f(x, z) \, dx,
\end{aligned}
\tag{8}
$$

Where

$$h(x, z) = \frac{\partial f(x, z)}{\partial z} / f(x, z) = \frac{\partial \log[f(x, z)]}{\partial z} \tag{9}$$

is called a "score function". Assuming this holds, we estimate the sensitivity by

$$\frac{\partial A(z)}{\partial z} \cong \frac{1}{L} \sum_{k=1}^{L} a\left(x\left(w^{(k)}, z\right)\right) h\left(w^{(k)}, z\right) \tag{10}$$

LRM has a wider range of application than does PDM because the pdf $f(x, z)$ is typically a smooth function with respect to the parameters z, whereas $a(x(w, z))$ is not. An exception that does not satisfy Eq. (8) will be discussed later in Sect. 2.3.

The restriction on the application of LRM to real systems is that it requires explicit knowledge of the pdf $f(x, z)$ to calculate the score function $h(x, z)$ from Eq. (9). This restriction might seem not to be a problem because we know the pdf of the random numbers $g(w)$ and the constructive algorithm, which computes the sample path $x = x(w, z)$ from w, and $f(x, z)$ can be calculated from these in theory. In fact, pdf $f(x, z)$ can be decomposed to the products of some conditional probability density functions for some systems, such as Markov chains, and discrete event systems without agent loop [2, 4]. Nevertheless, considering that we apply Monte Carlo simulation due to the lack of explicit knowledge on the pdf $f(x, z)$, the derivation of the score function from Eq. (9) is an intrinsically problematic approach.

In this study, we propose a "fixed-sample-path" method. Using this method, we can derive the score function not via the pdf $f(x, z)$ but directly from the constructive algorithm that computes the sample path $x = x(w, z)$ from the parameters z and the sample values w of the random numbers.

The paper is organized as follows. In Sect. 2, we describe the idea of the fixed-sample-path method and its formulation. In Sect. 3, the fixed-sample-path method is applied to two simple systems: a system with two-dimensional uniform random variables and the estimation of risk measures (Greeks) of option pricing in finance. Section 4 is a description of a more complicated example: a financial flow of funds network of 25 companies. Finally, Sect. 5 is the conclusion.

2 LRM with Fixed-Sample-Path Principle

2.1 Basic Idea

The reason that the calculation of the score function $h(x, z)$ requires explicit knowledge of the pdf $f(x, z)$ is that LRM in Eq. (8) is derived by differentiating Eq. (1), which depends on $f(x, z)$. In comparison, Eq. (2), which is used to derive PDM in Eq. (6), only depends on $g(w)$ and $x(w, z)$, which we know explicitly. Can we not derive LRM from Eq. (2) instead of Eq. (1)?

Let us consider sensitivity with respect to the ith parameter z_i. The key idea is, given a sample path $x = x(w, z)$ and a small variation Δz_i of the parameter z_i, to consider the small variation Δw of the sample values w of random variables that cancels the parameter variation, i.e., Δw satisfying.

$$x = x(w, z) = x(w + \Delta w, z + \Delta z_i). \qquad (11)$$

Since $w + \Delta w$ is, of course, not distributed with pdf $g(w)$, the expectation $A(z + \Delta z_i)$ is no longer calculable with simple expectation Eq. (3). However, using the importance sampling method, we can estimate $A(z + \Delta z_i)$ by averaging up $x(w, z) = x(w + \Delta w, z + \Delta z_i)$ with appropriate weights.

Concretely speaking, given a sample path $x = x(w, z)$, we consider a "fixed-sample-path" derivative of the random variables w with respect to the parameter z_i under the condition of fixing the sample path x:

$$\left. \frac{\partial w}{\partial z_i} \right|_{x = \text{const}} = - \frac{\frac{\partial x}{\partial z_i}}{\frac{\partial x}{\partial w}}. \qquad (12)$$

We note that the right-hand side of Eq. (12) is a formal expression because x might be a multidimensional vector. As discussed later in Sect. 2.3, the fixed-sample-path derivative $\left. \frac{\partial w}{\partial z_i} \right|_{x = \text{const}}$ can be calculated relatively easily from the constructive algorithm of the function $x = x(w, z)$. We use $\left. \frac{\partial w}{\partial z_i} \right|_{x = \text{const}}$ to calculate the score function $h(x, z)$.

2.2 Formulation of LRM

Let Δz_i be a small variation of the ith parameter z_i. Then, the expectation $A(z + \Delta z_i)$ under the parameter values $\mathbb{Z} + \Delta z_i = (z_1, z_2, \ldots, z_i + \Delta z_i, \ldots, z_N)$ is

$$
\begin{aligned}
&A(z + \Delta z_i) \\
&= \int_\Omega a(x(w, z + \Delta z_i)) g(w) \, dw \\
&= \int_{\Omega'} a(x(w' + \Delta w, z + \Delta z_i)) \frac{g(w' + \Delta w)}{\left| \frac{d_{w'}}{d_w} \right|} \, dw' \\
&= \int_{\Omega'} a(x) \frac{g(w' + \Delta w)}{\left| \frac{d_{w'}}{d_w} \right|} \, dw' \\
&= \int_{\Omega'} a(x) \frac{g(w + \Delta w)}{\left| i - \frac{d}{d_w} \left(\left. \frac{\partial w}{\partial z_i} \right|_{x = \text{const}} \right) \Delta z_i \right|} \, dw \\
&= \int_{\Omega'} a(x) \left\{ g(w) + \frac{\partial g}{\partial w} \cdot \left[\left(\left. \frac{\partial w}{\partial z_i} \right|_{x = \text{const}} \right) \right] \Delta z_i \right\} \left\{ 1 + \text{tr} \left[\frac{d}{d_w} \left(\left. \frac{\partial w}{\partial z_i} \right|_{x = \text{const}} \right) \right] \Delta z_i \right\} \, dw,
\end{aligned}
$$

$$\qquad (13)$$

where $\left.\frac{\partial w}{\partial z_i}\right|_{x=const}$ is the ratio between the small parameter variation Δz_i and the small variation Δw of the random variable, which keeps x constant, i.e., $x = x(w, z) = x(w + \Delta w, z + \Delta z_i)$ holds. Here, $\mathbb{I}$ is an identity matrix, tr denotes the trace of a matrix, and a centered dot " $\cdot$ " denotes the inner products of vectors. Also, $\left|\frac{dw'}{dw}\right|$ is the Jacobian determinant corresponding to the change of variables from w to $w' = w - \Delta W$, and Ω' is the image of Ω under this transform. To get from line 5 to line 6, we use the following relation. For a matrix $B(\varepsilon) = (b_{ij}) = \mathbb{I} - \varepsilon D$ with a matrix $D = (d_{ij})$ and a small real number $\varepsilon \ll 1$, the first-order Taylor approximation around $\varepsilon = 0$ leads to

$$
\begin{aligned}
\frac{1}{|\mathbb{I} - \varepsilon C|} &= \frac{1}{|B(\varepsilon)|} \\
&\cong \frac{1}{|B(0)|} - \varepsilon |B(0)| \sum_{i,j} \left(c_{ji}\big|_{\varepsilon=0} \right) \left(\frac{\partial b_{ij}}{\partial \varepsilon}\bigg|_{\varepsilon=0} \right) \\
&= \frac{1}{|\mathbb{I}|} - \varepsilon |\mathbb{I}| \sum_{i,j} \delta_{ji} \left(-d_{ij} \right) \\
&= 1 + \varepsilon\, \mathrm{tr}(C),
\end{aligned}
\tag{14}
$$

where (c_{ij}) is the inverse matrix of $B(\varepsilon)$ and (δ_{ij}) is an identity matrix.
Using Eq. (13), we obtain

$$
\begin{aligned}
\frac{\partial A}{\partial z_i} &= \lim_{\Delta z_i \to 0} \frac{A(z + \Delta z_i) - A(z)}{\Delta z_i} \\
&= \int_\Omega a(x) \sum_{j=1}^{M} \left[\frac{\partial}{\partial w_j} \left(\frac{\partial w_j}{\partial z_i}\bigg|_{x=const} \right) g(w) + \frac{\partial g}{\partial w_j} \left(\frac{\partial w_j}{\partial z_i}\bigg|_{x=const} \right) \right] - R_i \\
&= \int_\Omega a(x)\, h_i(w, z)\, g(w) d w - R_i \\
&= \mathbb{E}[a(X)\, h_i(w, z)] - R_i.
\end{aligned}
\tag{15}
$$

Therefore, the score function $h_i(w, z)$ can be written as

$$
h_i(w, z) = \sum_{j=1}^{M} \left[\frac{\partial}{\partial w_j} \left(\frac{\partial w_j}{\partial z_i}\bigg|_{x=const} \right) + \frac{1}{g(w)} \frac{\partial g}{\partial w_j} \left(\frac{\partial w_j}{\partial z_i}\bigg|_{x=const} \right) \right].
\tag{16}
$$

In addition, R_i is a term that is equivalent to the correction amount associated with the changing integration range Ω' of Eq. (13) to Ω. We call R_i a "boundary residual." From Eqs. (13) and (15), we obtain

$$R_i = \lim_{\Delta z_i \to 0} \frac{1}{\Delta z_i} \left\{ \int_{\Omega - \Omega'} a(x)\{1 + h_i(w, z)\Delta z_i\} \, g(w) \, dw \right.$$

$$\left. - \int_{\Omega - \Omega'} a(x)\{1 + h_i(w, z)\Delta z_i\} \, g(w) \, dw \right\}$$

$$= \int_{\partial\Omega} a(x) \, g(w) \left(\left.\frac{\partial w}{\partial z_i}\right|_{x = \text{const}} \right) \cdot n \, dw \tag{17}$$

$$= \int_{\Omega} \text{div} \left[a(x) \, g(w) \left(\left.\frac{\partial w}{\partial z_i}\right|_{x = \text{const}} \right) \right] dw$$

$$= \mathbb{E} \left[\frac{1}{g(w)} \text{div} \left[a(x) \, g(w) \left(\left.\frac{\partial w}{\partial z_i}\right|_{x = \text{const}} \right) \right] \right],$$

where $\Omega' - \Omega$ and $\Omega - \Omega'$ are the differential sets, $\partial\Omega$ is the boundary of Ω, n is the outward pointing unit vector of $\partial\Omega$, and div denotes the divergence with respect to w. From Eq. (10), if $\left.\frac{\partial w}{\partial z_i}\right|_{x = \text{const}}$ is zero (vector) on the boundary $\partial\Omega$, then

$$\frac{\partial A}{\partial z_i} = \int_{\Omega} a(x) \, h_i(w, z) \, g(w) \, dw \tag{18}$$

$$= \mathbb{E}[a(X) \, h_i(w, z)]$$

holds.

Here, we calculate the score function $h_i(w, z)$ for the typical distributions $g(w)$ of random variables with Eq. (16) for future convenience. For random variables w following the M-dimensional uniform distributions, considering $\frac{\partial g}{\partial w_j} = 0$, we obtain

$$h_i(w, z) = \sum_{j=1}^{M} \frac{\partial}{\partial w_j} \left(\left.\frac{\partial w_j}{\partial z_i}\right|_{x = \text{const}} \right) \tag{19}$$

For random variables w following the M-dimensional independent standard normal distributions, considering that pdf is written as $g(w) = \prod_{j=1}^{M} g_N(w_j)$, where $g_N(x) = \frac{1}{\sqrt{2\pi}} e^{-\frac{x^2}{2}}$ is the pdf of (one-dimensional) standard normal distributions and $\frac{g_N'(x)}{g_N(x)} = -w$ holds, we obtain

$$h_i(w, z) = \sum_{j=1}^{M} \left[\frac{\partial}{\partial w_j} \left(\left.\frac{\partial w_j}{\partial z_i}\right|_{x = \text{const}} \right) - w_j \left(\left.\frac{\partial w_j}{\partial z_i}\right|_{x = \text{const}} \right) \right]. \tag{20}$$

Once the score function $h_i(w, z)$ is calculated, we can estimate the sensitivity $\partial A/\partial z_i$ by Monte Carlo simulation using the LRM method

$$\frac{\partial A}{\partial z_i} = \mathbb{E}[a(X)\, h_i(\mathbb{W}, z)] - R_i$$

$$\cong \frac{1}{L} \sum_{k=1}^{L} a\left(x\left(w^k, z\right)\right) h_i\left(w^k, z\right) - R_i. \tag{21}$$

2.3   Discussion

The range in application of the fixed-sample-path method depends on the existence (and computability) of the fixed-sample-path derivative $\left.\frac{\partial w}{\partial z_i}\right|_{x=\text{const}}$. In other words, given a sample path x and a small variation Δz_i of the parameter z_i, the existence of the small variation Δw of the sample values w of random variables that satisfy Eq. (11) is the key to the fixed-sample-path method. This in general is not necessarily the case because the dimension of x, which we must keep fixed, can be bigger than the dimension of the random variables w. Nevertheless, for many applications, especially for the case in which the system behavior X is a time series $\mathbb{X}(t)$, the fixed-sample-path method is applicable. Here, we exemplify two cases. The first is the case in which the system behavior $\mathbb{X}(t)$ can be written as

$$\mathbb{X}(t) = f(t, y(w, z)), \tag{22}$$

i.e., $\mathbb{X}(t)$ follows a deterministic function $f(t, y)$ identified by a random variable y of which distribution is determined by the parameter z. Clearly,

$$\left.\frac{\partial w}{\partial z}\right|_{x(t)=\text{const}} = \left.\frac{\partial w}{\partial z}\right|_{y=\text{const}} \tag{23}$$

holds in this case. The second is the case in which the time evolution of the system behavior $\mathbb{X}(t)$ is determined by the relation

$$\mathbb{X}(t+1) = f\left(X(0), X(1), \ldots, X(t-1), \mathbb{W}^{(t)}, Z\right) \tag{24}$$

i.e., $\mathbb{X}(t+1)$ at $t+1$ is determined by the past history of $\mathbb{X}$ before t, a random variable $\mathbb{W}^{(t)}$ that is newly generated at each time t, and the parameter z. In this case, given a small variation Δz_i of the parameter, the small variation $\Delta \mathbb{W}^{(t)}$ of the random variables that cancels out Δz_i can be determined sequentially from the initial time $t =$ to the ending time $t = T$ from the relation

$$\begin{aligned} &f\left(X(0), X(1), \ldots, X(t-1), \mathbb{W}^{(t)}, z\right) \\ &= f\left(X(0), X(1), \ldots, X(t-1), \mathbb{W}^{(t)} + \Delta\mathbb{W}^{(t)}, z + \Delta z\right). \end{aligned} \tag{25}$$

A typical example of this case is Ito process, which is a well-known stochastic process in the financial field

$$dX_t = b(t, X_t)dt + \sigma(t, X_t)dB_t, \tag{26}$$

where B_t is a standard Brownian motion. We will show a brief example later in Sect. 3.2.

Another point we want to note is the boundary residual R_i. An exception that the conventional LRM with Eq. (8) is not applicable includes the case where the integral range of Eq. (1) depends on the parameter z, i.e., the distribution range of the system behavior X varies depending on the parameter z. The boundary residual R_i explicitly represents the correction amount associated with the change of the distribution range of X, and we can extend the range of application of LRM to this case using R_i as seen in Eq. (21). However, there would be considerable difficulty in numerical calculation of R_i by Eq. (17). Numerically efficient method for calculating R_i is one of the future works of this study.

3   Example Calculations

In this section, we apply the calculation method proposed in Sect. 2 to simple examples.

3.1   Two-Dimensional Uniform Distribution

As the first example, we consider a simple system consisting of a single parameter, $N = 1$, and two random variables, $M = 2$. Let $w = (w_1, w_2)$ be the two-dimensional uniform random numbers on $(0, 1) \times (0, 1)$, i.e., $g(w_1, w_2) = 1$ on $\Omega = \{(w_1, w_2); 0 < w_1, w_2 < 1\}$, and otherwise, $g(w_1, w_2) = 0$, and let z be a real-valued parameter. Assuming the system behavior as $x = (x_1, x_2) = \left(\sin(4z\, w_1), (2z + w_2)^2 + w_1 \right)$ and the behavior evaluation function as $a(x) = x_1 + x_2$, let us consider the sensitivity of the parameter value z to the expectation $A(z) = \mathbb{E}[a(x)]$. Note that since the distribution range of $x_2 = (2z + w_2)^2 + w_1$ depends on the parameter z, the boundary residual R is, as we look later, not zero for this example.

Direct Calculation: We can easily calculate the expectation $A(z)$ and its sensitivity $A'(z)$ by direct integration:

$$A(z) = \iint_{\Omega} \left\{ \sin(4z\, w_1) + z^2\, w_2 + w_1 \right\} dw_1 \, dw_2$$

$$= \frac{\sin^2(2z)}{2z} + 4z^2 + 2z + \frac{5}{6}, \tag{27}$$

$$A'(z) = \frac{4z \sin(4z) + \cos(4z) + 8z^2(4z+1) - 1}{4z^2}.$$

Calculation with Fixed-sample-Path Method: Let us calculate $A'(z)$ using LRM with the score function calculated by the fixed-sample-path method.

- **Step 1:** Generating random numbers
  Generate 2L number of random numbers under the uniform distribution on an interval $(0, 1)$, and represent them as $\mathbb{w}^{(k)} = \left(w_1^{(k)}, w_2^{(k)} \right)_{k=1...L}$.

- **Step 2:** Performing Monte Carlo simulation
  Calculate the system behavior $\mathbb{x}^{(k)} = \left(x_1^{(k)}, x_2^{(k)} \right)_{k=1...L}$ from $\left(w_1^{(k)}, w_2^{(k)} \right)_{k=1...L}$.
  Although this is quite easy for this simple example, this step might be computer-intensive for real-world problems.

- **Step 3:** Calculating $\frac{\partial \mathbb{w}}{\partial z}\big|_{\mathbb{x}=\text{const}}$
  For all $\left(w_1^{(k)}, w_2^{(k)} \right)_{k=1...L}$, calculate

$$\frac{\partial w_1}{\partial z}\bigg|_{\mathbb{x}=\text{const}} = -\frac{\partial x_1}{\partial z} \bigg/ \frac{\partial x_1}{\partial w_1} = -\frac{w_1}{z},$$

$$\frac{\partial w_2}{\partial z}\bigg|_{\mathbb{x}=\text{const}} = -\left(\frac{\partial x_2}{\partial z} + \frac{\partial x_2}{\partial w_1}\frac{\partial w_1}{\partial z}\bigg|_{\mathbb{x}=\text{const}} \right) \bigg/ \frac{\partial x_2}{\partial w_2} = \frac{w_1 - 4zw_2 - 8z^2}{4z^2 + 2zw_2}. \tag{28}$$

If an analytical calculation is impossible, we can adopt numerical approaches. For example, considering a small Δz (e.g., $\Delta z = 0.01$), find numerically $(\Delta w_1, \Delta w_2)$, which satisfies $\mathbb{x}((w_1, w_2), z) = \mathbb{x}((w_1 + \Delta w_1, w_2 + \Delta w_2), z + \Delta z)$, and approximate $\frac{\partial w_j}{\partial z}\big|_{\mathbb{x}=\text{const}}$ by $\frac{\Delta w_j}{\Delta z}$, $(j=1,2)$.
From Eq. (15), $\frac{\partial \mathbb{w}}{\partial z}\big|_{\mathbb{x}=\text{const}}$ turns out to be nonzero on the boundary $\partial\Omega$, and therefore, consideration of the boundary residual R is required.

- **Step 4:** Calculating $\frac{\partial}{\partial w_j}\left(\frac{\partial w_j}{\partial z}\big|_{\mathbb{x}=\text{const}} \right)$
  Calculate the derivatives of $\frac{\partial w_j}{\partial z}\big|_{\mathbb{x}=\text{const}}$ with respect to w_j:

$$\frac{\partial}{\partial w_1}\left(\frac{\partial w_1}{\partial z}\bigg|_{\mathbb{x}=\text{const}} \right) = -\frac{1}{z},$$

$$\frac{\partial}{\partial w_2}\left(\frac{\partial w_2}{\partial z}\bigg|_{\mathbb{x}=\text{const}} \right) = -\frac{w_1}{2z(2z+w_2)^2}. \tag{29}$$

We can also calculate them numerically by finding $\frac{\partial w_i}{\partial Z}\Big|_{x=\text{const}}$ under a small variation of w_j.

- **Step 5:** Calculating the score function $h(w, z)$

From Eq. (19), which is the case for uniformly distributed random variables, we obtain

$$h(w, z) = \sum_{j=1}^{2} \frac{\partial}{\partial w_j}\left(\frac{\partial w_j}{\partial z_i}\Big|_{x=\text{const}}\right) = -1 - \frac{w_1}{2z(w_2 + 2z)^2}. \tag{30}$$

Numerically, calculate the sum of $\frac{\partial}{\partial w_j}\left(\frac{\partial w_j}{\partial Z}\Big|_{x=\text{const}}\right)$.

- **Step 6:** Calculating the sensitivity $A'(z)$

To derive $A'(z)$, the boundary residual R is required. R is calculated using line 2 of Eq. (17):

$$\begin{aligned}
R - \int_0^1 a(x)\Big\{s(w, z)\big|_{w_2=1} - s(w, z)\big|_{w_2=0}\Big\}dw_1 \\
+ \int_0^1 a(x)\Big\{s(w, z)\big|_{w_1=1} - s(w, z)\big|_{w_1=0}\Big\}dw_2 \\
= \frac{4Z\sin(4Z) + \cos(4Z) + 8Z^2(4Z+1) - 1}{4Z^2},
\end{aligned} \tag{31}$$

which is the same result of direct calculation Eq. (27), as expected. Numerically, apply the Monte Carlo LRM method with the score function $s(w, z)$:

$$A'(z) \cong \frac{1}{L}\sum_{k=1}^{L} a\left(x^{(k)}\right)h\left(w^{(k)}, z\right) - R \tag{32}$$

3.2 Risk Measures (Greeks) in Finance

Currently, financial engineering is one of the most active fields of investigation that uses the Monte Carlo method and option pricing, and designing hedge strategies are especially important applications.

Let us calculate some typical risk measures (Greeks), Delta Δ, Vega ν, and Rho ρ for an Asian–European call option using LRM with the fixed-sample-path method. We suppose the underlying asset price $X(t)$ of the option follows a geometric Brownian motion (GBM) under a risk-neutral probability measure,

$$dX(t) = r\,X(t)dt + \sigma X(t)\,dB(t), \tag{33}$$

with the spot (initial) price $X(0) = X_0 > 0$, where r is the risk-free interest rate, σ is the volatility of the asset price, and $B(t)$ is the standard Brownian motion. Equation (33) has an explicit solution

$$X(t) = X(0)\exp\left[\left(r - \frac{\sigma^2}{2}\right)t + \sigma B(t)\right]. \tag{34}$$

The discounted value C_A of an Asian (average-price)–European call option derived from this asset with expiration date T and strike price K satisfies

$$C_A = e^{-rT}\mathbb{E}\left[\max\left(\frac{1}{T}\int_0^T X(t)dt - K, 0\right)\right], \tag{35}$$

where $\mathbb{E}[\cdot]$ denotes the expectation under the risk-neutral probability measure. Dividing T into M segments, we discretize Eq. (34) to

$$X_{j+1} = X_j \exp\left[\left(r - \frac{\sigma^2}{2}\right)\Delta t + \sigma\sqrt{\Delta t}w_{j+1}\right], \tag{36}$$

where $X_j = X(j\Delta t)$ is the system behavior, where $j = 1, \ldots, M$ and $\Delta t = T/M$, and $\{w_j\}_{j=1\ldots M}$ are independent standard normal random variables. Then, approximating continuous-time integral of Eq. (35) by discrete-time summation leads to

$$C_A = e^{-rT}\mathbb{E}\left[\max\left(\frac{1}{M}\sum_{j=1}^{M} X_j - K, 0\right)\right]$$
$$= e^{-rT}\mathbb{E}[a(X)] = e^{-rT}A(X_0, \sigma, \rho), \tag{37}$$

where we define the behavior evaluation function as $a(X) = \max\left(\frac{1}{M}\sum_{j=1}^M X_j - K, 0\right)$, which equals the payoff function of the option, and its expectation as $A(X_0, \sigma, \rho) = \mathbb{E}[a(X)]$.

On the basis of the above preparations, let us calculate three typical risk measures (Greeks), Delta Δ, Vega ν, and Rho ρ, defined as

$$\Delta = \frac{\partial C_A}{\partial X_0} = e^{-rT}\frac{\partial A}{\partial X_0}$$
$$\nu = \frac{\partial C_A}{\partial \sigma} = e^{-rT}\frac{\partial A}{\partial \sigma} \tag{38}$$
$$\rho = \frac{\partial C_A}{\partial r} = e^{-rT}\frac{\partial A}{\partial r} - TC_A.$$

We note that since the distribution range Ω of $\{W_j\}_{j=1...M}$ covers the whole $\mathbb{R}^M$, the boundary residual R_i equals zero.

$$\textbf{Delta} = \partial C_A / \partial X_0.$$

Delta Δ, which represents the sensitivity of option value C_A with respect to the spot price of the underlying asset, called "Delta Δ," is the most fundamental Greek in option trading. Considering the small variation of random variables $\{w_j\}_{j=1...M}$ that cancel out a given small variation of the spot price X_0, we easily obtain the fixed-sample-path derivative

$$\begin{cases} \left.\dfrac{\partial w_1}{\partial X_0}\right|_{x=const} = -\dfrac{\partial X_1}{\partial X_0} \Big/ \dfrac{\partial X_1}{\partial w_1} = -\dfrac{1}{\sigma\sqrt{\Delta t}X_0}, \\ \left.\dfrac{\partial w_j}{\partial X_0}\right|_{x=const} = 0 \; for \; j \geq 2. \end{cases} \tag{39}$$

The score function can be calculated by Eq. (20):

$$h_{X_0} = \frac{w_1}{\sigma\sqrt{\Delta t}X_0}. \tag{40}$$

Therefore,

$$\Delta = e^{-rT}\frac{\partial A}{\partial X_0} = e^{-rT}\,\mathbb{E}[a(\mathbb{X})h_{X_0}] = e^{-rT}\,\mathbb{E}\left[a(X)\frac{w_1}{\sigma\sqrt{\Delta t}X_0}\right] \tag{41}$$

holds. We can estimate Δ by Monte Carlo expectation (LRM) using Eq. (41) with a large number of sample paths generated by Eq. (36).

$$\textbf{Vega} = \partial C_A / \partial \sigma.$$

The sensitivity of option value C_A with respect to the volatility σ of the asset price is called "Vega v." The fixed-sample-path derivative with respect to σ is

$$\left.\frac{\partial w_j}{\partial \sigma}\right|_{x=const} = -\frac{\partial X_j}{\partial \sigma}\Big/\frac{\partial X_j}{\partial w_j} = \sqrt{\Delta t} - \frac{w_j}{\sigma}, \tag{42}$$

and the score function is

$$h_\sigma = \sum_{j=1}^{M}\left[\frac{(w_j)^2 - 1}{\sigma} - \sqrt{\Delta t}\,w_j\right] = \frac{\overline{(w_J)^2} - \sigma\sqrt{\Delta t}\overline{w_J} - M^2}{M\,\sigma}, \tag{43}$$

where we use the notations $\overline{w_j} = \frac{1}{M}\sum_{j=1}^{M}w_j$ and $\overline{(w_j)^2} = \frac{1}{M}\sum_{j=1}^{M}(w_j)^2$. Therefore, we obtain LRM estimator

$$\nu = e^{-rT}\frac{\partial A}{\partial \sigma} = e^{-rT}\,\mathbb{E}[a(X)h_\sigma] = e^{-rT}\,\mathbb{E}\left[a(X)\frac{\overline{(W_J)^2} - \sigma\sqrt{\Delta t}\overline{W_J} - M^2}{M\,\sigma}\right]. \quad (44)$$

$$\mathbf{Rho} = \partial C_A/\partial r.$$

The sensitivity of option value C_A with respect to the risk-free interest rate r is called "Rho ρ." Considering the small variation of random variables $\{w_j\}_{j=1\ldots M}$ that cancel out a given small variation of the risk-free interest r, the fixed-sample-path derivative is

$$\left.\frac{\partial w_j}{\partial r}\right|_{x=\mathrm{const}} = -\frac{\partial X_j}{\partial r}\bigg/\frac{\partial X_j}{\partial w_j} = -\frac{\sqrt{\Delta t}}{\sigma}. \quad (45)$$

The score function calculated by Eq. (20) is

$$h_r = \sum_{j=1}^{M}\frac{\sqrt{\Delta t}\,w_j}{\sigma} = \frac{\sqrt{\Delta t}}{M\,\sigma}\overline{W_J}. \quad (46)$$

Therefore, we obtain the LRM estimator

$$\rho = e^{-rT}\frac{\partial A}{\partial r} - TC_A = e^{-rT}\,\mathbb{E}[a(\mathbb{X})(h_r - T)] = e^{-rT}\,\mathbb{E}\left[a(X)\left(\frac{\sqrt{\Delta t}}{M\,\sigma}\overline{W_J} - T\right)\right]. \quad (47)$$

As might be expected, the score functions and LRM estimators of Delta Δ, Vega ν, and Rho ρ derived from the fixed-sample-path method in this section are the same as those derived from the conventional method by differentiating the probability density function [2, 8]. We note that the conventional method requires explicit knowledge of the relevant probability density function, whereas the fixed-sample-path method requires the knowledge of the time evolution of individual sample paths x only.

4   Analysis of Financial Flow of Funds Network

The calculation examples of Sect. 3 were aimed at pretty simple systems. In this section, we address a network model of the financial flow of funds among companies as an example of the relatively complicated system that shows the effectiveness of the fixed-sample-path method.

4.1 Outline of the Problem

Let us consider a network of the financial flow of funds among 25 companies, labeled 1–25, as shown in Fig. 1. While a network consisting 25 companies is not too complicated to understand and discuss the results, it is fairly complicated to perform sensitivity analysis with the conventional LRM method. In Fig. 1, the nodes represent each company, where the numbers written in the nodes represent the company's label. The edges represent the existence of the financial flow of funds along the edge directions. For simplicity, we suppose that the average amounts of fund transfers per unit period equal one for all edges. We suppose, in addition, that the assets of each company increase or decrease by an average amount per unit period denoted by parenthetical numbers beside each node, whereas the assets of the companies of which corresponding nodes have no parenthetical numbers do not change. This increase or decrease in assets represents the fund transfers from/to companies other than those of the 25 companies depicted in Fig. 1. As a result, the average net incomes and outgoings per unit period of each of the 25 companies are balanced.

We suppose the actual amounts of fund transfers through the edges to be random variables distributed around the above average amounts. The assets of each company increase or decrease depending on the variation of the difference between incomes and outgoings. As a result, there is the possibility for "company

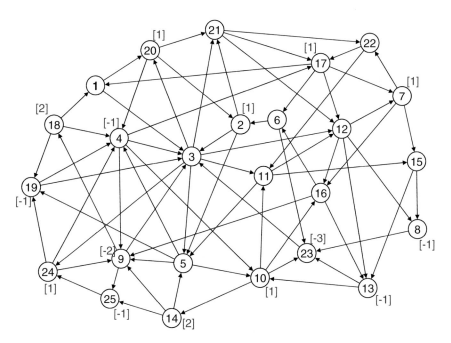

Fig. 1 Financial flow of funds network with 25 companies

bankruptcy," i.e., the assets of a certain company go negative at a certain time. Here, we suppose that companies in bankruptcy and the edges (funds transfer) related to them cease to exist. If company 1 in Fig. 1, for example, goes bankrupt at time t, we delete four edges: from Co. 1 to Co. 3, Co. 1 to Co. 20, Co. 17 to Co. 1, and Co. 18 to Co. 1. As a consequence, companies 3 and 20 become increasingly likely to go bankrupt because of an unfavorable balance without fund transfers from company 1, whereas companies 17 and 18 become less likely to go bankrupt because of a favorable balance. Bankruptcy of a company has an effect on the bankrupt probabilities of the other companies through the connection structure of the network in this way.

Now, we are interested in the relationship between the flow of funds of the edges and the bankrupt probabilities of the companies. If the average flows of each edge slightly change from 1, what happens in the bankrupt probability of company 1 or the average bankrupt probability of all 25 companies? Conversely, which edge is the most effective at reducing the bankrupt probability of company 1 if we change the average flow of funds? The edges linked directly from/to company 1 might naturally have a large influence, but is there a possibility that edges located away from company 1 have a large influence on its bankrupt probability by network effect? Given this awareness of the problems, the aim of this section is to estimate the sensitivities of the bankrupt probabilities of each company and the sensitivity of the average bankrupt probability of the all companies with respect to the average flow of funds of each edge by Monte Carlo simulation using LRM with the fixed-sample-path principle.

4.2 Formulation

Let us consider a network of the financial flow of funds among 25 companies, shown in Fig. 1. We call the "outside" of the network as "company 0" for notational convenience, i.e., the fund transfers from/to companies outside the network (denoted by parenthetical numbers beside each node) are considered to be the fund transfers from/to company 0. Let $X_i(t)$ be the total assets of company i (where $i = 1 \ldots 25$) at time t. We suppose the initial assets $X_i(0) = 25$ for all 25 companies. The existence function of company i is defined as

$$S_i(t) = \begin{cases} 1, & \text{if } X_i(t) \geq 0 \\ 0, & \text{otherwise,} \end{cases} \qquad (48)$$

i.e., $S_i(t)$ equals 1 if company i exists at time t, and $S_i(t)$ equals 0 if company i has been bankrupt. We define $S_0(t) = 1$ for all t for notational simplicity. Let $F_{ij}(t)$ be the amount of transfer of funds from company i to company j at time t. $F_{ij}(t)$ are random variables with mean $\mu_{ij} = 1$ for i, j (where $i = 1, \ldots, 25$ and $j = 1, \ldots, 25$) for which there exists an edge between company i and j, while μ_{ij} equals zero for i, j for which there exists no edge between them. In addition, $F_{0j}(t)$ and $F_{i0}(t)$, which denote the transfer of funds from/to the outside of the network, are random variables with mean μ_{0j}, $\mu_{i0} = 1 - 3$, shown in parentheses in the figure. Here, we suppose $F_{ij}(t)$ to be under log-normal distribution with mean μ_{ij}. and variance $\sqrt{\mu_{ij}}$. The assets $X_i(t)$ of the company (where $i = 1 \ldots 25$) satisfy the relation

$$X_i(t+1) - X_i(t) = \sum_{j=0}^{25} F_{ji}(t)S_j(t)\, S_i(t) - \sum_{j=0}^{25} F_{ij}(t)S_i(t)S_j(t). \qquad (49)$$

On the basis of the above premises, let us estimate the existence probabilities $S_i = \mathbb{E}[S_i(T)]$ of each company at $T = 100$ and the average existence probability $\overline{S_1} = \mathbb{E}\left[\frac{1}{25} \sum_{i=1}^{25} S_i(T)\right]$ of all 25 companies by Monte Carlo simulation. In addition, we estimate $\partial S_i/\partial \mu_{ij}$ and $\partial \overline{S_1}/\partial \mu_{ij}$, i.e., the sensitivity of S_i and $\overline{S_1}$ with respect to the average flow of funds of each edge, using the fixed-sample-path method. There exist 86 number of μ_{ij}, which are nonzero, i.e., 71 edges plus 15 parenthetical numbers. We can estimate $\partial S_i/\partial \mu_{ij}$ and $\partial \overline{S_1}/\partial \mu_{ij}$ for all 86 μ_{ij} simultaneously.

4.3   Derivation of Score Functions

To estimate $\partial S_i/\partial \mu_{ij}$ and $\partial \overline{S_1}/\partial \mu_{ij}$, the score functions $h_{(\mu_{ij})}$ are required. The log of a random variable under log-normal distribution with mean μ and variance σ is under normal distribution with mean and variance :

$$\begin{cases} m = \log\left(\dfrac{\mu^2}{\sqrt{\mu^2 + \sigma^2}}\right) \\[2mm] s = \sqrt{\log\left(1 + \dfrac{\sigma^2}{\mu^2}\right)}. \end{cases} \qquad (50)$$

Therefore, $F_{ij}(t)$, which is under log-normal distribution with mean μ_{ij} and variance $\sqrt{\mu_{ij}}$, can be written as

$$F_{ij}(t) = \exp\left(m_{ij} + s_{ij} w_{ij}^t\right) = \frac{\left(\mu_{ij}\right)^2 \sqrt{\log(1 + 1/\mu_{ij}} \exp\left(\sqrt{\log(1 + 1/\mu_{ij}} w_{ij}^t\right)}{\sqrt{\mu_{ij}(1 + \mu_{ij})}}, \qquad (51)$$

where w_{ij}^t is a random variable with the standard normal distribution.

Let us apply the fixed-sample-path method. Considering the relationship between a small variation of w_{ij}^t and a small variation of μ_{ij} under the condition of keeping $F_{ij}(t)$ fixed satisfies

$$\left.\frac{\partial w_{ij}^t}{\partial \mu_{ij}}\right|_{F_{ij}(t)=const} = -\frac{dF_{ij}(t)}{d\mu_{ij}}\bigg/\frac{\partial F_{ij}(t)}{\partial w_{ij}^t} = \frac{w_{ij}^t - \left(3+2\mu_{ij}\right)\sqrt{\log\left(1+\frac{1}{\mu_{ij}}\right)}}{2\,\mu_{ij}\left(1+\mu_{ij}\right)\log\left(1+\frac{1}{\mu_{ij}}\right)}, \tag{52}$$

and the fact that the system behavior $X_i(t)$ stays fixed if and only if all fund flows $F_{ij}(t)\,S_i(t)\,S_j(t)$ are fixed, we obtain the fixed-sample-path derivative

$$\left.\frac{\partial w_{ij}^t}{\partial \mu_{ij}}\right|_{x=const} = \begin{cases} \dfrac{w_{ij}^t - \left(3+2\mu_{ij}\right)\sqrt{\log\left(1+\frac{1}{\mu_{ij}}\right)}}{2\,\mu_{ij}\left(1+\mu_{ij}\right)\log\left(1+\frac{1}{\mu_{ij}}\right)} & \text{if } S_i(t)\,S_j(t)=1 \\ 0, & \text{otherwise,} \end{cases} \tag{53}$$

Therefore, from Eq. (20), the score function $h_{(\mu_{ij})}$ with respect to μ_{ij} is

$$h_{(\mu_{ij})} = \sum_{t=1}^{\min(\tau_i,\tau_j)}\left\{\frac{\partial}{\partial w_{ij}^t}\left(\left.\frac{\partial w_{ij}^t}{\partial \mu_{ij}}\right|_{x=const}\right) - w_{ij}^t\left(\left.\frac{\partial w_{ij}^t}{\partial \mu_{ij}}\right|_{x=const}\right)\right\}$$
$$= \sum_{t=1}^{\min(\tau_i,\tau_j)}\frac{1-\left(w_{ij}^t\right)^2 + w_{ij}^t\left(3+2\mu_{ij}\right)\sqrt{\log\left(1+1/\mu_{ij}\right)}}{2\,\mu_{ij}\left(1+\mu_{ij}\right)\log\left(1+1/\mu_{ij}\right)}, \tag{54}$$

where τ_i is the last time that company i exists:

$$\tau_i = \operatorname*{argmax}_{t\le T}[S_i(t)=1]. \tag{55}$$

4.4   Simulation Result

We did a Monte Carlo simulation with two million sample paths and estimated

$$\begin{aligned}
\mathcal{S}_i &= \mathbb{E}[S_i(100)] \cong \frac{1}{L}\sum_{k=1}^{L}S_i^{(k)}(100) \\
\overline{\mathcal{S}_1} &= \mathbb{E}\left[\frac{1}{25}\sum_{i=1}^{25}S_i(100)\right] \cong \frac{1}{L}\sum_{k=1}^{L}\left[\frac{1}{25}\sum_{i=1}^{25}S_i(100)\right] \\
\frac{\partial \mathcal{S}_i}{\partial \mu_{ij}} &= \mathbb{E}\left[S_i(100)h_{(\mu_{ij})}\right] \cong \frac{1}{L}\sum_{k=1}^{L}S_i^{(k)}(100)h_{(\mu_{ij})}^{(k)} \\
\frac{\partial \overline{\mathcal{S}_1}}{\partial \mu_{ij}} &= \mathbb{E}\left[\frac{1}{25}\sum_{i=1}^{25}S_i(100)h_{(\mu_{ij})}\right] \cong \frac{1}{L}\sum_{k=1}^{L}\left[\frac{1}{25}\sum_{i=1}^{25}S_i^{(k)}(100)h_{(\mu_{ij})}^{(k)}\right].
\end{aligned} \tag{56}$$

Figure 2 shows the simulation result. Figure 2a shows an over-drawn time series of 200 typical Monte Carlo sample paths of $\frac{1}{25}\sum_{i=1}^{25} S_i(t)$, the average existence probability of the 25 companies. Figure 2b–d show the convergence of the estimated values: (b) for S_i and $\overline{S_1}$, (c) for $\partial\overline{S_1}/\partial\mu_{ij}$, and (d) for $\partial S_1/\partial\mu_{ij}$. All of the estimated values are converged. As known, the convergence speeds of the sensitivities when using the LRM method are slower than those of the expectations themselves [2].

Table 1 shows the estimated values of $\overline{S_1}$ and S_i and their sensitivities $\partial\overline{S_1}/\partial\mu_{ij}$ and $\partial S_1/\partial\mu_{ij}$. The leftmost column of the table shows the estimated value of S_i (the average existence probability of the 25 companies) and the 25 estimated values of S_i (the existence probabilities of company i). The ten columns on the right side of the table show the sensitivities (differential coefficients) of $\overline{S_1}$ and S_i with respect to the average funds flow μ_{ij} of edges. Due to limited space, the sensitivities with respect to only ten edges arranged in descending order of their absolute values are shown, respectively, where the upper rows identify the edges, and the bottom rows show the estimated values of the differential coefficients.

From Table 1, for example, edge 23 → 3 (the edge from company 23 to company 3) turned out to have the largest sensitivity of 0.08596 to $\overline{S_1}$. Edge 0 → 7 (the flow of funds from the outside of the network to company 7) and edge 0 → 10 (from the outside to company 10) also had large sensitivities to $\overline{S_1}$. It is interesting that edge 23 → 3, which is an inner flow of the network, had larger sensitivity to the average existence probability $\overline{S_1}$ than did the inward flows from the outside of the network, which increased the total assets within the network. This would be explained by the fact that company 23 has four inward edges, while it has only one outward edge 23 → 3. An increasing flow of funds for 23 → 3, which clearly had an adverse effect on the survival of company 23, might be desirable for the survival of the many other companies in the network.

We turn attention to the existence probability S_1 of company 1. The top three edges having a large effect on S_1 were edge 1 → 3, edge 1 → 20, and edge 3 → 21 in descending order of the (absolute value of) sensitivities. We are convinced that edge 1 → 3 and edge 1 → 20, which are directly outward from node 1, had large and negative sensitivities to S_1. It is interesting that edge 3 → 21, which does not link to company 1 directly, had the third largest sensitivity. This would be explained if we note that edge 3 → 21 had the largest (and negative) effect on the survival of company 3 and edge 1 → 3 the largest (and negative) effect on the survival of company 1.

As seen above, we can estimate the sensitivity of $\overline{S_1}$ and S_i with respect to all 86 numbers of μ_{ij} by Monte Carlo simulation using the LRM method with the score functions derived using the fixed-sample-path principle. Although this example

Table 1 The estimated values of $\overline{S}_i$ and S_i and their sensitivities $\partial \overline{S}_i/\partial \mu_{ij}$ and $\partial S_1/\partial \mu_{ij}$

| | Existence probability | The estimated sensitivities with respect to each edge μ_{ij} | | | | | | | | | |
		#1	#2	#3	#4	#5	#6	#7	#8	#9	#10
$\overline{S}_i$	0.4769	23 → 3	0 → 7	0 → 10	0 → 18	10 → 23	0 → 17	0 → 20	0 → 2	3 → 12	3 → 24
		0.08596	0.07176	0.06742	0.06573	−0.06462	0.06146	0.05095	0.04989	−0.04944	−0.04906
S_1	0.6719	1 → 3	1 → 20	3 → 21	18 → 19	17 → 6	3 → 11	17 → 7	18 → 4	18 → 1	17 → 12
		−0.9747	−0.9134	0.4447	−0.3927	−0.3461	0.3424	−0.3414	−0.3368	0.3345	−0.3238
$S2$	0.7415	2 → 3	2 → 5	2 → 21	0 → 2	3 → 20	20 → 21	6 → 2	20 → 2	6 → 23	20 → 4
		−0.8100	−0.7856	−0.7707	0.6345	0.4148	−0.3370	0.3240	0.3066	−0.2995	−0.2838
S_3	0.2376	3 → 21	3 → 11	3 → 20	3 → 12	3 → 24	20 → 4	3 → 5	2 → 21	1 → 20	20 → 2
		−0.5485	−0.5407	−0.5334	−0.4134	−0.3956	0.3719	−0.3607	−0.2965	−0.2859	0.2798
S_4	0.4635	4 → 10	4 → 9	4 → 17	4 → 0	4 → 3	9 → 18	24 → 9	5 → 9	5 → 10	19 → 4
		−0.7510	−0.7464	−0.7339	−0.6120	−0.4982	0.4884	−0.4323	−0.3803	−0.3614	0.3497
S_5	0.5024	5 → 19	5 → 4	5 → 9	5 → 10	4 → 3	19 → 3	9 → 3	3 → 24	3 → 12	3 → 21
		−0.7143	−0.7053	−0.6733	−0.6124	0.4746	0.4496	0.4455	−0.4150	−0.3834	−0.3569
S_6	0.5322	6 → 23	6 → 2	16 → 13	23 → 3	17 → 12	16 → 9	17 → 1	23 → 0	0 → 17	12 → 16
		−1.0721	−1.0054	−0.4793	0.4554	−0.4080	−0.4065	−0.4058	0.3542	0.3474	0.3396
S_7	0.4482	7 → 16	7 → 15	7 → 22	0 → 7	22 → 17	12 → 16	17 → 1	17 → 6	16 → 9	0 → 17
		−1.0488	−1.0205	−0.9594	0.7789	0.5961	−0.4402	−0.4348	−0.3906	0.3893	0.3809
S_8	0.4284	8 → 23	8 → 0	23 → 3	15 → 13	12 → 13	23 → 0	12 → 16	15 → 8	3 → 5	12 → 8
		−1.0896	−0.8440	0.6284	−0.5632	−05147	04214	−04110	0.3514	−0.2901	0.2867
S_9	0.1940	9 → 25	9 → 18	9 → 0	25 → 24	9 → 3	14 → 25	18 → 4	24 → 19	14 → 9	5 → 19
		−0.5920	−0.5796	−0.4650	0.3994	−0.3581	−0.3199	03104	−0.2640	0.2539	−0.2515
S_{10}	0.4711	10 → 23	10 → 16	10 → 11	10 → 14	0 → 10	11 → 5	13 → 23	16 → 13	14 → 5	5 → 9
		−0.8638	−0.8049	−0.7756	−0.7606	0.6648	0.4750	−0.4707	0.4213	0.4069	−0.3430

(continued)

Table 1 (continued)

	Existence probability	The estimated sensitivities with respect to each edge μ_{ij}									
		#1	#2	#3	#4	#5	#6	#7	#8	#9	#10
S_{11}	0.4483	11 → 15	11 → 12	11 → 5	5 → 10	12 → 13	22 → 17	3 → 12	15 → 13	10 → 14	3 → 11
		−0.9037	−0.8759	−0.8294	0.4712	0.4209	−0.3898	−0.3650	0.3629	−0.3574	03.462
S_{12}	0.2472	12 → 7	12 → 16	12 → 8	12 → 13	7 → 22	17 → 7	16 → 9	0 → 7	8 → 23	3 → 5
		−0.6790	−0.5719	−0.5492	−0.5186	0.3617	−0.588	02915	−0.2793	0.2644	−0.2593
S_{13}	0.3799	13 → 10	13 → 23	13 → 0	10 → 16	10 → 11	12 → 3	23 → 3	15 → 8	16 → 6	3 → 5
		−0.8373	−0.8043	−0.6842	0.5073	0.4574	−0.4472	0.4331	−0.4145	−0.3261	−0.3106
S_{14}	0.7548	14 → 9	14 → 5	14 → 25	0 → 14	5 → 10	10 → 16	10 → 11	10 → 23	0 → 10	9 → 18
		−0.8365	−0.7869	−0.7464	0.6162	0.3702	−0.3328	−0.3267	−0.3026	0.2715	0.2418
S_{15}	0.5260	15 → 8	15 → 13	12 → 7	11 → 12	7 → 16	13 → 10	8 → 23	13 → 23	0 → 7	7 → 22
		−1.1379	−1.1117	0.5564	−0.5426	−0.5284	0.4491	0.4376	0.4156	0.4036	−0.3907
S_{16}	0.4141	16 → 6	16 → 13	16 → 9	13 → 10	12 → 13	10 → 14	9 → 3	10 → 11	9 → 18	0 → 10
		−0.8364	−0.8072	−0.8043	0.5246	−0.4697	−0.4008	0.3846	−0.3574	0.3417	0.3259
S_{17}	0.4926	17 → 12	17 → 6	17 → 1	17 → 7	0 → 17	21 → 22	12 → 8	22 → 17	12 → 13	7 → 22
		−0.9313	−0.8432	−0.8264	−0.7960	0.6776	−0.5995	0.3746	0.3744	0.3742	0.3387
S_{18}	0.6510	18 → 19	18 → 4	18 → 1	0 → 18	4 → 9	9 → 25	9 → 0	5 → 9	24 → 9	9 → 3
		−0.9755	−0.9174	−0.9007	0.7450	0.5324	−0.4179	0.3746	0.3700	0.3544	−0.3315
S_{19}	0.5141	19 → 4	19 → 0	19 → 3	18 → 4	24 → 4	5 → 4	18 → 19	24 → 19	4 → 17	5 → 19
		−1.1736	−0.8793	−0.7153	−0.6266	−0.6078	−0.5807	0.4290	0.4074	0.4007	0.3902
S_{20}	0.5610	20 → 21	20 → 2	20 → 4	0 → 20	21 → 17	3 → 21	2 → 3	4 → 3	4 → 17	1 → 20
		−0.9482	−0.9310	−0.8908	0.7290	0.4338	−0.4250	0.3988	0.3926	0.3468	0.3327
S_{21}	0.4799	21 → 17	21 → 22	21 → 12	3 → 12	17 → 1	2 → 21	17 → 6	12 → 16	20 → 4	12 → 13
		−0.9103	−0.8803	−0.8600	−0.4702	0.4002	0.3791	0.3728	0.3541	−0.3444	0.3119

(continued)

Table 1 (continued)

	Existence probability	The estimated sensitivities with respect to each edge μ_{ij}									
		#1	#2	#3	#4	#5	#6	#7	#8	#9	#10
S_{22}	0.5839	22 → 11	22 → 17	17 → 7	21 → 17	7 → 16	7 → 15	11 → 12	0 → 7	17 → 12	11 → 5
		−1.0411	−0.9813	0.6397	−0.5667	−0.5137	−0.4763	0.4563	0.4280	0.3974	0.3369
S_{23}	0.1888	23 → 0	23 → 3	3 → 12	6 → 2	13 → 0	10 → 14	8 → 0	6 → 23	0 → 10	10 → 23
		−0.5486	−0.4449	0.2889	−0.2859	−0.2788	−0.2577	−0.2489	0.2269	0.2241	0.2192
S_{24}	0.6154	24 → 19	24 → 4	0 → 24	24 → 9	19 → 3	4 → 3	3 → 5	3 → 20	3 → 11	3 → 21
		−0.8424	−0.7966	0.6685	−0.6268	0.5091	0.5016	−0.4815	−0.4246	−0.4240	−0.4064
S_{25}	0.3755	25 → 24	25 → 0	24 → 9	9 → 3	9 → 18	9 → 0	24 → 4	0 → 14	4 → 3	14 → 25
		−0.9278	−0.7619	0.5972	−0.5229	−0.4131	−0.3888	0.3871	0.3564	−0.3078	0.2978

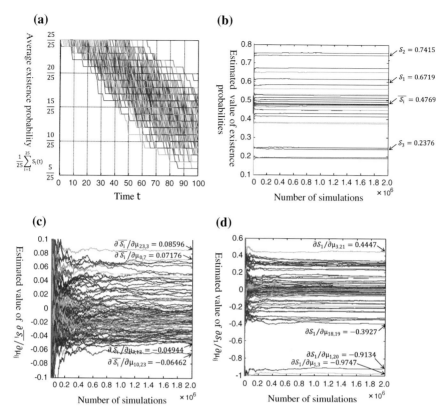

Fig. 2 Simulation result of the financial flow of funds network. **a** Over-drawn time series of the average existence probability by MonteCarlo simulation. **b** Estimated value of existence probabilities versus number of simulations. **c** Estimated value of versus number of $\partial \overline{S_i}/\partial \mu_{ij}$ simulations. **d** Estimated value of $\partial S_1/\partial \mu_{ij}$ versus number of simulations

network is pretty small, the LRM method with fixed-sample-path principle can be applicable and practical for much more complicated systems with numerous parameters, such as for systematic risk analysis of complicated financial networks, traffic flow on a complicated roadway network, and emerging "big data" analysis.

5 Conclusion

In this study, a fixed-sample-path method was proposed, which derives the score function of LRM not via the pdf $f(x, z)$. The key idea is to consider the fixed-sample-path derivative of the random variables w with respect to the parameter z_i under the condition of fixing the sample path x. The boundary residual R_i, which represents the correction associated with the change of the distribution range of the random variables in LRM, was also derived. Some examples including the estimation of risk measures (Greeks) of option and financial flow of funds networks showed the effectiveness of the fixed-sample-path method.

References

1. Glynn, P.W.: Optimization of stochastic systems via simulation. In: Proceedings of the Conference on Winter Simulation, pp. 90–105 (1989)
2. Glasserman, P.: Monte Carlo Methods in Financial Engineering (Stochastic Modelling and Applied Probability). Springer, New York (2003)
3. Ho, Y.C., Cao, X.R.: Perturbation Analysis of Discrete Event Dynamic Systems. Springer, Berlin (1991)
4. Bettonvil, B.: A formal description of discrete event dynamic systems including infinitesimal perturbation analysis. Eur. J. Oper. Res. **42**, 213–222 (1989)
5. Glynn, P.W.: Likelihood ratio gradient estimation: an overview. In: Proceedings of the Conference on Winter Simulation, pp. 366–374 (1987)
6. Rubinstein, R.Y., Kroese, D.P.: Simulation and the Monte Carlo Method. Wiley, New York (2011)
7. Reiman, M., Weiss, A.: Sensitivity analysis for simulations via likelihood ratios. Oper. Res. **37**, 830–844 (1989)
8. Broadie, M., Glasserman, P.: Estimating security price derivatives using simulation. Manage. Sci. **42**, 269–285 (1996)

A System Dynamics Simulator for Decision Support in Risk-Based IT Outsourcing Capabilities Management

Tarcio R. Bezerra, Antão Moura, Seth Bullock and Dietmar Pfahl

Abstract Organizations face important risks with IT Outsourcing (ITO)—the practice of delegating organizational IT functions to third parties. Here, we employ a system dynamics simulator to support ITO decision-making under risk, taking a dynamic and integrated view of both capabilities management and benefits management. After briefly presenting its functionality, we use the simulator to assess how deficits in two IT capabilities—Contract Monitoring (on the customer's side) and Service Delivery (on the supplier's side)—affect the earned values of service orders, the ITO budget, service completion deadlines, and damage to the customer-supplier relationship. Validation is ongoing at four institutions in Brazil, including a large, state tax collecting and finance agency. Initial results are encouraging and indicate that the simulator is useful for planning and managing ITO activities.

Keywords IT outsourcing · IT capabilities · Risk management · System dynamics simulation

T.R. Bezerra (✉) · A. Moura
Computing Systems Department, Federal University of Campina Grande,
Campina Grande, Brazil
e-mail: tarcio@copin.ufcg.edu.br

A. Moura
e-mail: antao@dsc.ufcg.edu.br

S. Bullock
Electronics and Computer Science, University of Southampton,
Southampton, UK
e-mail: sgb@ecs.soton.ac.uk

D. Pfahl
Institute of Computer Science, University of Tartu, Tartu, Estonia
e-mail: dietmar.pfahl@ut.ee

© Springer International Publishing Switzerland 2015
M.S. Obaidat et al. (eds.), *Simulation and Modeling Methodologies,*
Technologies and Applications, Advances in Intelligent Systems
and Computing 402, DOI 10.1007/978-3-319-26470-7_7

131

1 Introduction

Organizations use IT resources and capabilities as assets to create value in the form of goods and services. Capabilities represent the ability of an organization to coordinate, manage, and deploy resources to produce value [1]. When an organization lacks the internal IT capabilities required for the provision of all of its IT services, it must look for external organizations able to fill the gap [2]. This practice is called Information Technology Outsourcing—ITO.

Outsourcing is often used to transfer risk to third parties. However, this practice introduces new risks for customer organizations as well. The loss of internal technical skills is an important (new) risk factor for organizations embarking on an ITO initiative [3, 4]. The goal of ITO is not to lose control of IT, given the importance and centrality that it typically has for the outsourcing business. This concern should be reflected in the various phases of the outsourcing cycle. However, there exists no clear recipe for managers seeking to mitigate this risk in a rational and balanced way, without compromising the potential benefits of outsourcing [5].

The specialized literature offers many conceptual articles that identify lists of ITO risks or develop ITO risk models and empirical papers that address specific ITO risks, risk measurement [6], and risk management strategies [5]. However, there are still gaps to be filled by tools and models that help managers understand which capabilities to develop and/or maintain internal to their organizations, in which quantity or magnitude such capabilities are required, and how such capabilities behave in a dynamic scenario of constant interaction between internal IT and supplier teams.

Aiming to support risk-based decision-making on ITO and on IT capabilities management, we created and employed a simulation model [7] in the context of a Brazilian state tax and finance agency (SEFAZ). We have now extended the development and validation of this simulator by refining its structure and facilities to adopt earned value as a tracking indicator of service order (SO) evolution and to accommodate details of 38 projects at SEFAZ (20 already completed and 18 still running).

In this paper, we use the simulator to assess how deficits in two IT capabilities—Contract Monitoring (on the customer's side) and Service Delivery (on the supplier's side)—affect the earned values of SOs, the ITO budget, service completion deadlines, and damage to the customer-supplier relationship. Finally, we discuss ongoing validation efforts.

2 Related Work

This work relates to risk management in the context of IT Outsourcing. The term risk can take on different meanings depending on the setting. In ISO/DIS 31000 [8], risk is the effect of uncertainty on objectives, where an effect is a deviation from the expected outcome (positive and/or negative). In the scope of our work, we are interested in studying the risks of undesirable outcomes. Formally, the risk exposure (RE) is defined as $= P_{UO}L_{UO}$, where P_{UO} is the probability of an undesirable outcome, and L_{UO} is the loss due to an undesirable outcome.

Risk management in ITO is a topic that has been extensively studied for decades and is still a topic of great relevance. Within a recent review of ITO [9], 36 of the 164 publications cited specifically address ITO risk management. Among these articles, [10] develops risk models, [3, 11] incorporate strategies for risk mitigation contracts, [12] develops empirical research (longitudinal single case study), [6] focuses on risk measurement, and [13] identifies a list of risk factors.

Among the extensive list of risk factors identified in ITO by several authors, the lack of essential IT capabilities by customers and suppliers is almost ubiquitous [4, 5, 9]. The literature on ITO shows a strong relationship between the contracting organizations' capabilities and the expected outcomes of outsourcing initiatives [5, 9]: the capabilities to manage vendors and to negotiate contracts and technical/methodological capability in information systems development are strongly related to the ITO's success.

Authors in [4] review 97 articles focusing on ITO risk management. They summarize the main ITO risk factors and the impacts generated by them, categorize these factors and use them to build more complex risk and impact factors. The authors also associate these factors with related stages of a typical ITO life cycle.

In general, there are two methods that can be used to measure risk: quantitative and qualitative. In the quantitative method, the results depend heavily on the knowledge of the experts that assign values to the risk components. The use of purely quantitative approaches is extremely difficult and costly. The main challenge is the lack of data. This difficulty affects two components of risk: estimates concerning the likelihood and the impact of an event (e.g., in terms of cost or financial loss).

Our work uses a quantitative method to calculate risk exposure, based on a quantitative system dynamics simulation model of the contract monitoring process, where impact factors are calculated from differential equations and probabilities can be calculated from the results of multivariate sensitivity analyses. We believe that a quantitative approach, despite the difficulties in adopting it, has a more objective power to communicate risks to the decision makers as compared to the qualitative approach. This is of particular interest for business process managers wishing to make informed decisions based on quantitative (financial) values, especially in the case where risk treatment involves financial expenditures.

Software simulation modeling has been extensively used in risk management applied to various sectors of knowledge, industry, and services over the years. Some of these models use a static approach and others capture the dynamics of the processes to which they apply. Both approaches can stochastically generate values for risk factors as a strategy for representing uncertainty. Our adoption of system dynamics at the expense of other simulation techniques is that it is a holistic approach that is not limited to monocausal relationships, but allows one to represent a complex network of inter-dependencies among risk and impact factors, even when impact factors are fed back to the system as risk factors.

The model proposed by us aims to be a tool for supporting decision-making in ITO and in managing capabilities directly involved in the ITO process taking into account business benefits realization. We have sought inspiration from system dynamics simulation models applied to project management in general and in particular to software engineering projects [14–16], as well as to decision-making in people management [17]. One use case of our model is to support risk-based decision-making, considering the ITO risk factors and impacts that can be represented within the scope of the model. Risk assessment procedures to be applied to preexisting system dynamics models have been proposed in [18, 19] and influenced our work. There exist only few examples of simulation-based ITO risk management research. For example, authors of [20] present a model structure for risk analysis. Our work differs in that our approach is quantitative, and we focus on the risks related to the IT capabilities involved in the ITO process.

Further down the ITO lifecycle come the stages of evaluation and treatment of risks. As an example, [21] proposes the use of a decision tree to evaluate the outputs of a system dynamics model applied to project risk management. These later stages, however, are outside the scope of our model which focuses exclusively on ITO risk assessment.

3   A System Dynamics Simulator

Measuring IT capabilities quantitatively in order to properly allocate resources to better achieve planned results (e.g., project objectives) is a challenging problem, especially with regard to human resource skills and the impact of the tools and techniques used to support IT functions. However, there is a lack of tools and models that help managers make decisions about capabilities management.

We have developed a system dynamics simulator to support ITO decision-making. We modeled two IT capabilities: Contract Monitoring—a core capability in the context of outsourcing, which mediates all interactions between client and vendor capabilities; and Service Delivery—a generic single point of contact for IT services. The objective of the simulator is to assess the risks presented by deficits in these capabilities on the customer and supplier sides, including the risk of a premature contract termination. Due to space limitations, we identify risks without discussing ways of mitigating them and we describe details of the simulator's implementation only to the extent of informing on its main modules and output.

3.1   Architecture and Entities

The simulator's executable code (visualized as a stock and flow diagram) is segmented into views and its parameters are divided into four distinct categories: input, calibration, mediation, and output.

Input parameters characterize the benefits and performance metrics to be achieved, the IT resources available within the organization and the IT demand characteristics. Calibration parameters are used to tune the model's behavior to match the scenarios being simulated. Mediation parameters represent intermediate information obtained from the entries, from calibration and, in situations involving feedback loops, from output parameters, e.g., IT capabilities and second-level performance targets (desired workforce, desired skill level). Output parameters are values arising from the dynamic cause-effect relationships between model input, calibration, and mediator parameters. The model produces outputs that reflect the expected performance of IT resources (in terms of cost, quality, resource consumption, and earned value) in response to submitted inputs.

For clarity, maintainability, and reusability, the simulation model has been segmented into "views," reflecting the organization of policies captured in the modeling phase (financial management; demand management; capability forecasting and planning; sourcing management; insourced capabilities management; outsourced capabilities management; and contract monitoring of IT processes/functions).

Simulation Parameters: The most important input (I), calibration (C), and output (O) parameters are listed in Table 1 and will be detailed in Sect. 5.

Model Views: The views that highlight the core concepts of our risk assessment are briefly discussed below. Of the many dynamics diagrams implemented in the simulator, only the ones representing interactions between the capabilities of Contract Monitoring and Outsourced Service Delivery are illustrated since they are the focus of this paper. For additional details, please refer to [7].

Sourcing Management: In the sourcing management view, one can decide whether a particular IT capability will be fully executed by the internal team or completely or partially outsourced.

Table 1 Main simulation model parameters

Parameter	Unit	Type	Parameter	Unit	Type
Task (SU = Service Units)	SU	I	Contract Monitoring (CM) intangible effectiveness	–	C
Task conclusion time	Days	I	Time to adjust Service Delivery (SD) productivity	Days	C
Task budget	$	I	Cumulative cost of insourced CM capability	$	O
Initially available CM Workforce (WF)	Persons	I	Cumulative cost of outsourced SD capability	$	O
Initially average CM skill level	–	I	SD demand conclusion time	Day	O
SD SLA	–	I	Cost of rework	$	O
Minimum SD skill level	–	I	Penalties for rework	$	O
Time to adjust CM WF	Days	C	Cost performance index	–	O
Time to adjust CM skill level	Days	C	Schedule performance index	–	O
CM materials effectiveness	–	C	Supplier profitability index	–	O

Insourced Capabilities Management: This view contains the ITO contracting organization's side of the IT capabilities, among them the Contract Monitoring (CM) capability. Here, a capability is effectively a productivity rate, i.e., the number of service units (SU) processed per day. Therefore, the CM capability is given by the variable *Insourced CM Productivity*, in SU/Day, which is calculated based on the productivity of the resources involved (people, material resources, and intangible assets) using the following formula:

$$Insourced\ CM\ Productivity = Allocated\ Insourced\ CM\ Workforce$$
$$\times Maximum\ CM\ Rate\ per\ Person\ per\ Day \times Average\ CM\ Skill\ Level$$
$$\times CM\ Materials\ Effectiveness \times CM\ Intangibles\ Effectiveness$$

Allocated Insourced CM Workforce: This represents the number of people allocated to monitor the contract; *Maximum CM Rate per Person per Day* is a constant used to represent the number of service units that an "optimally skilled" workforce is able to process in a day. The Average *CM Skill Level* parameter takes values between 0 and 1 and represents the average fraction of the optimal skill level presented by the internal staff. As our work is focused on human resources, the constant *CM Intangibles Effectiveness* and *CM Materials Effectiveness* are just multipliers which represent the extent to which intangible and material resources empower staff productivity, respectively. The highlight of this view is the dynamic behavior of resources mobilized as capabilities governed by the need for productivity created by the SO to be processed (*Windowed Desired CM Productivity*) and subject to various operational delays (variables *Time to Adjust CM Workforce* and *Time to Adjust CM Average Skill Level*).

Contract Monitoring for Service Delivery: This view (Fig. 1) captures the specifics of the demands flow between the customer's IT organization and the ITO provider. This flow reflects the contract monitoring process and the interaction between this capability and the IT service delivery capability.

The *Actual Contract Monitoring Productivity* variable moves the streams of new SOs and those on warranty (rework) from the customer´s organization to the provider, as well as the flow of delivered services approval and defects detection.

The provider's capability to process the demands forwarded by the customer is represented by the variable *Outsourced SD Productivity*. *Outsourced SD Defect Injection Fraction* represents the error generation rate in service delivery.

Outsourced Capabilities Management: If all of the organization's own resources have been allocated and even so the internal generated capability is insufficient to meet the demand, then (if outsourcing is enabled and if there is available ITO budget) the simulator will adjust the provider's capability to the required levels subject to a required time for this adjustment. In our example, we use the Service Delivery (SD) generic capability.

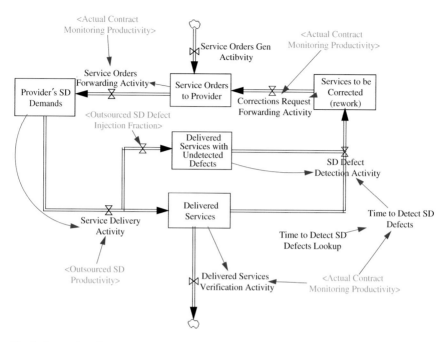

Fig. 1 Interactions between contract monitoring and outsourced service delivery capabilities

4 Illustration: ITO Contract Monitoring at SEFAZ

Following the Integrated Measurement, Modelling and Simulation (IMMoS) framework, integrating system dynamics model development with measurement-based quantitative modeling (GQM) [22], an initial version of the simulator was implemented, verified, and validated using tests of structure, behavior, and learning. This has been presented in [7]. Since then, we extended the collection to cover a greater number of organizations. Results of this new analysis led to adjustments in the simulator's structure (equations, causal relationships, and parameters) and to its (re)calibration. The model has been in use at the Finance and Revenue Agency of Alagoas State, Brazil (SEFAZ). SEFAZ has the largest IT department and the most important outsourcing activity in Alagoas, both in volume and in complexity [23]. Having gone through several generations of ITO, SEFAZ has experienced various contract formats and models.

For illustration, we consider the scope of an ITO contract that has been in operation for about a year. Its purpose was to provide design and implementation services for new information systems (projects) and to maintain those already in production (continuous services). We had access to records of SO performed during the contract. It is beyond the scope of this paper to analyze all of these SOs. We selected twenty projects fully completed by the supplier to capture the real behavior of all parameters involved in our model and thus perform part of its validation (tests of behavior). The validation methodology will be discussed in Sect. 4.8 below.

To illustrate the use of our model for risk assessment, while avoiding disclosing sensitive data, a fictitious project labeled SO FIS07 was synthetically generated based on real contract parameters and on simulation results. SO FIS07 has an expected workload of 5320 service units (SU), 174 days as expected conclusion time and its estimated cost is $256,211. Using this input set, the proposed model produced 163 days as simulated conclusion time and $267,560 as simulated cost, both indicators within a confidence interval of 10 %. The anticipated contract management cost is $124,609 and the cost of rework is $3,240. These two latter parameters were not originally registered by SEFAZ, but were estimated using simulation.

In what follows we study how the proposed simulation model was applied to ITO risk management at SEFAZ following a 5-step roadmap described in [19]: (1) Defining the risk factors; (2) Defining impacts; (3) Defining the variation of risk factors; (4) Conducting sensitivity analysis; and (5) Analyzing the results.

4.1 Defining the Risk Factors

A number of risk factors were selected based on important references in the ITO risk management literature [3, 4, 11, 13, 24–26]. It is important to emphasize that the focus of the proposed model is on human resources management related to IT capabilities, the Contract Monitoring Capability in particular. Risk factors associated with attributes of the contract itself and of the relationship between customer and supplier are outside the scope of this paper.

To relate the parameters of our model to the risk factors identified in the literature, we describe risk *scenarios*, as in [4, 26], which can be interpreted as complex risk factors. For illustration purposes, the following two risk scenarios (and related model parameters) were selected:

RS1—Insufficient Capability of the Contracting Organization in Monitoring ITO Contracts: In this scenario, contracts based on performance metrics (quality, cost, reward, penalties, revenue, etc.) are highly impacted as it becomes costly and inefficient to measure such metrics, which seriously compromises the results of the ITO initiative. Here, we consider that the contract monitoring process involves the following skills: the capability to estimate effort levels and timelines for completing tasks, to have sufficient knowledge of the outsourced function to check the delivered product or services, the ability to collect and record contract performance indicators and to negotiate with the supplier in the event of dispute.

Even when the available staff are highly experienced in all these skills, if there is an insufficient workforce performing contract management tasks, there will be a bottleneck in the work flow between customer and supplier. The lack of contract management tools may also lead to bottlenecks in managers' productivity, delay in acceptance of delivered services and the closure of invoices for payment, and difficulties in calculating penalties and in timely renegotiation and renewal of contracts. An incomplete or poorly detailed contract can generate dispute between customer and supplier about scope and quality levels of the contracted service, methodology for calculating the quality and cost indicators, penalties, and incentives. All these facts can lead, separately or in conjunction, to expected service conclusion time and cost misses; to acceptance of services with low quality level; and, to litigation with the supplier.

Contract monitoring capability is represented in our model by a productivity rate (CM Productivity), measured in service units per day (SU/Day) and calculated as a function of the parameters described below.

Associated simulation model parameters are: *Allocated CM Workforce* (in Number of Persons): Human resources allocated to perform tasks related to the ITO contract monitoring; *Initial Average CM Skill Level* (no measurement units): Initial average skill level of internal staff allocated to the ITO contract monitoring in this function; *Time to Adjust CM WF* (Day): Operating delay in adjusting the contract monitoring human resources; *Time to Adjust CM Skill Level* (Day): Time required to absorb and apply training and/or to gain experience on contract monitoring; *Time to Detect Defects* (Day): Time required for a defect in a delivered service to be detected by the contract monitoring team. The simulator models this parameter as a nonlinear function of the parameter CM Capability; so its behavior is endogenous.

RS2—Insufficient Capability of the Supplier to Deliver the Contracted Service:

Our work focuses on managing the contracting organization's resources and how to configure them to build IT capabilities. Therefore, we consider the supply-side capabilities in a consolidated basis (as a cloud). The supplier's service delivery capability involves the following skills: knowledge of the outsourced IT function and ability to deliver the product or service according to the performance parameters specified in the contract.

The less technical knowledge of the outsourced IT function the supplier has, the more they will fail to meet agreed performance requirements and this will directly affect the quality of the service delivered. Noncompliant delivered services will be resubmitted to the vendor for corrections, delaying the expected completion time for the service. The more rework is generated the more contract monitoring working hours will be consumed, rechecking delivered services. This will increase contract monitoring costs. Rework over the parameters agreed in the contract will also generate penalties and extra operational costs for the supplier, decreasing its profitability and causing it to reduce interest in the contract.

The service delivery capability is represented in our model by a productivity rate (*SD Productivity*), measured in service units per day (SU/Day).

Associated model parameters are: *Time to Adjust SD Productivity* (Day): Operating delay to adjust the service delivery capability; *SD SLA* (no measurement units): Service Level Agreement parameter is a real number in the range [0–1] that represents the minimum quality level of the delivered services. We say that a fraction (1–*SD SLA*) of the delivered service units will have defects and will need rework. This parameter does not influence penalties but influences the total cost of rework, which affects the supplier's profitability.

4.2 Defining the Impacts

The impact factors are attributes of the entities involved in IT services (client, provider, and service itself); usually representing their performance indicators such as cost, completion time, quality level, and satisfaction level. These indicators are affected by changes in risk factors. Based on the same rationale given in Sect. 4.1, here we describe impact *scenarios* as impact factors reach certain conditions.

"Earned value" offers a valuable approach for tracking performance against plans and controlling projects [27]. Earned value indicators compare planned values to actual values along the evolution of projects. For example, Cost Performance Index = (Task Budget × SO Completion Percentage)/Actual Cost.

To track the performance of SOs, we propose four earned value-based indicators: *SD Cost Performance Index*; *CM Cost Performance Index*; *Schedule Performance Index*; and *Supplier Profitability Index*. It is of interest to observe trends of earned value indicators by analyzing the slope of their curves. Interpretation of static performance positions may lead to less effective decisions.

The following impact scenarios are of interest.

IS1—Exceed ITO Budget: This impact scenario arises when the expected cost for SOs is exceeded. The associated model parameter is the earned value indicator *SD Cost Performance Index* (No Unit*)*, calculated based on *Task Budget* ($)—the estimated cost for the SO, based on its workload and on contract formulas; on *SD Conclusion Fraction* (No Unit)—the actual conclusion percentage of all service units from an SO; and on *Cumulative Cost of SD Capability* ($)—the cost of the capability (internal and outsourced) used to process all service units from an SO.

IS2—Exceed the Expected Service Conclusion Time: This impact scenario arises when the expected conclusion time for SOs is exceeded. The associated model parameter is the earned value indicator *SD Schedule Performance Index* (No Unit), calculated based on *Task Expected Conclusion Time* (Day)—the estimated conclusion time for the SO, based on its workload and on contract formulas; on *SD Conclusion Fraction* (No Unit)—the actual conclusion percentage of all service units from an SO; and on *Elapsed SD Time* (Day)—the number of elapsed days that a supplier effectively spent so far to process a SO's service units.

IS3—High Contract Management Cost: The costs of internal resources are usually neglected or not computed in public sector outsourcing processes, where salaries of career employees are not considered as part of the project's budget [28]. The effort (and cost) involved in managing contracts in Brazil typically represent between 30 and 40 % of the related service cost [28]. Exceeding this threshold means incurring additional management costs.

The associated model parameter is the earned value indicator **CM Cost Performance Index** (No Unit), calculated based on **Cumulative Cost of CM Capability** (\$): Cost of the capability used for monitoring ITO contracts along the SO execution; and on **Cumulative Cost of SD Capability** (\$).

IS4—Premature Contract Termination and Service Discontinuity or Debasement: This impact scenario is more subjective. From the customer's point of view one can monitor indications that the supplier is losing money or is not achieving the profitability projected at the beginning of the contract. Therefore, the supplier has reduced interest in continuing the relationship. Thus, in a possible replacement scenario, services may be discontinued or have their quality compromised by the lack of resources for their proper functioning.

Associated model parameter is the earned value indicator **Supplier Profitability Index** (No Unit) which, in the case where SOs have fixed prices, based on an initial agreed effort estimation, indicates if extra costs (penalties, cost of rework) are eroding the profitability of SOs. It is calculated based on **Task Budget** (\$); on **SD Conclusion Fraction** (No Unit); on **Cumulative Cost of SD Capability** (\$); and on **Cost of Penalties for Rework** (\$) which is the total cost of penalties issued to the supplier upon reaching a contractually agreed rework index.

All "expected values" mentioned in the description of impact scenarios are established relative to a baseline. This baseline can be elicited from empirical data, interviews with experts, or generated synthetically using simulation.

4.3   Relationships Between Risk and Impact Scenarios

Figure 2 summarizes the cause-and-effect relationships between risk and impact scenarios within the model. These relationships were established based on [13, 24–26], and on interviews with experts from SEFAZ. Figure 2 also illustrates

Fig. 2 Cause-and-effect relationships between risk and impact scenarios

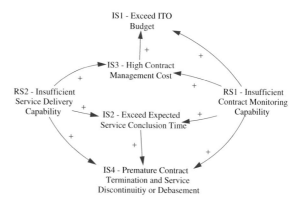

Table 2 Risk factor variation

Parameters	Estimated values			Parameters	Estimated values		
	Min	Exp	Max		Min	Exp	Max
Customer conditions				Supplier conditions			
Initial available CM workforce	2	5	7	Time to adjust SD productivity	5	15	30
Initial average CM skill level	0.4	0.7	1				
Time to adjust CM WF	5	15	30	SD SLA (minimum level of service)	0.85	0.9	0.98
Time to adjust CM skill level	10	30	40				

hypotheses (e.g., higher contract management costs should increase the chance of exceeding ITO budget) to be explored in the sensitivity analysis in Sect. 4.5.

4.4 Variation of Risk Factors

The range of risk factor values reflects the uncertainty with which decision makers predict impacts. Such uncertainties are generated stochastically by varying the simulation input variables (risk factors) according to probability distribution functions. These functions are constructed based on empirical data and goodness-of-fit tests or the triangular probability distribution function is used with parameters estimated by experts. Here, we use observed data at SEFAZ and estimates from experts. It is important to emphasize that the subjective estimation of numerical parameters made by experts based on their experience and knowledge does not violate the quantitative nature of our approach. Also note that history (information in logs), conditions (such as physical, temporal or financial limitations) and guidelines (such as those established in corporate policies) may reduce the "subjectivity" in providing estimates.

To better understand the impacts caused by variation in risk factors, these variations will be divided into (a) Customer conditions and; (b) Supplier conditions, as laid out in Table 2.

4.5 Sensitivity of Impact Factors

The sensitivity charts generated by the Vensim DSS simulation environment [29] allow an intuitive visual analysis of the magnitude of the impacts caused by the realization of risk conditions at different confidence intervals. For instance, one can observe the cumulative probability of an impact factor exceeding an expected value.

Fig. 3 Cost of service
(peak = $267,560)

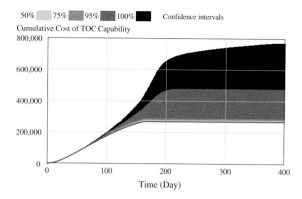

Fig. 4 Service cost
performance index
(peak = 0.96)

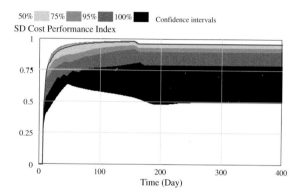

Fig. 5 Histogram of contract
monitoring cost performance
index (peak = 0.46) with
adjusted Weibull distribution

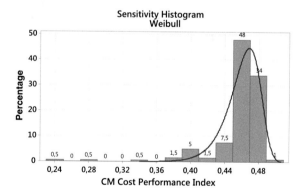

Figure 3 shows how variability in risk factors influences the *Cumulative Cost of SD Capability* over time. Figure 4 shows the variation of the impact factor *SD Cost Performance Index* over time. Figure 5 shows the histogram of the variation of the impact factor *CM Cost Performance Index*.

In all 200 simulations performed for the sensitivity analysis, the Latin Hyper-cube sampling technique with a default noise seed (1234) and triangular probability distributions with the ranges in Table 2 were used to generate the multivariate random sample of risk factors. The Latin Hypercube sampling ensures that the full range of each parameter being varied is explored more evenly and consistently in the simulations.

In Figs. 3 and 4, the solid red line (peak) is the simulation result for the scenario in which all risk factors simultaneously assume their expected values. It is the baseline for the risk analysis. The shaded areas represent the confidence intervals adopted for the sensitivity analysis, where 50 % (light gray), 75 % (gray), 95 % (dark gray), and 100 % (black) of simulated impact factors are located. The limits of the black area represent the situation of the maximum and minimum impacts on service performance.

Each simulation runs for 400 days. This end-of-simulation condition was adopted because, this time interval holds more than double the estimated simulated SO completion time and it is possible to observe all behaviors of interest.

Besides the sensitivity charts, the simulator generates histogram charts and the main statistical estimators (mean, median, standard deviation, minimum value, maximum value, and normalized standard deviation). This information enables various statistical analyses on the impact factors, including the identification of their probability distribution functions.

4.6 Analysis of Simulation Results

Following the risk management cycle, the information gathered from the sensitivity analysis (the risk assessment) can be used to prioritize risks, invest in risk prevention, risk control, and risk mitigation activities. It is possible to calculate potential financial losses and to quantify indicators that can be used to support qualitative or subjective management decisions.

To better understand the impacts of risk factors, we divided the sensitivity analysis into two subsections. First, we recorded the effects of uncertainty about the contracting organization's conditions on the impact factors. Then, we recorded the impacts caused by uncertainty about the supplier's conditions.

The impacts will arise when the expected values (peak line) for the impact factors are exceeded. The polarity of the relationship between risk factors and impact factors will define in which region of the graph undesirable values will be located. For instance, if x is the expected cost of an SO to the supplier and $F(x)$ the probability distribution function associated with this cost, the probability of service budget overrun is $1 - F(x \leq \text{expected cost})$. F can be identified as the Chi-Square function from the frequency distribution of outputs using goodness-of-fit tests. Other possible tests for identifying F are Kolmogorov–Smirnov and Anderson–Darling [19, 27]. Here, we use Chi-Square and Anderson-Darling from the statistical software Minitab® 17.1.0 [30].

Table 3 Confidence bounds for the cost of service

Conf. Bound	Min cost	Max cost	Conf. Bound	Min cost	Max cost
50 %	$270,690	$277,245	95 %	$265,900	$476,938
75 %	$266,430	$288,535	100 %	$265,900	$769,672
All peak	$267,560				

Varying the Customer's Conditions: In total, 200 simulations were performed in which all model parameters related to the contract monitoring capability of the ITO customer varied simultaneously.

Impact on SO Cost: Fig. 3 shows the cost accumulation of the capability employed to execute the SO. The peak line shows a slight exponential growth in the initial phase of the simulation, during which the service delivery capability is being allocated and used. The inflection point indicates the time when this capability begins to be demobilized and its accumulated costs tend to stabilize (stop growing). This fact indicates that the SO has been fully executed. In terms of sensitivity analysis, the location of the peak line in relation to confidence intervals reveals a very unfavorable prognosis for the execution of the SO within the planned cost. Visually, one can estimate that between 50 and 75 % of the simulation results exceed the planned cost for the SO. Table 3 shows the confidence bounds for the cost of the SO.

Figure 4 shows the dynamic changes in the earned value indicator SD Cost Performance Index that express the ratio between the expected and the actual cost of the SO.

The peak line shows that the evolution of the actual cost of service is very close to the planned cost since the beginning of the project. The relative position between the peak line and the confidence intervals reveals them in more than 75 % of the simulations; the cost performance was above 0.8. This will not financially impact the SEFAZ, since according to the contract, the client organization will only pay the cost calculated based on the effort initially estimated for the service. In interviews with vendor project managers, it was said that a cost performance of not less than 0.8 is considered satisfactory. This implies that the simulated scenario of uncertainty about the customer conditions presented impacts only for the supplier, but within a tolerable cost range, which is good. Hence, no treatment is required for the risk factors involved. That is because the value considered typical of the input parameter *CM Initial Workforce* (five persons—see Table 2) supported the demand well.

Examining the histogram for the SO execution cost (not shown here), it was observed that the distribution resembles a negative exponential distribution. Therefore, given that the average of 200 simulations of the SO execution cost is $327,993 with standard deviation of $97,248, the cumulative probability of a budget overrun is approximately $[1 - P (x \leq \$267,560)] = 44.33$ %.

Impact on SO conclusion time: Table 4 shows that the SO conclusion time is lower than expected in more than 90 % of the 200 simulations. Observing the raw data

Table 4 Confidence bounds
for schedule performance
index

Conf. Bound	Min	Max	Conf. Bound	Min	Max
50 %	1.067	1.067	95 %	0.97	1.067
75 %	1.06	1.067	100 %	0.46	1.067
All peak	1.067				

(not shown here), the schedule performance index is less than one in only five simulations. This implies that the majority of variations in risk factors associated with customer were favorable.

In order to illustrate what could happen with this impact if SEFAZ were to allocate a smaller team to monitor contracts, a less favorable scenario was set for a new round of simulation runs. In this new scenario, the uncertainty range of human resources for the contract management team was made worse by adopting the values min = 1, peak = 2, and max = 3 (as opposed to 2, 5, and 7 in Table 2). In this new scenario, in more than 95 % of the simulations, the schedule performance index was around 0.45 (mean = 0.455, median = 0.451, standard deviation = 0.034). This implies that in a more pessimistic scenario, the supplier is likely to face schedule delays (due to lower throughput by the customer in "approving deliverables") and will thus have to allocate more human resources in order to try to catch up and avoid missing deadlines.

Impact on Contract Monitoring Cost: Examining the histogram for the earned value indicator *CM Performance Index* (Fig. 5) and using the Anderson–Darling test, it was observed that the distribution resembles a Weibull distribution with estimators Shape = 28.58 and Scale = 0.4695.

Assuming F(x) is a Weibull probability distribution function, the cumulative probability of contract monitoring costs being higher than anticipated (greater than 40 % of the SO cost, using this threshold from [28]) $[1 - F(x \leq 0.4)] = 0.9$, i.e., risk materializes in 90 % of the simulations.

Impact on Supplier Profitability: For all contracts analyzed within this model, the supplier is financially penalized in the following situations: (a) in rework, because it bears the costs of penalties and the operating cost of reworking; (b) in delays verifying delivered services and in closure of invoices for payment.

Variations in contract monitoring capability will not impact the amount of generated defects but, rather, will influence the time to detect them. In 95.5 % of 200 simulations, the cost of penalties varied less than 10 % from the baseline value (peak = 12,810; mean = 12,861; median = 12,810; standard deviation = 1,144). However, low levels of this capability will cause bottlenecks in the approval of delivered services. In tasks with strong interdependence, the supplier cannot go ahead with the work but will have to bear costs for idle capacity. In addition, the delay caused by the contract management team will increase pressure on the project schedule. The supplier may have to mobilize more capacity for the project in an attempt to keep the initial deadline, increasing its costs. The variation in contract monitoring capability caused more variation in the cost of service, as can be observed in Fig. 4.

For the simulated scenario, the sensitivity histogram for the earned value indicator *Supplier Profitability Index* resembles a normal distribution with the estimators mean = 0.89 and standard deviation = 0.064. Assuming F(x) as a normal probability distribution function, the cumulative probability of the supplier profitability index being lower than 0.8 (again, a level considered satisfactory by interviewed supplier's project managers) [F(x ≤ 0.8)] = 0.079, i.e., risk materializes in 7.9 % of the simulations.

Again, a new round of simulations was run assuming a less favorable scenario where the uncertainty range of human resources for the contract management team was made worse by adopting the values min = 1, peak = 2, and max = 3 (as opposed to 2, 5, and 7 in Table 2). The histogram of supplier profitability index for this new scenario shows that in 33 % of the simulations this earned value indicator was below 0.8. This implies that in a more pessimistic scenario, the supplier is likely to have losses with this contract and the customer needs to allocate more human resources to contract monitoring in order to reduce the chance of premature contract termination.

We conclude this section by emphasizing that the simulated uncertainty levels in the customer's conditions initially cause direct impacts on service conclusion time in the most pessimistic scenario (IS2) and in contract management cost, which is an endogenous risk factor related to exceeding the budget for the service (IS1). Service conclusion time overruns can bring indirect impacts to the customer, depending on how the outsourced service relates to the business layer. The high cost of contract management related tasks (IS3) is often overlooked by public organizations in Brazil since they do not include wages of the internal team responsible for this task in IT project budgets.

Univariate simulations which vary the customer's risk factors one at a time were also undertaken in order to identify factors that had the most impact. The *Initial Contract Monitoring Workforce* factor is responsible for the greatest variation in the impact factors. In a more unfavorable scenario of contract monitoring human resources, we found that the cost of services and supplier profitability were impacted (IS1 and IS4) the most. Therefore, the model indicates that a more effective action to control or mitigate the risks of insufficient contract monitoring capability is to ensure that sufficient human resources are allocated to this activity. Other components that comprise this capability are also important and should be considered when mitigating this risk. Learning curve delays have also been shown to be important factors in other reference models and in empirical investigations [14–17] suggesting that it may be important to improve the treatment of knowledge acquisition in future versions of the model. Investing in training and contract management tools, and using methodologies and expertise to more accurately estimate the effort and cost of IT projects are actions that can mitigate risks associated with costs overrun and litigations with suppliers.

Varying the Supplier's Conditions: 200 simulations were performed simultaneously varying all model parameters related to the supplier's service delivery capability, namely: *Time to Adjust SD productivity, Minimum SD Skill Level,* and *SD SLA* (Service Level Agreement).

Changes imposed on the supplier's conditions cause significant impact in the cost of the service, in the cost of contract monitoring, and in the service conclusion time. The earned value indicator *SD Cost Performance Index* varied from 0.72 to 0.8 in 30 % of the simulations, in the risky region (bellow 0.8 as commented earlier). *CM Cost Performance Index* varied from 0.42 to 0.47, assuming the median = 0.46 in 43.5 % of the simulations. *Schedule Performance Index* varied from 0.82 to 1.08, assuming the median = 1.06 in 57.5 % of the simulations.

The most significant impacts were on the earned value indicator Supplier Profitability Index, which varied from 0.7 to 0.77 in 33.5 % of simulations (the risky range) and from 0.86 to 0.92 in 66.5 % of simulations. The risk factor to which impact factors were most sensitive was the *Service Level Agreement,* associated with the overall quality of service provided in relation to the percentage of defects generated.

The impact scenario IS4 (premature contract termination caused by low profitability for the supplier) has a high probability of arising in this scenario based on low *Supplier Profitability Index.*

The simulations performed can provide multiple insights for decision makers regarding prevention and control of premature contract termination, which may compromise the quality of services and the achievement of planned benefits. The effects of a supplier's low service delivery capability go beyond the obvious and immediate delay in projects. They can compromise the quality of the relationship and the profitability of the contract, affecting the supplier itself, which could not withstand such impacts for long.

It is important that the customer monitors its suppliers' level of satisfaction in order to anticipate scenarios where switching supplier is needed—typically a complex and costly process in Brazilian public organizations by red tape and inefficiency in its flow. This monitoring can be achieved using the *Supplier Profitability Index* indicator.

SEFAZ recently faced this situation in its information systems development and maintenance outsourcing contract signed in 2010 with a budget of tens of millions of dollars over a multiple year contract. This contract was prematurely terminated (not renewed) by the supplier after 24 months. The supplier justified their lack of interest in renewing the contract, during their annual renegotiation, claiming the financial infeasibility of the relationship. According to the SEFAZ contract manager, the company presented an unsatisfactory performance throughout the contract, undergoing several fines for SLA violations. Despite the low quality delivered by the supplier, SEFAZ tried to keep the contract because the bureaucracy involved in a change of supplier would be very time-consuming, negatively affecting all related system projects with no assurance that performance issues would be resolved by a new supplier in a satisfactory period of time. Moreover, delays caused by rework did not increase SO's direct costs to the customer. By not being calculated, the extra costs and any losses caused by non-deployed systems were ignored by SEFAZ. These facts gave the false impression that SEFAZ was not suffering financial loss.

For the supplier, on the other hand, fines and operating costs caused by the rework were eroding the profitability of the contract. After being purchased by a global IT provider, the new supplier's managers chose not to continue the contract with SEFAZ.

Interviews with the replacement supplier of the referred service and the analysis of the records of 20 projects executed in this recent contract revealed that the low contract monitoring capability presented by the SEFAZ was affecting the supplier's cash flow due to delays in approval of the services delivered for further payment. Customer and supplier have been working together to improve this process through mutual learning.

4.7 Implications of the Proposed Model to Risk Management at SEFAZ

Risk management at SEFAZ is currently carried out with a tool that uses a quali tative approach based on compliance. In this tool, a governance map is built associating objects in three different layers: business processes in the strategic layer, IT processes in the tactical layer, and IT asset (material resources, systems, and human resources) in the operational layer. Qualitative levels of importance (low, medium, and high) are assigned to each of the connections between objects. A set of controls is associated with each category of IT assets. These controls represent risk factors to which each asset is subject. The process of risk assessment with this tool consists in informing whether or not the controls are implemented. At the end of this process, qualitative risk indices (very low, low, medium, high, and very high) are calculated for each asset and propagated to the strategic layer through the links defined by the governance map.

We have interviewed four users of this tool (an information security officer, an IT manager, a business process manager, and a software project manager). They judge that the way risks are currently measured/reported suffices to prioritize them based on the indices. On the other hand, the qualitative nature of such indices does not allow decision makers to estimate the actual extent of impacts and thus pre cludes trade off analyses of investments when addressing those risks. The inter viewees believe the proposed model will facilitate these estimation and analyses.

4.8 Validation Efforts at SEFAZ

The validation efforts of the base model used in this study for the purpose of risk assessment have led to its improvement, as follows:

1. The production and execution of a goal-oriented measurement plan (GQM plan), part of the system dynamics development framework IMMoS [22],

deepened insight into the model parameters and into the availability of records on project performance in organizations surveyed;

2. The volume of empirical data about SEFAZ projects available for analysis has improved estimates of calibration parameters, uncertainty levels faced by managers and confirmed the dynamic hypothesis incorporated in the model;

3. The lack of detailed records regarding the customer, reflecting the difficulty of the contract management team in maintaining outsourcing contracts performance indicators, led us to gather data directly from the supplier. Consequently, more knowledge about the supplier was acquired, including better understanding of the impacts suffered by him, which improved the analysis of IS4.

5 Conclusions and Outlook

In this paper, we showed how our model, to support decision-making in ITO and in IT capabilities, can be used to analyze and prioritize risks. Following the risk assessment procedure in [19] applied to the context of a Brazilian state tax and finance agency (SEFAZ), we analyzed the impact of two kinds of risks (lack of contract monitoring capability in contracting organizations and lack of service delivery capability in suppliers) on the ITO budget, on the deadline for completion of services and on the relationship between customer and supplier.

Our experiments indicate that a lack of contract monitoring capability in ITO contracting organizations directly impacts service cost and service conclusion time, and influences the cost of contract management, which is an endogenous risk factor related to exceeding the service budget. The bottleneck produced by low contract monitoring capability in approvals of deliveries increases the schedule pressure, inducing the supplier to increase its internal capability level to compensate for delays. Another effect of this bottleneck is the maintenance of idle capacity at the supplier, awaiting the approval of required artifacts for the continuation of projects. In both situations, the supplier's costs increase. This has been confirmed by the analysis of empirical data collected during the execution of the GQM plan. It was also found that low levels of service delivery capability in the supplier most significantly impact the earned value indicator *Supplier Profitability Index*. These may induce early termination of the contract.

The observation of earned value dynamics offers model user's capability to prioritize risks based on these behavioral tendencies.

The base model used in this study underwent a complete validation cycle (see [7]). The learned lessons and the results of the initial model validation as well as the analysis of new empirical data collected during the execution of a goal-oriented measurement plan are being used to guide structural enhancement and calibration of the model.

However, in order to complete validation of its utility for supporting risk-based decision-making for ITO, the model needs to undergo new tests of learning with the same interviewed group of users, as well as a more comprehensive comparative study between our approach and the current risk assessment approach used at SEFAZ.

References

1. Office of Government Commerce (OGC): ITIL Core Books, Service Strategy, TSO, UK (2007)
2. Barney, B.: Firm resources and sustained competitive advantage. J. Manag. **17**, 99–120 (1991)
3. Ngwenyama, O.K., Sullivan, W.E.: Outsourcing contracts as instruments of risk management: insights from two successful public contracts'. J. Enterp. Inf. Manage. **20**(6), 615–640 (2007)
4. Martens, B., Teuteberg, F.: Why risk management matters in IT outsourcing: a literature review and elements of a research agenda. In: Proceedings of the 17th European Conference on Information Systems, p. 1–13 (2009)
5. Lacity, M.C., Khan, S.A., Willcocks, L.P.: A review of the IT outsourcing literature: Insights for practice. J. Strateg. Inf. Syst. **18**(3), 130–146 (2009)
6. Bahli, B., Rivard, S.: Validating measures of information technology outsourcing risk factors. Omega **332**, 175–187 (2005)
7. Bezerra, T.R., Bullock, S., Moura, A.: A simulation model for risk management support in IT outsourcing. In: Proceedings of the 4th International Conference on Simulation and Modeling Methodologies, Technologies and Applications SIMULTECH 2014—Vienna, Austria, 28–30 August, pp. 339-351 (2014). ISBN 978-989-758-038-3
8. International Standards Organization: ISO 31000:2008—Risk Management: Principles and Guidelines on Implementation (2008)
9. Lacity, M., Khan, S., Yan, A., Willcocks, L.: A review of the IT outsourcing empirical literature and future research directions. J. Inf. Technol. **25**, 395–433 (2010)
10. Osei-Bryson, K.M., Ngwenyama, O.K.: Managing risks in information systems outsourcing: an approach to analyzing outsourcing risks and structuring incentive contracts. Eur. J. Oper. Res. **174**(1), 245–264 (2006)
11. Ngwenyama, O.K., Technology, I., Sullivan, W.E., Patricia, B.: 'Secrets of a successful outsourcing contract: A risk analysis framework for analyzing risk factors. Technology **416**, 1–12 (2006)
12. Willcocks, L.P., Lacity, M.C., Kern, T.: Risk mitigation in IT outsourcing strategy revisited : Longitudinal case research at LISA. Inf. Syst. **8**(1999), 285–314 (2000)
13. Earl, M.J.: The risk of outsourcing IT. Sloan Manage. Rev. Spring (1996)
14. Abdel-Hamid, T.K., Madnick, S.E.: Software project dynamics: an integrated approach. Prentice-Hall, Englewood Cliffs (1991)
15. Lin, C.Y., Abdel-Hamid, T.K., Sherif, J.S.: Software-engineering process simulation model (SEPS). J. Syst. Softw. **38**, 263–277 (1997)
16. Garousi, V., Khosrovian, K., Pfahl, D.: A customizable pattern-based software process simulation model: design, calibration and application. Softw. Process Improv. Pract. **14**(3), 165–180 (2009)
17. Costa, M.D., Braga, J.L., Abrantes, L.A., Ambrósio, B.G.: Support to the decision making process in human resources management in software projects based on simulation models. Rev. Eletrônica de Sistemas de Informação **12**, (1), jan–mai 2013, artigo 51 (2013). doi:10. 5329/RESI.2013.1201005
18. Houston, D.X., Mackulak, G.T., Collofello, J.S.: Stochastic simulation of risk factor potential effects for software development risk management. J. Syst. Softw. **59**(3), 247–257 (2001)

19. Pfahl, D.: ProSim/RA—software process simulation in support of risk assessment. In: Biffl, S., Aurum, A., Boehm, B., Erdogmus, H., Grünbacher, P. (eds.) Value-based software engineering, pp. 263–286. Springer Press, Berlin (2005)
20. Gui-sem, W., Xiang-yang, L.: The risk analysis on IT service outsourcing of enterprise with system dynamics. Int. Conf. Serv. Sci. (2010). doi:10.1109/ICSS.2010.47
21. Tan, B., Anderson, E.G., Dyer, J.S., Parker, G.G.: Evaluating system dynamics models of risky projects using decision trees: alternative energy projects as an illustrative example. Syst. Dyn. Rev. 26(1), 1–17 (2010)
22. Pfahl D.: An integrated approach to simulation-based learning in support of strategic and project management in software organisations. Ph.D. thesis, Universität Kaiserslautern, Germany
23. Cunha, M.C.: Aspectos e Fatores da Terceirização de Sistemas de Informação no Setor Púbico: Um Estudo em Instituições Públicas de Alagoas. Universidade Federal de Pernambuco', Tese de Doutorado (2011)
24. Rivard, S., Albert, B., Patry, M.: Assessing the risk of IT outsourcing. IEEE (1998)
25. Rivard, S., Albert, B., Patry, M., Dussault, S.: Managing the risk of IT outsourcing. IEEE, Montreal (1998)
26. Bahli, B., Rivard, S.: The information technology outsourcing risk: A transaction cost and agency theory-based perspective. J. Inf. Technol. 18, 211–221 (2003)
27. Madachy, R.J.: Software process dynamics. Wiley, Hoboken (2007)
28. Carvalho, S.: Um processo para gestão de contratos de aquisição de serviços de desenvolvimento de software na administração pública. Masters Dissertation, UFPE, Recife, Brazil (2009)
29. The Ventana Simulation Environment—Vensim DSS. http://www.vensim.com
30. http://www.minitab.com

Analysis of Fractional-order Point Reactor Kinetics Model with Adiabatic Temperature Feedback for Nuclear Reactor with Subdiffusive Neutron Transport

Vishwesh A. Vyawahare and P.S.V. Nataraj

Abstract In this paper, a fractional-order nonlinear model is developed for the nuclear reactor with subdiffusive neutron transport. The proposed fractional-order point reactor kinetics model is a system of three coupled, nonlinear differential equations. The model represents subprompt critical condition. The nonlinearity in the model is due to the adiabatic temperature feedback of reactivity. This model originates from the fact that neutron transport inside the reactor core is subdiffusion and should be better modeled using fractional-order differential equations. The proposed fractional-order model is analyzed for step and sinusoidal reactivity inputs. The stiff system of differential equations is solved numerically with Adams-Bashforth-Moulton method. The proposed model is stable with self-limitting power excursions. The issue of convergence of this method for the proposed model for different values of fractional derivative order is also discussed.

Keywords Nuclear reactor · Point reactor kinetics model · Dynamic model · Fractional-order modeling

1 Introduction

The heart of a nuclear power plant is the nuclear reactor. In this, the heat energy is generated by carrying out a controlled fission of nuclei of fissile radioactive materials with the help of neutrons. Fission reactions are a result of neutrons moving inside the reactor core and colliding with the nuclei of core material. Due to the use of radioactive materials and high probability of this fission chain reaction becoming

V.A. Vyawahare (✉)
Department of Electronics Engineering, Ramrao Adik Institute of Technology,
Nerul, Navi Mumbai, India
e-mail: vishwesh.vyawahare@rait.ac.in

P.S.V. Nataraj
IDP in Systems and Control Engineering, Indian Institute of Technology Bombay,
Powai, Mumbai, India
e-mail: nataraj@sc.iitb.ac.in

© Springer International Publishing Switzerland 2015
M.S. Obaidat et al. (eds.), *Simulation and Modeling Methodologies,
Technologies and Applications*, Advances in Intelligent Systems
and Computing 402, DOI 10.1007/978-3-319-26470-7_8

uncontrollable, utmost care has to be taken to design, construct, maintain, and oper-
ate/control a nuclear reactor. In view of this, the mathematical modeling of nuclear
reactor is a key step in designing an efficient and safe reactor. As given in [1, 2],
this reactor model is fundamentally based on the model of neutron transport in reactor core. Thus, the validity and applicability of reactor model will depend on how
perfectly one models the neutron transport in its core.

In the classical analysis, the diffusion approximation of neutron transport is used
widely. The integer-order (IO) neutron diffusion equation (in one-dimension), based
on the Fick's constitutive law, is

$$\frac{1}{v}\frac{\partial \phi(x,t)}{\partial t} + (\Sigma_a - v\Sigma_f)\phi(x,t) = D\frac{\partial^2 \phi(x,t)}{\partial x^2}, \tag{1}$$

where v is the neutron velocity, $\phi(x,t)$ is the neutron flux at location x at time instant t,
D is the diffusion coefficient, v is the average number of neutrons emitted per fission
reaction, and Σ_a, Σ_f are the respective macroscopic cross-sections of absorption and
fission reactions. But the concept of modeling neutron transport as diffusion has
some problems, viz., the diffusion model is applicable mainly in the moderator of
the core and should not be used to model the neutron movements near the regions
with strong absorption, next, it predicts infinite speed of propagation of neutrons
[3–5]. In an attempt to rectify these shortcomings and achieve a better representation
of neutron movements, an FO neutron telegraph equation model was proposed in
[6, 7] (see (2)). It was developed using the stochastic technique of continuous-time
random walk (CTRW) as given in [8] for a slab reactor.

$$\frac{\tau^\alpha}{v^\alpha}\frac{\partial^{2\alpha} \phi(x,t)}{\partial t^{2\alpha}} + M_1\frac{\partial^\alpha \phi(x,t)}{\partial t^\alpha} + M_2\phi(x,t) = D\frac{\partial^2 \phi(x,t)}{\partial x^2}, \tag{2}$$

where $0 < \alpha < 1$, and $M_1 = \tau^\alpha(\Sigma_a - v\Sigma_f) + 1/v^\alpha$ and $M_2 = \Sigma_a - v\Sigma_f$. The terms
like $\frac{\partial^\alpha}{\partial t^\alpha}$ denote the Caputo fractional time-derivatives of order α (see (4)). The FO
point reactor kinetics (FPRK) model was reported in [9], which is a system on cou-
pled nonlinear ordinary differential equations (ODEs):

$$\frac{d^\alpha}{dt^\alpha}P(t) = \frac{\rho(t) - \beta}{\Lambda}P(t) + \lambda C(t),$$

$$\frac{d}{dt}C(t) = \frac{\beta}{\Lambda}P(t) - \lambda C(t), \tag{3}$$

with one delayed neutron group (1G). Here $P(t)$ is the reactor power, $\rho(t)$ is the
reactivity, Λ is the mean generation time between the birth of neutron subsequence
absorption inducing fission, $C(t)$ is the average concentration of the delayed neutrons,
and λ is the average decay constant of the delayed neutrons. Various versions of the
FPRK model were reported in [10]. The FPRK model (or the PRK model in general)
forms a set of coupled, nonlinear differential equations in reactor power $P(t)$ and
delayed neutron concentration $C(t)$.

Note that the Caputo fractional time-derivative definition is considered in models (2) and (3) which is defined as follows [11]. The Caputo fractional derivative (FD) of order $\alpha \in \mathbb{R}^+$ of a causal function $f(t)$ is given by

$$_0D_t^\alpha f(t) = \frac{1}{\Gamma(n - \alpha)} \int_0^t \frac{f^n(\tau)}{(t - \tau)^{\alpha - n + 1}} d\tau, \tag{4}$$

with $n \in \mathbb{N}, n - 1 < \alpha < n$, and $f^n(\tau)$ is the nth-order derivative of the function $f(t)$. It is seen that this definition requires $f(t)$ to be n-times differentiable and furthermore this derivative has to be integrable. This condition makes definition (4) quite restrictive. Nevertheless, it is preferred by engineers and physicists because FO differential equations (FDEs) with Caputo derivatives have same initial conditions (ICs) (which have well-defined physical meanings) as that for the integer-order differential equations. Recently, fractional derivatives have been extensively used for modeling a variety of systems and processes [12, 13], and also in control [14]. One of the major applications of FDEs is in modeling of anomalous diffusion occurring in complex systems [8]. These FDE models are found to be more realistic and compact than their counterparts, the classical integer-order models. For a detailed history and an exhaustive bibliography of fractional calculus and its applications, see [15].

The point reactor kinetics model establishes dependence of the neutron flux or power in reactor core on reactivity. The remarkable feature about reactor mechanism is that *the reactivity also depends on the power.* So there is an *inherent* feedback (negative, in fact) present in the reactor [1, 16]. There is a kind of 'cyclic' mechanism related to neutron flux and reactivity: reactivity affecting power, which in turn affects the reactivity. The justifications explaining the dependence of reactivity on the power can be summarized simplistically as

1. Reactor power depends on the reactivity.
2. Core temperature depends on the reactor power.
3. Reactivity depends on the core temperature.

In this paper, we consider this feedback mechanism to develop FO point reactor kinetics (FPRK) model with temperature feedback of reactivity, which mimics the situation of a subprompt critical reactor subjected to a small positive reactivity ($\rho_0 < \beta$). Literature survey reveals that there have been only two attempts in which the analysis of FPRK model with reactivity feedback is carried out. The first of these references [17] analyzes the FPRK model with Newtonian reactivity feedback and uses the FPRK model developed in [18]. The second contribution [19] reports the analysis of FO Nordheim-Fuchs model with adiabatic temperature feedback of reactivity. To the best of our knowledge, this is for the first time that the development of a subprompt critical nonlinear fractional-order point reactor kinetics model using adiabatic temperature feedback of reactivity for a nuclear reactor is reported.

The proposed FPRK model is based on the fundamental assumption of considering the neutron transport as anomalous diffusion, particularly subdiffusion [20, 21]. The literature survey reveals that there have been attempts to develop other types of fractional-order models of the neutron transport and the nuclear reactor, [5, 18, 22–26]. For a detailed and rigorous review on PRK models in general, see [18].

The paper is organized as follows. Next section discusses in brief the inherent reactivity feedback mechanism present in nuclear reactor. In Sect. 3, development of the proposed FPRK model is presented. Analysis of the proposed FO model with step and sinusoidal reactivity inputs is given in Sects. 4 and 5, respectively. A comparison of the results with the IO point reactor kinetics (IPRK) model in terms of time evolution of power, reactivity and reactor core temperature is presented. Issues related to the use of various solution methods (both analytical and numerical) for solving the stiff FPRK model are discussed in detail. Conclusion is given in Sect. 6. An appendix at the end gives the numerical algorithm used for solving fractional-order model.

2   Reactivity Feedback Mechanism in Nuclear Reactor

The feedback mechanism in the reactor can be represented using the block diagram in Fig. 1 (see [1, 16]). Let the reactor be represented by the PRK model. We start with the assumption that the reactor is operating at a steady-state equilibrium power level P_0. Now there will be two types of reactivities present in the reactor,

$$\rho_f[P_0] \equiv \text{feedback reactivity due to } P_0,$$
$$\rho_0 \equiv \text{external reactivity.} \tag{5}$$

The feedback reactivity $\rho_f[P_0]$ mostly corresponds to a negative reactivity, trying to reduce the neutron flux, and ultimately the power. Hence the feedback shown in Fig. 1 is to be considered as a negative feedback, and is also known as the power defect in reactivity [1]. If we allow this process to continue, it will result into the gradual reduction in the number of fission reactions and so in the number of neutrons. The reactor will gradually become more and more subcritical and a time will come when the reactor will eventually shut down. In order to keep the reactor running and maintain its criticality, an external (positive) reactivity ρ_0 must be applied (like withdrawal of control rods) to balance the negative reactivity such that

$$\rho(t) = \rho_0 + \rho_f[P_0] = 0. \tag{6}$$

Fig. 1 Closed-loop configuration with reactivity feedback

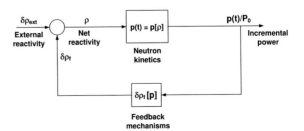

Now let the power change to a new value $P(t)$. The incremental power is defined as the deviation of the power from the equilibrium value,

$$p(t) \equiv P(t) - P_0. \tag{7}$$

The corresponding incremental changes in the reactivities can be expressed as

$$\delta\rho_{ext}(t) = \rho_{ext}(t) - \rho_0,$$
$$\delta\rho_f[p] = \rho_f[P] - \rho_f[P_0], \tag{8}$$

where, $\rho_f[P] \equiv$ feedback reactivity due to $P(t)$, and $\rho_{ext}(t) \equiv$ external reactivity to counterbalance $\rho_f[P]$. As a result, the net reactivity input to the reactor is comprised of two components,

$$\rho(t) = \delta\rho_{ext}(t) + \delta\rho_f[p], \tag{9}$$

which is depicted in the block diagram of Fig. 1.

3 FPRK Model with Reactivity Feedback

In this section, we develop the FPRK model with feedback of reactivity. Note that we consider the situation of one delayed group (1G) only, although extending this model to the six delayed group case is quite trivial and straightforward.

First, we consider the IPRK model with reactivity feedback [16], which is a system of three nonlinear IO ordinary differential equations (IDEs):

$$\frac{d}{dt}P(t) = \frac{\rho(t) - \beta}{\Lambda}P(t) + \lambda C(t),$$
$$\frac{d}{dt}C(t) = \frac{\beta}{\Lambda}P(t) - \lambda C(t), \tag{10}$$
$$\frac{d}{dt}\rho(t) = \frac{d}{dt}\gamma(t) - K_C\alpha_T P(t),$$

where, $\gamma(t)$ is the impressed reactivity, α_T is the temperature coefficient of reactivity and K_C is the reciprocal of the reactor heat capacity. The ICs are $P_0 = P(0)$, $C_0 = C(0)$, and $\rho_0 = \rho(0)$.

Now we derive the FPRK model with reactivity feedback. The 1G FPRK model (3) is rewritten:

$$\frac{d^{\alpha}}{dt^{\alpha}}P(t) = \frac{\rho(t) - \beta}{\Lambda}P(t) + \lambda C(t),$$
$$\frac{d}{dt}C(t) = \frac{\beta}{\Lambda}P(t) - \lambda C(t). \tag{11}$$

We now append to this system the reactivity feedback equation. Temperature feedback for time-varying reactivity is given by [16]

$$\rho(t) = \gamma(t) - \alpha_T(T(t) - T_0), \tag{12}$$

where $\gamma(t)$ is the impressed time-varying reactivity and $T(t)$ is the reactor core temperature with T_0 being the initial temperature at $t = 0$. Next the adiabatic model is:

$$\frac{d}{dt}T(t) = K_C P(t). \tag{13}$$

Differentiating (12) with respect to t, and using (13), we get the ODE for reactivity, as

$$\frac{d}{dt}\rho(t) = \frac{d}{dt}\gamma(t) - \alpha_T K_C P(t). \tag{14}$$

Thus we have a system of three nonlinear differential equations: an FDE for $P(t)$ and two ODEs for $C(t)$ and $\rho(t)$ as

$$\begin{aligned}
\frac{d^\alpha}{dt^\alpha}P(t) &= \frac{\rho(t) - \beta}{\Lambda}P(t) + \lambda C(t), \\
\frac{d}{dt}C(t) &= \frac{\beta}{\Lambda}P(t) - \lambda C(t), \\
\frac{d}{dt}\rho(t) &= \frac{d}{dt}\gamma(t) - K_C \alpha_T P(t).
\end{aligned} \tag{15}$$

This is the FPRK model with reactivity feedback. It is also appended with three ICs P_0, C_0, and ρ_0.

In the next section, we analyze and compare the IO and FO models (10) and (15) for two types of reactivity insertions, step and sinusoidal. This will give a clear picture of how the reactor power varies for a change in reactivity input with the negative temperature feedback when subdiffusive neutron transport framework is used.

4 Analysis of the Proposed Model with Step Reactivity Insertion

In this section, a thorough analysis of the proposed FPRK model (15) is carried out. We make the reactor below prompt critical, that is a step reactivity variation of much smaller magnitude ($\rho_0 < \beta$) is impressed. It will be seen that the negative temperature feedback of reactivity limits the power rise, eventually bringing it back to zero [1, 16, 27]. Since this nonlinear model is very difficult to solve analytically, we go for the numerical solution. The Adams-Bashforth-Moulton method, used widely in the field of fractional calculus and FO control was used to solve these models. The variations in power, precursor concentration, temperature, and reactivity are obtained for different values of fractional power α. Some issues related to the convergence of the numerical method are reported and discussed.

A certain amount ρ_0 of positive reactivity is suddenly inserted into the reactor. We need to keep the reactor below prompt critical [16] by choosing $\rho_0 < \beta$. Thus, $\gamma(t) = \rho_0 \Rightarrow \frac{d}{dt}\gamma(t) = 0$. So the ODE for reactivity in IPRK and FPRK models (10), (15) becomes

$$\frac{d}{dt}\rho(t) = -\alpha_T K_C P(t). \tag{16}$$

We use the data from [1]: $\beta = 0.0075$, $\lambda = 0.08\text{s}^{-1}$, $\Lambda = 10^{-3}$ s. The ICs chosen are $P_0 = 1$ w, $C_0 = 93.75$, and $\rho_0 = 0.0025$. We take $K_C = 0.05$ K/MW s and $\alpha_T = 5 \times 10^{-5} \text{K}^{-1}$. Variation in the reactor temperature $T(t)$ is also studied. It is obtained using (12) (with $\gamma(t) = \rho_0$) as

$$T(t) = \frac{\rho_0 - \rho(t)}{\alpha_T} + T_0. \tag{17}$$

As a convention it is assumed that $T_0 = 0\,°\text{C}$.

The IPRK system (10) was solved using the MATLAB ODE solver `ode15s` suitable for the stiff ODEs [28]. This particular solver was chosen because the PRK models incorporating temperature feedback of reactivity, in general, form a stiff system of nonlinear ODEs [18, 29]. The step-size was $h = 1 \times 10^{-3}$ s. Time-variation of power, delayed neutron precursor concentration, reactivity, and temperature are shown in Fig. 2.

As we see, the power starts rising due to the insertion of positive reactivity. However, its rate of increase is much slower. This peculiar behaviour is due to the presence of delayed neutrons as they help in slowing down the dynamics of the reactor. This increase in power causes the reactor temperature to rise. The adiabatic negative temperature feedback shows its effect and reactivity starts decreasing. The power attains a peak value of $P_{max} = 14.63$ w at the instant $t = 101.75$ s. The reactivity and the precursor concentration at this instant are 0.0004523 and 1289.13 respectively. Finally, the power reduces to zero with the reactivity settling at -0.002592. The final core temperature is $101.83\,°\text{C}$.

Next we solve and analyze the FPRK model (15). The same data and ICs are considered. As it is impossible to obtain a closed-form solution for the nonlinear FDE system, we opt for other techniques for its solution. To mention explicitly, the Adomian Decomposition Method (ADM) [30], and Variational Iteration Method (VIM) [31] were tried. These methods as such don't come under the category of numerical methods, because they provide solutions in the form of a power series with easily computable terms. These methods are claimed to have many advantages over the classical numerical methods, viz., no discretization, high accuracy, minimal calculations, to name a few. We tried to implement these methods for our problem using Mathematica [32]. But due to inherent stiff nature of the FPRK model, these methods did not work and a convergent solution could not be achieved A similar observation

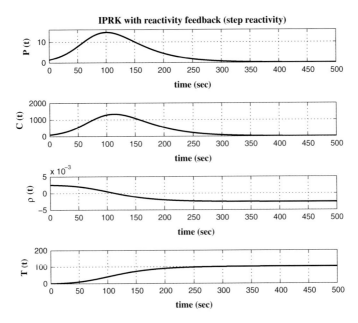

Fig. 2 IPRK model with reactivity feedback: plots for $P(t)$, $C(t)$, $\rho(t)$, and $T(t)$ for a step reactivity input

about the convergence of these methods for FO Nordheim-Fuchs model has been reported in [19].

It was then decided to use an improved version of the Adams-Bashforth-Moulton (ABM) algorithm [33] which is based on the Predictor-Corrector scheme for the FDE system [34]. This method worked perfectly for the given FDE system. It should be noted that the order of convergence for the ABM method is a non-decreasing function of the fractional order α. Only two values of fractional order α, 0.7, and 0.9 are considered as we could not make the algorithm converge for smaller values of α. Salient steps in the ABM algorithm are given in the Appendix.

The FPRK model (15) is solved for two values of α and analysis is carried out in detail.

1. $\alpha = 0.7$

   The step-size used for the ABM method was $h = 0.05$ s. The plots for $P(t)$, $C(t)$, $\rho(t)$, and $T(t)$ are shown in Fig. 3.

   We notice that the behaviour of this FPRK model is in line with the reactor dynamics. The impressed step reactivity causes the power to shoot up, albeit at a slower rate. The power reaches to its peak value, $P_{max} = 14.2031$ w (which is less than the P_{max} of IPRK model) at $t = 103.75$ s. The reactivity at this peak value of power is 0.000454 and the corresponding precursor concentration is 1255.09. The rest of the dynamics is similar to the IPRK model. Negative reac-

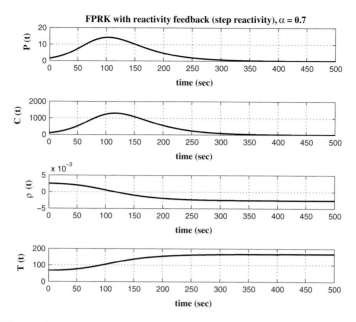

Fig. 3 FPRK model with reactivity feedback ($\alpha = 0.7$): plots for $P(t)$, $C(t)$, $\rho(t)$, and $T(t)$ for a step reactivity input

tivity required to bring power to zero is -0.002585. The temperature ultimately settles at $101.7043\,°C$.

2. $\boldsymbol{\alpha = 0.9}$

Same step-size $h = 0.05$ s. is used to solve the nonlinear FPRK model (15) with $\alpha = 0.9$. As said earlier, the behaviour of this FPRK model is very similar to that of the IPRK model. The effect of imposing a positive reactivity of $\rho_0 = 0.0025$ on power, precursor concentration, core temperature, and the reactivity itself is depicted in Fig. 4.

The power rise as a result of achieving the subprompt criticality is slow. The peak power is 14.5434 w and occurs at $t = 101.75$ s. The reactivity and precursor concentration at P_{max} are 0.0004513 and 1282.094 respectively. The power excursion is controlled by reduction in reactivity due to negative temperature feedback. Finally, a negative reactivity of -0.002591 brings the power to zero with reactor temperature settling at $T = 101.8262\,°C$.

To carry out a comparative study for the IO and FO models, these observations are compiled in Tables 1 and 2. Survey of these two tables and the plots in Figs. 2, 3, and 4 bring following observations to our notice:

1. The power overshoot in all models is small. In case of FPRK models, for both $\alpha = 0.7$ and 0.9, the peak power attained was nearly equal to that in the IPRK model. These maximum values in power occur almost at the same instant for all

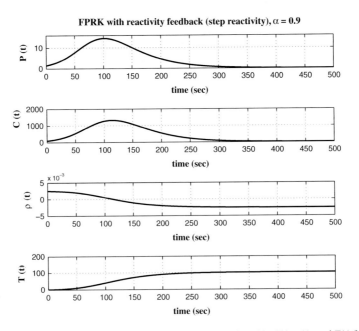

Fig. 4 FPRK model with reactivity feedback ($\alpha = 0.9$): plots for $P(t)$, $C(t)$, $\rho(t)$, and $T(t)$ for a step reactivity input

Table 1 IPRK and FPRK models with reactivity feedback: values of various quantities at P_{max} for step reactivity input

α	P_{max} (watts)	P_{max} at (sec)	$\rho(t)$ at P_{max}	$C(t)$ at P_{max}
0.7	14.2	103.7	4.5e-4	1255
0.9	14.5	101.7	4.5e-4	1282
IPRK	14.6	101.1	4.5e-4	1289

Table 2 IPRK and FPRK models with reactivity feedback: values of various quantities at steady-state for step reactivity input

α	$P(t)$ (watts)	$C(t)$	$\rho(t)$	$T(t)$ °C
0.7	0.0258	2.9	−0.002585	101.7
0.9	0.0131	1.6	−0.002591	101.8
IPRK	0.0111	1.4	−0.002592	101.8

    these models. Also, values of positive reactivity and precursor concentration at P_{max} are almost equal.

2. The inherent negative feedback of temperature helps in limiting the power overshoot and eventually brings it down to a negligible value. This dynamics is

observed in both IO and FO models. The delayed neutron dynamics also works for this cause.

3. After the power excursion dies out, it settles to a very low value (almost equal for IPRK and FPRK models). A very little negative reactivity is required to bring this power to zero. The reactor temperature settles at around 101 °C for all the models.

4. However a different scenario is observed in the dynamics of precursor concentration. The values of $C(t)$ when power reaches to its peak value are different for IPRK and FPRK models. Furthermore, the values also depend on the fractional derivative power α. It is seen that as the derivative order increases, precursor concentration at P_{max} increases (lowest for FPRK model with $\alpha = 0.7$ and highest for IPRK model). However, exactly opposite situation is observed during the steady-state. Now, the final value at which $C(t)$ settles after the power excursion has died out is largest for FPRK model with derivative order 0.7. It gradually decreases and is lowest for the IPRK model.

One may deduce that for the reactivity feedback case, the behaviours of IPRK and FPRK models (with $\alpha = 0.7, 0.9$) nearly coincide. However, it should be noted that the nuclear reactor is a critical and complex system and its safe operation is a crucial issue. A slight variation in any parameter may have severe effect on performance of the system. In view of this, the results predicted by the proposed FPRK model yield a better understanding of the nuclear reactor dynamics.

Furthermore, we think that the effect of considering subdiffusive neutron transport would be more saliently visible for lower values of α. But as mentioned earlier, we found it almost impossible to get the ABM method converged for smaller values of fractional differentiation order. Nevertheless, the results presented here confirm the validity of the developed FPRK model. It faithfully captures the behaviour of a reactor subjected to a subprompt step in reactivity under the influence of adiabatic temperature feedback.

5 Analysis of the Proposed Model with Sinusoidal Reactivity Insertion

In this section, we try to explain how the reactor behaves when it is subjected to the sinusoidally varying reactivity when the negative feedback of temperature is present, in the framework of subdiffusive neutron transport. Such type of variations in reactivity can be produced by sinusoidal oscillations of the control rods. We study how neutron population responds to this type of variation [35]. It is well known from the control theory fundamentals [36] that the process of excitation of system by sinusoidal input is used to determine system parameters. So this technique, popularly known as Rod Oscillator Method, is used to determine a number of neutron kinetics parameters. Here we vary the reactivity sinusoidally to study its effect on power, precursor concentration, and temperature.

The reactivity variation is given as [16]

$$\gamma(t) = \rho_0 + \rho_1 \sin \omega t, \tag{18}$$

where, ω is the angular velocity in rad/sec., ρ_1 is the amplitude, and

$$\frac{d}{dt}\gamma(t) = \omega\rho_1 \cos \omega t. \tag{19}$$

Using this, (14) becomes

$$\frac{d}{dt}\rho(t) = \omega\rho_1 \cos \omega t - \alpha_T K_C P(t), \tag{20}$$

which replaces the ODE for reactivity in IPRK and FPRK models (10), (15). We wish to analyze and compare various features of the IPRK and FPRK models. Again the same data, used for the step reactivity input, is considered. The initial values of power, precursor concentration, and reactivity, (P_0, C_0, ρ_0) remain the same as in the case of step reactivity. We take $\rho_1 = 0.005$ and $\omega = 1$ rad/s. Note that the value of amplitude ρ_1 is selected such that at $\omega t = \pi/2$, $\rho(t) = \rho_0 + \rho_1 = 0.0075 = \beta$, making the reactor prompt-critical.

Before we attempt to analyze and solve the FDE system, we consider the IO model. The nonlinear system (10) of differential equations is solved using the MAT-LAB ODE solver `ode15s`. A step-size of $h = 0.01$ s. was used. Plots for power, delayed neutron precursor concentration, reactivity, and the core temperature are shown in Fig. 5. Here, when the sinusoidal reactivity is introduced in the reactor, the power starts increasing. But unlike the step reactivity case, this increase in power is not monotonous. Oscillations with gradually increasing amplitude are produced in power. At $t = 33.36$ s., we have the peak power. The precursor concentration plot also shows very small oscillations during its ascend. The value of reactivity at P_{max} is 0.005187. The general dynamics of the reactor does not change. Positive reactivity increases power, which in turn increases the core temperature. This has a negative effect on the reactivity, as a result it decreases and consequently brings down the power to zero. Finally, the reactivity settles into a sinusoidal variation oscillating between the peak values -0.01061 and -5.75×10^{-4}. Thus at steady-state, the reactivity applied to the reactor is negative. The final core temperature is $161.5429\,^\circ C$.

Next we consider the FPRK model (15). Again we could solve this nonlinear system of ODEs for $\alpha = 0.7$ and 0.9 only, due to the diverging behaviour of the ABM method for lower values. The step-size used was $h = 0.025$ s. The response of the FPRK model to the sinusoidally varying reactivity for these two values of α can be understood very well by examining the variation in power, precursor concentration, and temperature as shown in Figs. 6 and 7. The reactor dynamics exhibited by these FPRK models is quite similar to that of the IPRK model. Here also, the insertion of sinusoidal reactivity forces the power to increase. It rises with oscillations whose amplitude grows gradually. Power reaches to a peak value. In the mean time,

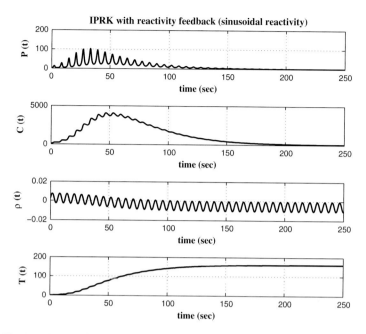

Fig. 5 IPRK model with reactivity feedback: plots for $P(t)$, $C(t)$, $\rho(t)$, and $T(t)$ for a sinusoidal reactivity input

increase in power heats up the reactor. The negative effect of temperature on reactivity becomes predominant, thereby reducing it. This forces the power to reduce and finally it settles to a very low value. Throughout this period, the reactivity continues to oscillate sinusoidally. But as seen, this sine wave gets shifted slightly downward. At the steady-state, it oscillates between two negative values. The delayed neutron concentration increases to a peak and then finally settles to a very low value. Core temperature increases and settles to a final value. Various observations are compiled in Tables 3 and 4. We summarize the results related to sinusoidal reactivity insertion as follows:

1. As expected, the delayed neutron precursor dynamics helps in limiting the magnitude of power overshoot.
2. The values of P_{max} as predicted by IPRK and FPRK models are slightly different. Also, there is a noticeable difference in the precursor concentration at the peaking of power. The instant at which power reaches to its maximum is almost same for all the models and so are the corresponding values of reactivity required to achieve this peak power
3. At steady-state, each model shows that power reaches to a very low value. The final values of precursor concentration and core temperature predicted by these models are almost equal. The reactivity continues to be in the sinusoidal variation. For all models, it settles into a sinusoid oscillating between two negative values.

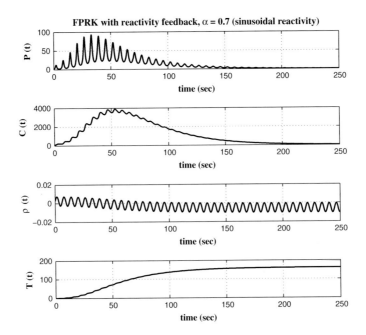

Fig. 6 FPRK model with reactivity feedback ($\alpha = 0.7$): plots for $P(t)$, $C(t)$, $\rho(t)$, and $T(t)$ for a sinusoidal reactivity input

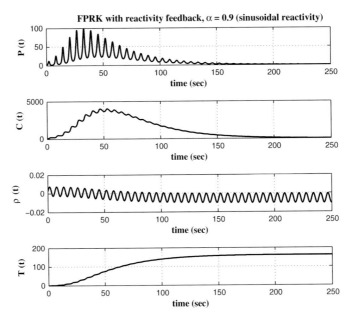

Fig. 7 FPRK model with reactivity feedback ($\alpha = 0.9$): plots for $P(t)$, $C(t)$, $\rho(t)$, and $T(t)$ for a sinusoidal reactivity input

Table 3 IPRK and FPRK models with reactivity feedback: values of various quantities at P_{max} for sinusoidal reactivity input

α	P_{max} (watts)	P_{max} occurs at (sec)	Reactivity at P_{max}	Precursor concentration at P_{max}
0.7	92.8448	33.25	0.005541	2689.2
0.9	99.2303	33.325	0.005307	2880.6
1 (IPRK)	102.9028	33.36	0.005187	2978.5

Table 4 IPRK and FPRK models with reactivity feedback: values of various quantities at steady-state for sinusoidal reactivity input

α	$P(t)$ (watts)	$C(t)$	$\rho(t)$ oscillates between min–max	T(t) °C
0.7	0.1297	23.1327	-0.01505 to -5.0765×10^{-4}	160.1719
0.9	0.0999	22.0859	-0.01058 to -5.7974×10^{-4}	161.6061
1 (IPRK)	0.0938	21.0333	-0.01508 to -5.7507×10^{-4}	161.5428

Thus, in this and the previous section we analyzed the FPRK model with reactivity feedback. As the aim was to study the reactor with below prompt critical situation, the dynamics of precursor concentration was also considered in the model. The derived model was analyzed in detail. As a part of the study, we examined the behaviour of FPRK model when subjected to step and sinusoidal reactivity. The reactivity insertion was chosen in such a way so as to keep the reactor subprompt critical. This exercise confirms that the developed fractional-order nonlinear model with reactivity feedback represents the reactor dynamics. A comparison of this FO model with the classical IO model is also carried out and various results are presented.

6 Conclusions

To get a more reliable and realistic model of the nuclear reactor, it is necessary to consider processes by which the reactivity is affected by the neutron flux or power. Using this fact, in this paper, we proposed a new version of the fractional-order point reactor kinetics model considering the effect of temperature feedback on the reactivity. The adiabatic model of temperature dependence on power is used.

The proposed FPRK model is used to analyze small perturbations (below prompt critical) in the power. This model is a system of three coupled, nonlinear ODEs (one FDE + two IDEs). The model was subjected to step and sinusoidal reactivity inputs. Various standard methods like VIM, ADM, were tried. But it was found that the stiff nature of the PRK model in general proved as a big hurdle in achieving the convergence for these methods. Finally, the ABM method was used to numerically solve the FO model. 'It becomes more and more difficult to make a numerical method applied to an FDE converge for smaller values of α': this frequently observed phenomenon was experienced for the model under consideration.

The behaviour predicted by the FO model was found to be in line with reactor physics. Each time, the power excursion was found to be self-limiting and therefore stable. Thus this paper presents a major step in the development and analysis of fractional-order model for a nuclear reactor under the consideration of reactivity feedback. The developed FO model very faithfully mimics the actual behaviour of the reactor in these situations. Also the developed FO model has broader applicability, and is easy to derive and solve. The classical integer-order model forms a special case of the proposed FO model.

The analysis carried out in this paper can be made more exhaustive by studying the FPRK models with other types of reactivity feedback mechanisms and carrying out a comparative study of these models. Further, it has been proved in literature that the heat transfer mechanism is better represented using fractional dynamics. Hence a more detailed study of the reactivity feedback in nuclear reactor can be achieved by additionally considering a fractional-order model for the temperature dependence of reactivity.

Appendix: Fractional Second-order Adams-Bashforth-Moulton Method

Here the fractional second-order Adams-Bashforth-Moulton (ABM) method which is used in Sect. 4 is explained in brief. The main computational steps involved in the algorithm are presented here for the equispaced grid points. For details, refer to [33, 37, 38]. It is an extension of the classical ABM method used to numerically solve the first-order ODEs. It comes in the category of the so-called PECE (Predict, Evaluate, Correct, Evaluate) type since it involves calculation of the predictor value which is in turn used to compute the corrector value. This method and its variants are popular in the field of fractional calculus and applied areas [39, 40]. The algorithm explained below is for a single fractional differential equation. However, it can be easily modified to handle a system of FDEs.

Consider the single term FDE with Caputo FD

$$_0D_t^\alpha y(t) = f(t, y(t)), \tag{21}$$

where $\alpha \in \mathbb{R}^+$ and with the appropriate initial conditions:

$$D_t^k y(0) = y_0^{(k)}, \quad k = 0, 1, \dots, m - 1, \tag{22}$$

where, $m = \lceil \alpha \rceil$ is the ceil function. The equivalent Volterra integral equation is

$$y(t) = \sum_{k=0}^{m-1} \frac{t^k}{k!} D_t^k y(0) + \frac{1}{\Gamma(\alpha)} \int_0^t (t-\tau)^{\alpha-1} f(\tau, y(\tau)) d\tau. \tag{23}$$

The integration limits from 0 to t imply the nonlocal structure of the fractional derivatives.

The next step is to use the product trapezoidal quadrature formula to replace the integral in (23). We approximate the following integral

$$\int_0^{t_{k+1}} (t_{k+1} - \tau)^{\alpha-1} g(\tau) d\tau, \tag{24}$$

as

$$\int_0^{t_{k+1}} (t_{k+1} - \tau)^{\alpha-1} g_{k+1}(\tau) d\tau, \tag{25}$$

where $\tilde{g}_{k+1} \equiv$ piecewise linear interpolation for $g(t)$ with grid points at t_j, $j = 0, 1, 2, \ldots, k + 1$. Thus we can write the integral (25) as

$$\int_0^{t_{k+1}} (t_{k+1} - \tau)^{\alpha-1} g_{k+1}(\tau) d\tau = \sum_{j=0}^{k+1} a_{j,k+1} g(t_j), \tag{26}$$

for the equispaced nodes ($t_j = jh$ with some fixed step-size h). The values of $a_{j,k+1}$ are given for $j = 0$ as

$$\frac{h^\alpha}{\alpha(\alpha + 1)} \left(k^{\alpha+1} - (k - \alpha)(k + 1)^\alpha \right),$$

for $1 \leq j \leq k$ as

$$\left(\frac{h^\alpha}{\alpha(\alpha + 1)} \right) (d),$$

where

$$d = (k - j + 2)^{\alpha+1} + (k - j)^{\alpha+1} - 2(k - j + 1)^{\alpha+1},$$

and for $j = k + 1$ as

$$\frac{h^\alpha}{\alpha(\alpha + 1)}.$$

So the corrector formula is

$$
\begin{aligned}
y_{k+1} = & \sum_{j=0}^{m-1} \frac{t_{k+1}^j}{j!} y_0^{(j)} \\
& + \frac{1}{\Gamma(\alpha)} \sum_{j=0}^{k} a_{j,k+1} f(t_j, y_j) \\
& + \frac{1}{\Gamma(\alpha)} \left(a_{k+1,k+1} f(t_{k+1}, y_{k+1}^P) \right),
\end{aligned}
\tag{27}
$$

where now the predictor y_{k+1}^P is evaluated as

$$
y_{k+1}^P = \sum_{j=0}^{m-1} \frac{t_{k+1}^j}{j!} y_0^{(j)} + \frac{1}{\Gamma(\alpha)} \sum_{j=0}^{k} b_{j,k+1} f(t_j, y_j),
\tag{28}
$$

with

$$
b_{j,k+1} = \frac{h^\alpha}{\alpha} \left((k+1-j)^\alpha - (k-j)^\alpha \right).
\tag{29}
$$

For $0 < \alpha < 1$, the predictor and corrector expressions get modified as

$$
y_{k+1}^P = y_0 + \frac{1}{\Gamma(\alpha)} \sum_{j=0}^{k} b_{j,k+1} f(t_j, y_j),
\tag{30}
$$

and

$$
\begin{aligned}
y_{k+1} = & y_0 + \frac{1}{\Gamma(\alpha)} \sum_{j=0}^{k} a_{j,k+1} f(t_j, y_j) \\
& + \frac{1}{\Gamma(\alpha)} \left(a_{k+1,k+1} f(t_{k+1}, y_{k+1}^P) \right).
\end{aligned}
\tag{31}
$$

As already mentioned in Sect. 4, the convergence of this algorithm deteriorates as $\alpha \to 0$. This algorithm was coded in MATLAB.

References

1. Duderstadt, J.J., Hamilton, L.J.: Nuclear Reactor Analysis. Wiley, New York (1976)
2. Glasstone, S., Sesonske, A.: Nuclear Reactor Engineering: Vol. 1. CBS Publishers & Distributors, Chennai (2002)
3. Beckurts, K.H., Wirtz, K.: Neutron Physics. Springer, New York (1964)

4. Meghreblian, R.V., Holmes, D.K.: Reactor Analysis. McGraw-Hill Book Company, New York (1960)
5. Espinosa-Paredes, G., Morales-Sandoval, J.B., Vázquez-Rodríguez, R., Espinosa-Martínez, E.G.: Constitutive laws for the neutron transport current. Ann. Nucl. Energy **35**, 1963–1967 (2008)
6. Vyawahare, V.A., Nataraj, P.S.V.: Modeling neutron transport in a nuclear reactor as subdiffusion: The neutron fractional-order telegraph equation. In: The 4th IFAC Workshop on Fractional Differentiation and its Applications, Badajoz, Spain (2010)
7. Vyawahare, V.A., Nataraj, P.S.V.: Fractional-order modeling of neutron transport in a nuclear reactor. Appl. Math. Model. **37**, 9747–9767 (2013)
8. Compte, A., Metzler, R.: The generalized Cattaneo equation for the description of anomalous transport processes. J. Phys. A Math. Gen. **30**, 7277–7289 (1997)
9. Vyawahare, V.A., Nataraj, P.S.V.: Development and analysis of the fractional point reactor kinetics model for a nuclear reactor with slab geometry. In: The 5th IFAC Workshop on Fractional Differentiation and its Applications, Nanjing, China (2012)
10. Vyawahare, V.A., Nataraj, P.S.V.: Development and analysis of some versions of the fractional-order point reactor kinetics model for a nuclear reactor with slab geometry. Commun. Nonlinear Sci. Numer. Simul. **18**, 1840–1856 (2013)
11. Samko, S.G., Kilbas, A.A., Marichev, O.I.: Fractional Integrals and Derivatives. Gordon and Breach Science Publishers, Philadelphia (1997)
12. Das, S.: Functional Fractional Calculus for System Identification and Controls. Springer, New York (2011)
13. Magin, R.L.: Fractional Calculus in Bioengineering. Begell House Publishers, Redding (2006)
14. Monje, C.A., Chen, Y.Q., Vinagre, B.M., Xue, D., Feliu, V.: Fractional-order Systems and Control: Fundamentals and Applications. Springer, London (2010)
15. Machado, J.T., Kiryakova, V., Mainardi, F.: Recent history of fractional calculus. Commun. Nonlinear Sci. Numer. Simul. **16**, 1140–1153 (2011)
16. Hetrick, D.L.: Dynamics of Nuclear Reactors. American Nuclear Society, Illinois (1993)
17. Espinosa-Paredes, G., del Valle-Gallegos, E., Núñez-Carrera, A., Polo-Labarrios, M.A., Espinosa-Martínez, E.G., Vázquez-Rodríguez, R.: Fractional neutron point kinetics equation with newtonian temperature feedback effects. Prog. Nucl. Energy **73**, 96–101 (2014)
18. Espinosa-Paredes, G., Polo-Labarrios, M.A., Espinosa-Martínez, E.G., del Valle-Gallegos, E.: Fractional neutron point kinetics equations for nuclear reactor dynamics. Ann. Nucl. Energy **38**, 307–330 (2011)
19. Vyawahare, V.A., Nataraj, P.S.V.: Development and analysis of fractional-order Nordheim-Fuchs model for nuclear reactor. In: Daftardar-Gejji, V. (ed.) Fractional Calculus: Theory and Applications. Narosa Publishing House, Chennai (2014)
20. Klages, R., Radons, G., Sokolov, I.M. (eds.): Anomalous Transport. WILEY-VCH Verlag GmbH & Co., New York (2008)
21. Metzler, R., Klafter, J.: The random walk's guide to anomalous diffusion: a fractional dynamics approach. Phys. Rep. **339**, 1–77 (2000)
22. Das, S., Biswas, B.B.: Fractional divergence for neutron flux profile in nuclear reactor. Int. J. Nucl. Energy Sci. Technol. **3**, 139–159 (2007)
23. Sardar, T., Ray, S.S., Bera, R., Biswas, B., Das, S.: The solution of coupled fractional neutron diffusion equations with delayed neutrons. Int. J. Nucl. Energy Sci. Technol. **5**, 105–113 (2010)
24. Das, S., Das, S., Gupta, A.: Fractional order modeling of a PHWR under step-back condition and control of its global power with a robust $PI^\lambda D^\mu$ controller. IEEE Trans. Nucl. Sci. **58**, 2431–2441 (2011)
25. Kadem, A.: The fractional transport equation: an analytical solution and a spectral approximation by Chebyshev polynomials. Appl. Sci. **11**, 78–90 (2009)
26. Kadem, A., Baleanu, D.: Analytical method based on Walsh function combined with orthogonal polynomial for fractional transport equation. Commun. Nonlinear Sci. Numer. Simul. **15**, 491–501 (2010)

27. Nahla, A.A.: An analytical solution for the point reactor kinetics equations with one group of delayed neutrons and the adiabatic feedback model. Prog. Nucl. Energy **51**, 124–128 (2009)
28. Mathworks: MATLAB Manual. The Mathworks Inc., MATLAB version 7.1 (R14), USA (2005)
29. Aboanber, A.E., Nahla, A.A.: On pade' approximations to the exponential function and application to the point kinetics equations. Prog. Nucl. Energy **44**, 347–368 (2004)
30. Daftardar-Gejji, V., Jafari, H.: Adomian decomposition: a tool for solving a system of fractional differential equations. J. Math. Anal. Appl. **301**, 508–518 (2005)
31. Odibat, Z., Momani, S.: Application of Variational Iteration Method to nonlinear differential equations of fractional order. Int. J. Nonlinear Sci. Numer. Simul. **1**, 15–27 (2006)
32. Ruskeepaa, H.: Mathematica Navigator: Mathematics Statistics and Graphics. Academic Press, Amsterdam (2009)
33. Diethelm, K.: The Analysis of Fractional Differential Equations: An Application-Oriented Exposition Using Differential Operators of Caputo Type. Springer, Berlin (2010)
34. Diethelm, K., Ford, N.J., Freed, A.D.: A Predictor-Corrector approach for the numerical solution of fractional differential equations. Nonlinear Dyn. **29**, 3–22 (2002)
35. Stacey, W.M.: Nuclear Reactor Physics. WILEY-VCH Verlag GmbH & Co., Weinheim (2007)
36. Ogata, K.: Modern Control Engineering. Prentice-Hall, Englewood Cliffs (2002)
37. Diethelm, K., Ford, N.J., Freed, A.D., Luchko, Y.: Algorithms for the fractional calculus: a selection of numerical methods. Comput. Methods Appl. Mech. Eng. **194**, 743–773 (2005)
38. Connolly, J.A.: The numerical solution of fractional and distributed order differential equations. PhD thesis, University of Liverpool, UK (2004)
39. Li, C., Peng, G.: Chaos in Chen's system with a fractional order. Chaos Solitons Fractals **22**, 443–450 (2004)
40. Tavazoei, M.S., Haeri, M.: A necessary condition for double scroll attractor existence in fractional-order systems. Phys. Lett. A **367**, 102–113 (2007)

Analysis of Model Predictive Control for Fractional-Order System

Mandar M. Joshi, Vishwesh A. Vyawahare and Mukesh D. Patil

Abstract This paper attempts to analyze the performance of model predictive control (MPC) strategy for a fractional-order system. MPC is a popular control technique that is extensively used for the control of industrial processes. Here MPC is applied to a system with fractional dynamics. The original fractional-order model of the system is considered as 'plant' and its integer-order approximation is considered as the 'model'. The effect of approximation of fractional-order model on the MPC is studied. Also a study of the effect of uncertainties in the plant parameters and the non-integer derivative order is carried out.

Keywords Predictive control · Fractional calculus · Fractional-order systems · Integer-order approximation · Internal model control

1 Introduction

Model predictive control (MPC) is an optimal control theory based on numerical optimization. Future control efforts and future plant responses are predicted using a system model and optimized at regular intervals with respect to a performance index. It is a computational technique for improving control performance in applications and processes in chemical and petrochemical industries. Predictive control has become arguably the most widespread advanced control methodology currently in use in industry [1–5].

M.M. Joshi (✉)
Department of Electrical Engineering, College of Engineering, Pune, India
e-mail: mandarjoshi137@gmail.com

V.A. Vyawahare · M.D. Patil
Department of Electronics Engineering, Ramrao Adik Institute of Technology,
Nerul, Navi Mumbai, India
e-mail: vishwesh.vyawahare@rait.ac.in

M.D. Patil
e-mail: mukesh.rait@gmail.com

© Springer International Publishing Switzerland 2015
M.S. Obaidat et al. (eds.), *Simulation and Modeling Methodologies,
Technologies and Applications*, Advances in Intelligent Systems
and Computing 402, DOI 10.1007/978-3-319-26470-7_9

The basic MPC concept can be explained as follows. In MPC, model of the system is used to predict the future output and control efforts required to attain the reference trajectory. Hence, the accuracy of the model determines the control and delivers precise future input trajectory in order to follow the reference signal. This is the basic philosophy of any MPC. MPC is more a methodology and not a single technique and it is known by many different nomenclatures such as Model Predictive Control (MPC), Model based predictive control (MBPC), Receding horizon control (RHC), Moving horizon control (MHC), Internal Model Control (IMC), etc.

The literature survey reveals that MPC has made outstanding contributions for solving control problems in industry [6]. It has so far been applied mainly in the petrochemical industry, but is currently being increasingly applied in other sectors of the process industry. The main advantages of using MPC in these applications are:

1. Inherent handling of multivariable control problems.
2. Actuator limitations are considered.
3. Operation closer to constraints is possible for more profitability.
4. Applications with relatively low control update rates provides sufficient time for the necessary online computations.

Fractional calculus is used for modeling real-world and man-made systems more compactly and faithfully. Fractional calculus has been known from 1695, when Leibnitz introduced the symbol for the nth derivative $d^n y / dx^n$, and L'Hopital discussed the possibility that 'n' could be $1/2$. Nonetheless, this branch of mathematical analysis was really developed in the 19th century by Liouville, Riemann, Letnikov and others [7]. One of the most popular examples of fractional-order modeling of a physical system is the semi-infinite lossy transmission line [8]. Other systems with fractional-order dynamics are viscoelasticity, dielectric polarization and electromagnetic waves [8]. Fractional-order control can be regarded as the generalization of the conventional integer-order control theory. Fractional-order models represent a system in more accurate ways [8]. Hence it is worth exploring how the MPC strategy will work for systems with fractional dynamics. There are some systems which are of fractional nature and hence a fractional-order model can be used to represent such systems instead of integer-order models.

For implementation of MPC, a state-space model is required and approximation method helps to obtain the same. Similar work can be found in [5]. A fractional-order Transfer Function (FOTF) can not be converted to a state-space model through simulation softwares like MATLAB. Therefore, we need to get an approximation that is, an integer-order equivalent (transfer function) which can represent that FOTF and also can be converted to state-space model using MATLAB. Moreover, an approximation has to be made for a fractional operator s^α or $s^{-\alpha}$ with $0 < \alpha < 1$.

In this paper, MPC is applied to an FO system with different fractional-order derivative order, α for which Oustaloup's recursive approximation has been used to obtain respective models. Obtained models are then converted to acquire strictly proper natured discrete state-space models by adding a pole at far away from the origin. Authenticity of these models were checked with the original one through frequency and step response. Finally, these models were used for prediction purposes in MPC. In order to make MPC realistic, analytically calculated output equation for FO systems has been utilized to represent process output. Furthermore, model predictive control strategy was tested on the fractional-order system in the presence of uncertainty in the plant parameters. Design of model predictive control for a fractional order system with approximation method and by using fractional-order plant itself to deliver output state to track the reference even in the presence of parametric uncertainty in the plant becomes the contribution of this work.

The paper is organized as follows. In Sect. 2, a brief introduction to fractional calculus is given and dynamics of FO systems is reviewed; Sect. 3 presents the Oustaloup's recursive approximation method and its application for fractional operators. Philosophy behind MPC and formulation of MPC problem is covered in Sect. 4. Section 5 presents the results obtained for MPC with FO systems and that with parametric uncertainty. Section 6 concludes this work with some observations.

2   Fractional Calculus

Fractional Calculus can be defined as integration and differentiation of non-integer order. In the development of the theory of fractional calculus, there are different definitions available for fractional-order integration and differentiation such as Riemann-Liouville fractional differentiation and integration, Caputo fractional differentiation, Grunwald-Letnikov fractional differentiation [9]. Caputo fractional differentiation is defined as [10]

$$
{}_{0}^{C}D_{t}^{\alpha}f(t) = \frac{1}{\Gamma(m-\alpha)} \int_{0}^{t} (t-\tau)^{(m-\alpha-1)} D_{\tau}^{m}f(\tau)d\tau, \tag{1}
$$

where α is the order of the differentiation and m is the nearest integer on higher side of α.

2.1   Fractional-Order Systems

Fractional-order systems are represented by fractional-order models. Also, for integer-order systems, fractional-order models give better predictions [8, 11, 12]. Fractional-order systems are theoretically of infinite order [13]. A fractional-order system can be represented mathematically by following fractional-order differential equation (FDE) [10]

$$a_n D^{\alpha_n} y(t) + a_{n-1} D^{\alpha_{n-1}} y(t) + \cdots$$
$$+ a_1 D^{\alpha_1} y(t) + a_0 y(t)$$
$$= b_m D^{\beta_m} u(t) + b_{m-1} D^{\beta_{m-1}} u(t) + \cdots$$
$$+ b_1 D^{\beta_1} u(t) + b_0 u(t), \tag{2}$$

where D^{α_i} and D^{β_i} represents Caputo or Riemann Liouville Fractional Derivative (RLFD).

Using Laplace transform, one can obtain the transfer function for fractional system. The order of the fractional-order Transfer Function (FOTF) is theoretically infinite since a fractional-order (FO) system is basically infinite dimensional system [10]. But for convention we assume that the order of a FOTF is equal to the highest degree of denominator which will be a real number and not necessarily an integer. In an FOTF, the numerator and denominator are not polynomials but are pseudo-polynomials.

3 Approximation Method

Approximation methods are based on different techniques. In continuous domain there are three types of techniques on which approximations are based. Curve fitting/frequency response matching, continued fraction expansion (CFE) and regular Newton process for iterative approximation [14]. Only Curve fitting/frequency response matching is briefly explained below since it is the principle on which Oustaloup's recursive approximation is based.

3.1 Curve Fitting/Frequency Response Matching

In this technique, frequency domain identification is used to obtain the rational function (integer-order function) of a given fractional operator. For example, we will try to get an integer-order function of s^α by matching their frequency response. This can be written as an optimization problem, where the cost function will be of the form [14]

$$J = \int W(w) |G(w) - \hat{G}(w)|^2 dw, \tag{3}$$

where $W(w)$ is the weighting function and $G(w)$ represents the frequency response of original function which is in this case is s^α. $\hat{G}(w)$ represents the frequency response of integer-order approximation of s^α. We will minimize the cost function in order to achieve the best approximation. This is the basic idea behind approximation methods based on curve fitting [14].

3.2 Oustaloup Recursive Approximation

It gives the approximation for a fractional-order differentiator s^α and is widely used. This approximation is given by following equations [7, 15]:

$$s^\alpha \approx K \prod_{k=1}^{N} \frac{s + w'_k}{s + w_k}, \qquad (4)$$

where the poles, zeros and gain can be evaluated as [7]:

$$w'_k = w_l w_u^{(2k-1-\alpha)/N}, \qquad (5)$$

$$w_k = w_l w_u^{(2k-1+\alpha)/N}, \qquad (6)$$

$$K = w_h^\alpha, \qquad (7)$$

$$w_u = \sqrt{w_h/w_l}, \qquad (8)$$

where w_l and w_h are lower and higher frequency values. It should be noted that small values of N obviously result in low-order approximations which are simple in nature but with ripples in both Bode gain and phase plots. To practically eliminate such a ripple, N has to be increased. A higher value of N makes the computation heavier [8].

3.3 Comparison of FO System and Its Approximation

This section contains bode plots and step response for the FO system for which the main results of this paper are shown. The FO plant given below is used for five different values of $\alpha = 0.1, 0.3, 0.5, 0.7$ and 0.9; $(a = b = 1)$ thus giving five different FO systems.

$$G(s) = \frac{b}{s^\alpha + a} \qquad a, b \in \mathbb{R}. \qquad (9)$$

The FO system shown above is used here since it represents model of some physical systems and has been used as a benchmark system in [16, 17]. Few examples of physical systems represented by (9) are: model of beam heating process [18], model of thermal systems [19] and an explicit systems model for electrical networks [12]. Above FO system (9) was also used for approximation of high-order integer systems [20].

In the plots (shown below) the original FO system (9) is compared with its obtained approximation via frequency as well as step response. FOTF represents the original FO system and Oustaloup represents approximation obtained for the FO system. To obtain frequency response of original system (9), s is replaced by jw and then the system is simulated for a fixed frequency range (in this case 10^{-2} to 10^6 rad/s) to obtain the magnitude and phase vectors. The bode plot is obtained by multiplying suitable factors to the magnitude and phase vector. To obtain the approximation of fractional operator s^α in (9), Oustaloup recursive approximation (4) was used. An example of approximation for one of the FO systems is given below.

Approximation of $\frac{1}{s^{0.5}+1}$. Approximation was carried out for $s^{0.5}$ and following results were obtained:

$$s^{0.5} = \frac{N(s)}{D(s)}, \tag{10}$$

where

$$N(s) = 1000s^{10} + 2.985e^8 s^9 + 1.219e^{13} s^8 + 7.722e^{16} s^7 + 7.727e^{19} s^6 + 1.225e^{22} s^5$$
$$+ 3.076e^{23} s^4 + 1.224e^{24} s^3 + 7.691e^{23} s^2 + 7.498e^{22} s + 1e^{21},$$
$$D(s) = s^{10} + 7.498e^5 s^9 + 7.691e^{10} s^8 + 1.224e^{15} s^7 + 3.076e^{18} s^6 + 1.225e^{21} s^5$$
$$+ 7.727e^{22} s^4 + 7.722e^{23} s^3 + 1.219e^{24} s^2 + 2.985e^{23} s + 1e^{22}.$$

Adding one to $s^{0.5}$ and then inverting it will lead to an expression for $1/(s^{0.5} + 1)$, which is:

$$\frac{1}{s^{0.5} + 1} = \frac{N1(s)}{D1(s)}, \tag{11}$$

where

$$N_1(s) = s^{10} + 7.498e^5 s^9 + 7.691e^{10} s^8 + 1.224e^{15} s^7 + 3.076e^{18} s^6 + 1.225e^{21} s^5$$
$$+ 7.727e^{22} s^4 + 7.722e^{23} s^3 + 1.219e^{24} s^2 + 2.985e^{23} s + 1e^{22},$$
$$D_1(s) = 1001s^{10} + 2.992e^8 s^9 + 1.227e^{13} s^8 + 7.844e^{16} s^7 + 8.034e^{19} s^6 + 1.347e^{22} s^5$$
$$+ 3.849e^{23} s^4 + 1.996e^{24} s^3 + 1.988e^{24} s^2 + 3.735e^{23} s + 1.1e^{22}.$$

A pole at far end in the left plane of imaginary axis was added to convert the above obtained system into strictly proper form; explanation for this conversion is given in Sect. 4.4. The expression for $1/(s^{0.5} + 1)$ becomes:

$$\frac{1}{s^{0.5} + 1} = \frac{N2(s)}{D2(s)}, \tag{12}$$

with

$$N_2(s) = s^{10} + 7.498e^5 s^9 + 7.691e^{10} s^8 + 1.224e^{15} s^7 + 3.076e^{18} s^6 + 1.225e^{21} s^5$$
$$+ 7.727e^{22} s^4 + 7.722e^{23} s^3 + 1.219e^{24} s^2 + 2.985e^{23} s + 1e^{22},$$
$$D_2(s) = 1.586e^{-13} s^{11} + 1001s^{10} + 2.992e^8 s^9 + 1.227e^{13} s^8 + 7.844e^{16} s^7 + 8.034e^{19} s^6$$
$$+ 1.347e^{22} s^5 + 3.849e^{23} s^4 + 1.996e^{24} s^3 + 1.988e^{24} s^2 + 3.735e^{23} s + 1.1e^{22}.$$

Similarly approximation for other four FO plants was evaluated. Once, an integer-order equivalent (transfer function) is obtained for FO plant, traditional methods can be used to plot frequency and step response [21].

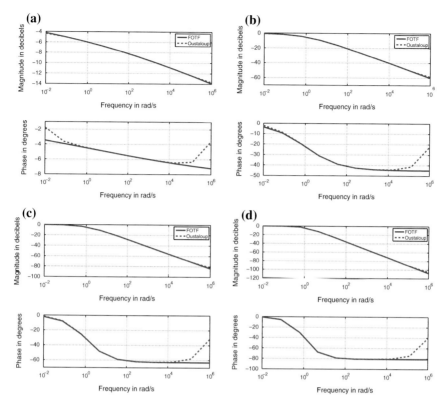

Fig. 1 Frequency response of $\frac{1}{s^{\alpha}+1}$, **a** bode plot for $\frac{1}{s^{0.1}+1}$, **b** bode plot for $\frac{1}{s^{0.5}+1}$, **c** bode plot for $\frac{1}{s^{0.7}+1}$, **d** bode plot for $\frac{1}{s^{0.9}+1}$

Frequency Response of $\frac{1}{s^{\alpha}+1}$. Figure 1 shows the frequency domain validation of the IO approximations for different values of α (plot for $\alpha = 0.3$ is not shown due to space constraints).

It can be observed that for the frequency range of 10^0 to 10^4 rad/s we get an exact match for the frequency response, which advocates that the obtained approximation can replace the original FOTF for the said frequency range.

Step Response of $\frac{1}{s^{\alpha}+1}$. To obtain the expression for step response of FO plant, the transfer function (9) is rewritten such that an expression can be obtained in terms of $Y(s)$ and $U(s)$, where the first is output and the later is the input.

$$Y(s) = \frac{U(s)}{s^{\alpha} + 1}.$$

(13)

Now for step input $U(s) = 1/s$ and then applying inverse Laplace transformation [7], gives:

$$Y(t) = 1 - E_{\alpha}(-t^{\alpha}),$$

(14)

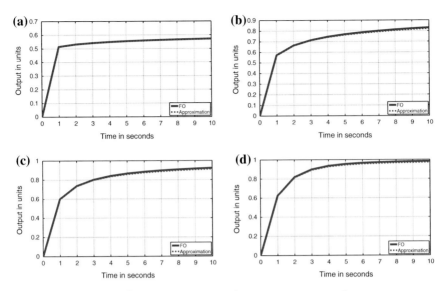

Fig. 2 Step response of $\frac{1}{s^a+1}$, **a** step response of $\frac{1}{s^{0.1}+1}$, **b** step response of $\frac{1}{s^{0.5}+1}$, **c** step response of $\frac{1}{s^{0.7}+1}$, **d** step response of $\frac{1}{s^{0.9}+1}$

where $E_\alpha(-t^\alpha)$ is one parameter Mittag-Leffler Function (MLF) [22]. Figure 2 shows the time domain validation of the IO approximations for different values of α (plot for $\alpha = 0.3$ is not shown due to space constraints).

From both frequency and step response it is observed that the dynamics of original FO system is retained in the obtained IO approximation and hence using this approximation for analysis and simulation in model predictive control as a finite-dimensional model of original FO system is justified.

4 Model Predictive Control

In this section, the theory of MPC is discussed in detail.

4.1 Principle of Model Predictive Control

For a system, the reference trajectory r(k) is defined as the ideal trajectory along which a plant should return to the set-point trajectory w(k), at any instant k. The current error between the output signal y(k) and the setpoint is an error represented as:

$$err(k) = w(k) - y(k). \tag{15}$$

Then the reference trajectory is chosen such that if the output followed it exactly, then the error after i steps would be

$$err(k + i) = exp^{-iT_s/T_{ref}} err(k), \tag{16}$$

where T_{ref} is the time constant of exponential assuming that reference trajectory approaches the set-point exponentially and T_s is the sampling interval. Therefore, the reference trajectory is defined as

$$r(k + i|k) = w(k + i) - y(k + i). \tag{17}$$

The notation $r(k + i|k)$ implies that the reference trajectory depends on the conditions at time k [6, 8].

Now, the predictive controller uses an internal model to estimate the future behavior of the plant. This process starts at the current time k and carries over a future prediction horizon H_p. This predicted behavior depends on the assumed control effort trajectory $\hat{u}$ that is to be applied over the prediction horizon and we have to select the input which promises the best predicted behavior. The notation $\hat{u}$ represents the predicted value input at time $k + i$ at instant k; the actual input $u(k + i)$ may be different than $\hat{u}$. After a future trajectory is decided, only the first element of the trajectory is required as the control input to the process. Hence $u(k)$ becomes $\hat{u}(k|k)$. The tasks of prediction, measurement of output and determination of input trajectory are repeated at every sampling instant. This is known as the receding horizon strategy, since prediction horizon slides along by one sampling interval at each step while its length remains the same.

Formulation of MPC problem needs few basic elements such as model, cost function, prediction equations and control law. These are covered in the next section.

4.2 Problem Formulation

In MPC, the first element of paramount importance is the model of the process/plant, which is required for prediction of the future states. This model must show the dependance of the output on the current measured variable and the current inputs. A precise model will deliver accurate predictions, but for MPC we do not always need precise model though they are always appreciated. Since the decisions that is, the optimal control effort is updated regularly, any model uncertainty can be dealt within a fair range [23]. Hence, accuracy of the model used for the prediction can be compensated with regular updating of states. In this paper, FO system is used as a plant/process. Transfer function of the FO plant is:

$$G(s) = \frac{b}{s^\alpha + a}, \quad a, b \in \mathbb{R}, \tag{18}$$

where α is varied between 0 and 1.

Second element is the cost function, which is required to evaluate the input trajectory such that it minimizes/maximizes the cost function. Selection of the cost function is a crucial issue and involves engineering and theocratical expertise. The cost function should be as simple as one can get away with for the desired performance. Choice of cost function affects the complexity of the implied optimization and hence it should be taken into consideration that the cost function must be simple enough to have a straightforward optimization [23]. For this reason 2-norm measures are popular and the cost function used in this paper is also a 2-norm measure.

$$J = \sum_{i=1}^{n_y} \|r_{k+i} - y_{k+i}\|_2^2 + \lambda \sum_{i=0}^{n_u-1} \|\Delta u_{k+i}\|_2^2, \tag{19a}$$

$$= \sum_{i=1}^{n_y} \|e_{k+i}\|_2^2 + \lambda \sum_{i=0}^{n_u-1} \|\Delta u_{k+i}\|_2^2, \tag{19b}$$

where λ is the weighting scalar; first term represents the sum of squares of the predicted tracking errors from an initial horizon to an output horizon n_y and the second term represents the sum of squares of the control over the control horizon n_u. It is assumed that control increments are zero beyond the control horizon, that is

$$\Delta u_{k+i|k} = 0, i \geq n_u.$$

The third element is the optimization, which is used to evaluate the control law. The control law is computed from the minimization of the cost J (19a) with respect to n_u future control moves, that is $\Delta \underline{u}$. It is denoted as

$$\min_{\underset{\rightarrow}{\Delta u}} J = \|\underline{r} - \underline{y}\|_2^2 + \lambda \|\Delta \underline{u}\|_2^2, \tag{20a}$$

$$= \|\underline{e}\|_2^2 + \lambda \|\Delta \underline{u}\|_2^2, \tag{20b}$$

where

$$\underline{r} = \left[r(k+1) \ r(k+2) \ \cdots \right]^T. \tag{21}$$

$$\underline{y} = \left[y(k+1) \ y(k+2) \ \cdots \right]^T. \tag{22}$$

$$\Delta \underline{u} = \left[u(k+1) \ u(k+2) \ \cdots \right]^T. \tag{23}$$

Once the model and the cost function are defined, next step is to get prediction equations and these are obtained from discrete state-space model of the plant.

4.3   Prediction with State-Space Models

Prediction with a state-space model is straightforward, provided the nature of the model should be strictly proper. Proof for prediction equation is as follows [1]: Consider the state-space model which gives the one step ahead prediction:

$$x_{k+1} = Ax_k + Bu_k; \quad y_{k+1} = Cx_{k+1}, \tag{24a}$$

and at instant $k + 2$

$$x_{k+2} = Ax_{k+1} + Bu_{k+1}; \quad y_{k+2} = Cx_{k+2} \tag{24b}$$

Substitute (24a) into (24b) to eliminate x_{k+1}, we get

$$x_{k+2} = A^2 x_k + ABu_k + Bu_{k+1}; \quad y_{k+2} = Cx_{k+2}, \tag{24c}$$

at instant $k + 3$ using (24c)

$$x_{k+3} = A^2 x_{k+1} + ABu_{k+1} + Bu_{k+2}; \quad y_{k+3} = Cx_{k+3}. \tag{24d}$$

Substitute (24a) to eliminate x_{k+1}

$$x_{k+3} = A^2 [Ax_k + Bu_k] + ABu_{k+1} + Bu_{k+2}; \quad y_{k+3} = Cx_{k+3}. \tag{24e}$$

Hence a generalized equation for n-step ahead predictions will be

$$x_{k+n} = A^n x_k + A^{n-1} Bu_k + A^{n-2} Bu_{k+1} + \dots + Bu_{k+n-1}, \tag{24f}$$

$$y_{k+n} = C[A^n x_k + A^{n-1} Bu_k + A^{n-2} Bu_{k+1} + \dots + Bu_{k+n-1}]. \tag{24g}$$

Hence, one can form the whole vector of future predictions up to a horizon n_y as follows:

$$\begin{bmatrix} x_{k+1} \\ x_{k+2} \\ x_{k+3} \\ \vdots \\ x_{k+n_y} \end{bmatrix} = \begin{bmatrix} A \\ A^2 \\ A^3 \\ \vdots \\ A^{n_y} \end{bmatrix} x_k + H_x \begin{bmatrix} u_k \\ u_{k+1} \\ u_{k+2} \\ \vdots \\ u_{k+n_y-1} \end{bmatrix}, \tag{25a}$$

$$\begin{bmatrix} y_{k+1} \\ y_{k+2} \\ y_{k+3} \\ \vdots \\ y_{k+n_y} \end{bmatrix} = \begin{bmatrix} CA \\ CA^2 \\ CA^3 \\ \vdots \\ CA^{n_y} \end{bmatrix} x_k + H \begin{bmatrix} u_k \\ u_{k+1} \\ u_{k+2} \\ \vdots \\ u_{k+n_y-1} \end{bmatrix}, \tag{25b}$$

$$H_x = \begin{bmatrix} B & 0 & 0 & \dots \\ AB & B & 0 & \dots \\ A^2B & AB & B & \dots \\ \vdots & \vdots & \vdots & \vdots \\ A^{n_y-1}B & A^{n_y-2}B & A^{n_y-3}B & \dots \end{bmatrix}, \tag{25c}$$

$$H = \begin{bmatrix} CB & 0 & 0 & \dots \\ CAB & CB & 0 & \dots \\ CA^2B & CAB & CB & \dots \\ \vdots & \vdots & \vdots & \vdots \\ CA^{n_y-1}B & CA^{n_y-2}B & CA^{n_y-3}B & \dots \end{bmatrix}. \tag{25d}$$

4.4 Discrete State-Space Model for FO Plants

In this paper, discrete state-space model is used in MPC for tracking problem. Discrete state-space model is obtained using Oustaloup recursive approximation. As discussed in the above section, the state-space model considered for prediction equations was of strictly proper nature. However, the state-space model obtained for five FO systems were not of strictly proper nature and hence, a pole was added at a far-away location in the left half of s-plane to make the system strictly proper in nature. A state-space model thus obtained was verified using frequency and step response (refer Sect. 3.3) and it was observed that the addition of pole affects the frequency response at higher frequencies. Thus, for a specific region of frequency plot (from 10^0 to 10^4 rad/s) the FO system retains its original form. Finally, obtained state-space models were discretized at a step size of 1 second. One example of discrete state-space model obtained using approximation method is given below.

Discrete State-Space Model for $\frac{1}{s^{0.5}+1}$. Expression (12) was used to obtain the discrete state-space model (using MATLAB) with step size of one second. The obtained A, B, C and D matrices are as follows:

$$A = \begin{bmatrix} A1 & -6.934e^{34} \\ I & 0 \end{bmatrix},$$

where, I is 10x10 identity matrix and A1 is single row of 10 elements given by:

$$A1 = [-6.31e^{15}, -1.886e^{21}, -7.732e^{25}, -4.944e^{29},$$
$$-5.064e^{32}, -8.492e^{34}, -2.426e^{36}, -1.258e^{37},$$
$$-1.253e^{37}, -2.354e^{36}],$$
$$B = [1, 0, 0, 0, 0, 0, 0, 0, 0, 0]^T,$$
$$C = [6.303e^{12}, 4.726e^{18}, 4.848e^{23}, 7.714e^{27}, 1.939e^{31},$$
$$7.72e^{33}, 4.87e^{35}, 4.867e^{36}, 7.683e^{36}, 1.882e^{36}, 6.303^e 34],$$
$$D = 0.$$

Similarly, discrete state-space models for other four FO plants were obtained. Once again the frequency and step response of obtained state-space model were compared with that of the original FOTF and it was observed that conversion of transfer function to state-space does not affect frequency and step response. Hence, use of state-space model (obtained through approximation) instead of FO system is further justified.

Once the model, cost function and the prediction equations are defined for MPC; the next step is to evaluate the control law.

4.5 Control Law

As stated above, the control law is determined from the minimization of a 2-norm measure of predicted performance (20). True optimal control uses $n_y = n_u = \infty$ in performance index [23]; but there is a limitation on MPC algorithms and hence this assumption cannot be validated, making MPC suboptimal. Though an upper limit on n_y and n_u cannot be infinity but the limits should be selected such that they are always higher than the settling time of the system. For large n_y and n_u MPC gives near identical control to an optimal control law with the same weights. While on the other hand, very small n_y and n_u will result in severely suboptimal control law [23, 24]. Hence, proper choice of n_y and n_u is required for MPC. Selection of n_y and n_u will affect the control law, since they are calculated ones and used each time when the control law is updated. The control law for discrete state-space model and unconstrained condition is [23]:

$$\Delta u_k = P_r \underset{\rightarrow}{r} - Kx - P_r[y_k - \hat{y}_k], \tag{27}$$

where y_k is the process output (real output) and $\hat{y}_k$ is the model output. P_r and K are given by

$$P_r = (H^T H + \lambda I)^{-1} H^T M,$$
$$K = (H^T H + \lambda I)^{-1} H^T P,$$
$$P = \begin{bmatrix} CA \\ CA^2 \\ CA^3 \\ \vdots \\ CA^{n_y} \end{bmatrix},$$

and M is a weighting vector.

Expression (27) is used to evaluate optimal control effort u^* to be given to the plant/process and to the model. At each instant the optimal control effort u^* is updated by evaluating the control law (27) by considering revised process output

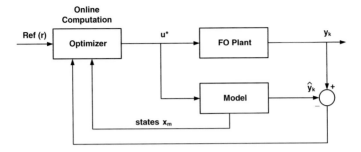

Fig. 3 Block diagram for MPC

y_k, model output $\hat{y}_k$ and the states x. Figure (3) represents block diagram of MPC with FO plant and a model, obtained using approximation method (Fig. 3).

4.6 Process Output Evaluation

When MPC is applied to a real process, its output y_k is measured in real-time at every instant for evaluation of optimal control effort. In case of simulations, generally the model of the process/plant is used to evaluate the process output. Hence, the bias $[y_k - \hat{y}_k]$ is redundant as y_k is equal to $\hat{y}_k$ at each instant. In this paper, the process output y_k has been evaluated by formulating an output equation in time domain for the input given at each instant to the FO plant mentioned in (18) for respective α values. Expression of y_k in time domain was obtained by applying inverse Laplace transformation ([7]) to the expression obtained using transfer function of FO plant (9) for respective α values. Laplace transform of Caputo fractional differentiation (1) and a special function known as Mittag-Leffler function (MLF) is used to evaluate the expression of process output in time domain. For the evaluation of process output expression, definitions of 1-parameter MLF, 2-parameter MLF and Laplace transform of Caputo derivative is required and they are as follows:
Definition for Laplace transform of Caputo fractional differentiation is [25]

$$L\left[{}^{C}_{0}D^{\alpha}_{t}f(t)\right] = s^{\alpha}F(s) - \sum_{k=0}^{m-1} s^{\alpha-k-1}\left[\frac{d^k}{dt^k}f(t)\right]_{t=0}, \tag{28}$$

where $0 < \alpha < 1$, $m - 1 < \alpha \le m$ and $m \in \mathbb{N}$.
Definition for 1-parameter MLF is [22]

$$E_{\alpha}(t) = \sum_{k=0}^{\infty} \frac{t^k}{\Gamma(\alpha k + 1)}, \tag{29}$$

where $\alpha \in \mathbb{R}^{+}$.

Definition for 2-parameter MLF is [22]

$$E_{\alpha,\beta}(t) = \sum_{k=0}^{\infty} \frac{t^k}{\Gamma(\alpha k + \beta)}, \tag{30}$$

where $\alpha, \beta \in \mathbb{R}^+$.

Analytical expression for process output y_k is obtained using inverse Laplace transformation on the expression by rewriting transfer function (9) in terms of $Y(s)$ and $U(s)$ using definition of Caputo fractional differentiation (1) and using Laplace transform of Caputo fractional differentiation (28) as follows:

$$\frac{Y(s)}{U(s)} = \frac{b}{s^\alpha + a}. \tag{31a}$$

Above expression is a general form of (9)

$$s^\alpha Y(s) + aY(s) = bU(s). \tag{31b}$$

The actual FDE with Caputo differentiation is

$$\,_0^C D_t^\alpha y(t) + ay(t) = bu(t), \tag{31c}$$

for $0 < \alpha < 1$. Taking Laplace transform as given in (28) for Caputo differentiation, we get

$$s^\alpha Y(s) - Y(0) + aY(s) = bU(s), \tag{31d}$$

$$Y(s) = \frac{bU(s)}{s^\alpha + a} + \frac{Y(0)}{s^\alpha + a}. \tag{31e}$$

Taking inverse Laplace transform we get

$$L^{-1}\left(\frac{bU(s)}{s^\alpha + a}\right) = L^{-1}\left(\frac{1}{s^\alpha + a}\right) * L^{-1}(bU(s)). \tag{31f}$$

Note that $u(t)$ is a constant value at any kth instant and hence $U(s) = \frac{u_{k-1}}{s}$ whereas u_{k-1} is the constant value obtained as the optimal control effort at the previous instant. Using the standard inverse Laplace transform expressions as given in [7], (31f) becomes

$$L^{-1}\left(\frac{bU(s)}{s^\alpha + a}\right) = b \times u_{k-1}[1 - E_\alpha(-at^\alpha)]. \tag{31g}$$

Again using [7] for Laplace inverse of second term

$$L^{-1}\left(\frac{Y(0)}{s^\alpha + a}\right) = y(o) \times [(t^{\alpha-1})E_{\alpha,\alpha}(-at^\alpha)]. \tag{31h}$$

Hence, y_k is obtained for using (31g) and (31h) as:

$$y_k = b \times u_{k-1}[1 - E_{\alpha,1}(-at_k^\alpha)] + y_{k-1}[t_k^{\alpha-1}E_{\alpha,\alpha}(-at_k^\alpha)]. \tag{32}$$

Now, by substituting $a = 1$ and $b = 1$ in (32), an expression for y_k can be obtained for FO plant described by (9) and this expression takes into account intial/current conditions of plant:

$$y_k = u_{k-1}[1 - E_{\alpha,1}(-t_k^\alpha)] + y_{k-1}[t_k^{\alpha-1}E_{\alpha,\alpha}(-t_k^\alpha)] \tag{33}$$

In the following section, results obtained for simulation of unconstrained MPC with FO plant with different α values are shown.

5 Results

In the previous sections of this paper, the control law and the process output has been evaluated which were used to obtain these results. MPC is applied to an FO plant defined at (18) for five different values of $\alpha = 0.1, 0.3, 0.5, 0, 7$ and 0.9. Simulations are carried out to obtain output and input values for unconstrained condition. Output plot (Fig. 4) shows reference, process output y_k and model output $\hat{y}_k$ for respective FO plants. Subsequent plot shows the optimal control effort u^* at each instant for respective plants. For all plots y-axis represents amplitude in units and x-axis represents sampling instances. Sampling time of 1 second is used for these results. Though here results of simulations for one sampling time are shown, it was observed that the simulation of MPC for these FO plants is also possible for other step sizes. The range of step size for which simulations can be carried out was found to be from 0.5 to 1 seconds. This becomes one of the advantages of using continuous approximation methods for MPC of FO systems.

Process output and model output are seen to have different values initially because both are calculated using different equations, though same initial conditions are fed to both the models . Process output is calculated using (33) with the help of MAT-LAB routine mlf() developed by Podlubny. Model output is calculated using state-space model using (24) obtained through Oustaloup recursive approximation. It was observed that the difference between process output and model output eventually becomes negligible. Results clearly show that the optimal control effort is obtained so that the process output can track the reference signal. In real scenario, there will be an actual plant with MPC applied to it. It will be possible to measure real output at every instant and that will be fed to the optimizer along with the calculated output of internal model (see Fig. (3)). In most of the cases there will be a small amount of difference in the actual output and the model output at initial stages due to the errors

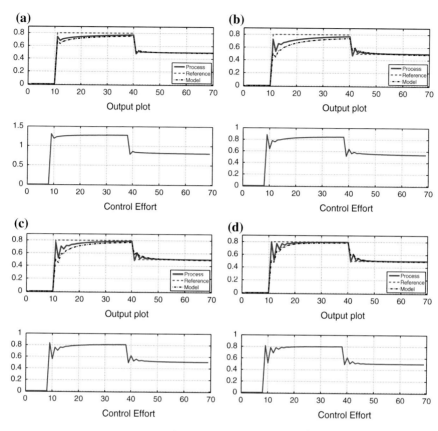

Fig. 4 Output and control effort for $\frac{1}{s^a+1}$, **a** output and control effort for $\frac{1}{s^{0.1}+1}$, **b** output and control effort for $\frac{1}{s^{0.5}+1}$, **c** output and control effort for $\frac{1}{s^{0.7}+1}$, **d** output and control effort for $\frac{1}{s^{0.9}+1}$

in the model of the process. Here in simulations, the same scenario was created by using process output expression instead of using the same state-space model which was used as an internal model (Fig. 4).

Few observations are noted from these results. Magnitude of control effort goes on reducing as the fractional-order α goes on increasing. Also, the process output becomes more oscillatory with increasing α value, when the reference signal changes. There are two different reasons which may cause these phenomena. As stated earlier, control effort is obtained so that the process output can track the reference signal and hence process output expression is the reason behind this reduction. Process output expression uses two MLFs for its computation (see (33), (29) and (30)). MLF takes a long time for computation and the nature of MLF changes very rapidly for small α values. The other reasons behind irregularity can be the infinite memory nature of the FO system. For small α values a larger amount of past values are considered for computation.

Both these observations are not seen as soon as constraints are applied on the input as well as output. In this paper, unconstrained case was considered since for an optimal solution one needs to first confirm that solution exists for unconstrained case before going to constrained situation. In optimal theory, it is said that if the control law does not deliver optimal solution and robustness in the unconstrained case then it will not give optimal solution in the constrained case [23]. Unfortunately the converse does not hold true and hence it is worthwhile to check the performance of the system for unconstrained case first.

From these results it is evident that FO system's model can be used in its true form using approximation method for MPC. Furthermore, the obtained process and model output depicts the behaviour of the original FO system in terms of nature of plot and the required settling time. Systems with small values of α tend to take large time to settle. The reference trajectory is followed faithfully. The reference signal used for the simulation contains both positive and negative steps and hence it can be inferred that MPC for FO systems can work for all types of limiting reference signals.

5.1 Effects of Parametric Uncertainty

Parametric uncertainty in the form of plant parameters a, b and α has been introduced to test the model predictive control strategy for single fractional operator plant. To authenticate the working of approximation method with the addition of parametric uncertainty, frequency and step response of the original plant with parametric uncertainty and that of approximated function have been compared. Figure 5 shows the same for $\alpha = 0.5$. Figure 5a, c show the matching of frequency and step response respectively between original function and approximated function (using Oustaloup's approximation) for the parametric uncertainty of $\pm 10\%$ of the nominal value introduced in a and b. The step and Bode plots are compared in Fig. 5b, d for an additional $\pm 10\%$ uncertainty in α.

The first plot of MPC results (Fig. 6a) shows the simulated result of MPC on the FO plant and subsequent plots show the simulated results when uncertainties were added in the plant parameters. The next plot (Fig. 6b) shows the results for uncertainty added in terms of plant parameters a, b and α by $\pm 10\%$ in the plant only and the model used for prediction was of the original plant (with zero uncertainty). It is observed that the MPC strategy works satisfactorily in this situation though some disturbances and an overshoot are visible in the output state. The use of the original model (without uncertainties) does not significantly affect the performance of the MPC. This is due to the use of process output equation instead of the model obtained through approximation method to measure the output state. However, the use of process output equation is not the only reason for the functionality of MPC strategy in this particular case. The model which was derived for the plant without any uncertainty was accurate enough to deliver the correct (albeit not precise) pre-

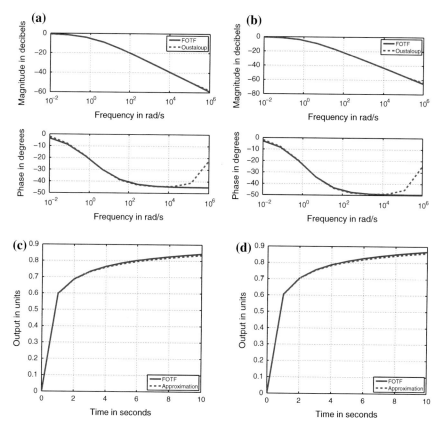

Fig. 5 Frequency and step response with parametric uncertainty, **a** bode plot for $\frac{1.1}{s^{0.5}+1.1}$, **b** bode plot for $\frac{1.1}{s^{0.55}+1.1}$, **c** step response for $\frac{1.1}{s^{0.5}+1.1}$, **d** step response for $\frac{1.1}{s^{0.55}+1.1}$

dictions to lead optimization on the right track, resulting in follow through of the output state with the reference signal.

The next plot (Fig. 6c) presents the output state and the control effort with parametric uncertainty added in the plant parameters a and b in both model and plant. The model used for prediction in this case was derived from the plant with the parametric uncertainties added in it. Hence, the model used in this situation is more accurate than the one used to obtain the results in the previous case (Fig. 6b). From the plot (Fig. 6c), it is evident that the MPC strategy is working well in tracking the reference signal. The difference in the results of both cases in the form less disturbances and overshoot in the later one is clearly seen. The state space model used for prediction in these two cases was different and hence the reduction in the steady-state error is seen for the later case (Fig. 6c).

Figure 6d depicts the output state and the control effort with parametric uncertainty added in the plant parameters a, b and α in both model and plant. It is seen

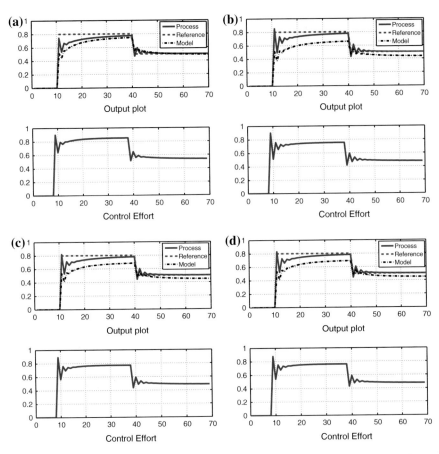

Fig. 6 Output and control effort with parametric uncertainty, **a** output and control effort with zero parametric uncertainty, **b** output and control effort with parametric uncertainty added in the plant only, **c** output and control effort with parametric uncertainty added in the form of a and b in both the model and the plant, **d** output and control effort with parametric uncertainty added in the form of a, b and α in both the model and the plant

that the presence of uncertainty in the order of the derivative α does not significantly change the output as well as in the control effort. This result is very much similar to the one obtained in the previous case (Fig. 6c). This result is quite interesting.

6 Conclusions

The paper reports the use of model predictive control theory for fractional-order system. The discrete integer-order state-space model approximation was used. Control law was obtained for unconstrained MPC problem and was used as input for both the fractional-order plant and its integer-order approximation model.

It was observed that MPC can be utilized for FO plants once the model of the FO plant is obtained and a unique analytical expression can be used to depict process output for FO plants. Approximation method was used to obtain model of the FO plant and its authenticity was verified using frequency and step response. Obtained results reveal that MPC can be used for FO plants and the model used works fine. Moreover, use of systematic expression obtained for process output makes it realistic. Hence, classical servo problem can be solved for a class of linear FO systems with the help of approximation method using MPC.

This work also tries to address the issue of model predictive control strategy for fractional-order plant in the presence of parametric uncertainties. From the results it is evident that the MPC can handle the small parametric uncertainty in both the fractional-order model as well as its integer-order approximation.

References

1. Muske, K.R., Rawlings, J.B.: Model predictive control with linear models. Process Syst. Eng. **39**, 262–287 (1993)
2. Rawlings, J.B.: Tutorial overview of model predictive control. IEEE Control Syst. Mag. **20**, 30–52 (2000)
3. Morari, M., Lee, J.H.: Model predictive control: past, present and future. J. Comput. Chem. Eng. **23**, 667–682 (1999)
4. Bemporad, A.: Model predictive control design: New trends and tools. In: IEEE Confernce on Decision Control (2006)
5. Rhouma, A., Bouani, F., Bouzouita, B., Ksouri, M.: Model predictive control of fractional order systems. J. Comput. Nonlinear Dyn. **9**, 031011 (2014)
6. Maciejowski, J.M.: Predictive Control with Constraints. Prentice Hall, London (2002)
7. Monje, C.A., Chen, Y.Q., Vinagre, B.M., Xue, D., Feliu, V.: Fractional Order Systems and Control. Springer, Berlin (2010)
8. Boudjehem, D., Boudjehem, B.: A fractional model predictive control for fractional order systems. Fractional Dynamics and Control, pp. 59–71. Springer, New York (2012)
9. YaLi, H., Ruikun, G.: Application of fractional order model reference adaptive control on industry boiler burning system. In: International Conference on intelligent Computation Technology and Automation (2010)
10. Podlubny, I.: Fractional differential equations: an introduction to fractional derivatives, fractional differential equations, to methods of their solution and some of their applications. Volume 198. Academic press, New York (1998)
11. Hortelano, M.R., de Madrid y Pablo, A.P., Hierro, C., Berlinches, R.H.: Generalized predictive control of arbitrary real order. In: New Trends in Nanotechnology and Fractional Calculus Applications. Springer, Berlin (2010)
12. Romero, M., de Madrid, A.P., Maoso, C., Vinagre, B.M.: Fractional order generalized predictive control: formulation and some properties. In: International Conference Control, Automation, Robotics and Vision Singapore (2010)
13. Caponetto, R., Dongola, G., Fortuna, L.: Fractional Order Systems Modeling and Control Applications. Second edn. Volume 72. World Scientific Publishing Co., Pvt. Ltd
14. Vinagre, B.M., Podlubny, I., Hernandez, A., Feliu, V.: Some approximation of fractional order operators used in control theory and applications. Fract. Calc. Appl. Anal. **3**, 210–213 (2000)

15. Valerio, D., Costa, J.: Variable order fractional order derivatives and their numerical approximations. J. Signal Process. **91**, 470–483 (2011)
16. Poinot, T., Trigeassou, J.C.: A method for modelling and simulation of fractional systems. Signal process. **83**, 2319–2333 (2003)
17. Poinot, T., Trigeassou, J.C.: Identification of fractional systems using an output-error technique. Nonlinear Dyn. **38**, 133–154 (2004)
18. Dzieliński, A., Sierociuk, D.: Fractional order model of beam heating process and its experimental verification. In: New Trends in Nanotechnology and Fractional Calculus Applications, pp. 287–294. Springer, Berlin (2010)
19. Gabano, J.D., Poinot, T.: Fractional modelling and identification of thermal systems. Signal Process. **91**, 531–541 (2011)
20. Mansouri, R., Bettayeb, M., Djennoune, S.: Approximation of high order integer systems by fractional order reduced-parameters models. Math. Comput. Model. **51**, 53–62 (2010)
21. Nise, N.S.: Control Systems Engineering. Wiley, New York (2007)
22. Podlubny, I., Petras, I., Vinagre, B.M., O'leary, P., Dorcak, L.: Analogue realizations of fractional-order controllers. J. Nonlinear Dyn. **29**, 281–296 (2002)
23. Rossiter, J.A.: Model—Based Predictive Control: A Practicle Approach. CRC Press, Boca Raton (2003)
24. Camacho, E., Bordons, C.: Model Predictive control. Volume 2. Springer, London (2004)
25. Podlubny, I., Dorcak, L., Kostial, I.: On fractional derivatives, fractional-order dynamic systems and $PI^{\lambda}D^{\mu}$-controllers. Volume 5. In: 36th IEEE Conference on Decision and Control, 4985–4990 (1997)

CFD Modeling of a Mixed Mode Boosted GDI Engine and Performance Optimization for the Avoidance of Knocking

Michela Costa, Ugo Sorge, Paolo Sementa and Bianca Maria Vaglieco

Abstract The paper applies simulation techniques for the prediction and optimization of the thermo-fluid-dynamic phenomena characterizing the energy conversion process in a GDI engine. The 3D CFD model validation is realized on the ground of experimental measurements of in-cylinder pressure cycles and optical images collected within the combustion chamber. The model comprehends properly developed submodels for the spray dynamics and its impingement over walls. This last is particularly important due to the nature of the mixture formation mode, being wall-guided. Both homogeneous stoichiometric and lean stratified charge operations are considered. In the case of stoichiometric mixture, the possible occurrence of knocking is also accounted for by means of a submodel able to reproduce the preflame chemical activity. The CFD tool is finally included in a properly formulated optimization problem aimed at minimizing the engine-specific fuel consumption with the avoidance of knocking through a non-evolutionary algorithm.

Keywords CFD optimization · 3D engine model · GDI spark ignition engine · Control · Knocking

M. Costa (✉) · U. Sorge · P. Sementa · B.M. Vaglieco
Istituto Motori, CNR, Via Marconi, 4, 80125 Naples, Italy
e-mail: m.costa@im.cnr.it

U. Sorge
e-mail: u.sorge@im.cnr.it

P. Sementa
e-mail: p.sementa@im.cnr.it

B.M. Vaglieco
e-mail: b.m.vaglieco@im.cnr.it

© Springer International Publishing Switzerland 2015
M.S. Obaidat et al. (eds.), *Simulation and Modeling Methodologies,*
Technologies and Applications, Advances in Intelligent Systems
and Computing 402, DOI 10.1007/978-3-319-26470-7_10

195

1 Introduction

This work has the primary purpose of showing how a properly developed simulation model may be of importance for the prediction of the behavior of a complex system as a gasoline direct injection (GDI) spark ignition engine, hence for the choice of the optimal control parameters of its actual operation.

The well-established role of computational fluid dynamics (CFD) as a tool for the analysis of thermo-fluid-dynamic systems is further confirmed by its application in the design phase of energy conversion systems, and, in particular, of internal combustion engines. Simulation analyses allow running a virtual prototype of a certain propulsion system and testing various geometric configurations or control parameters within times and costs absolutely negligible if compared with the corresponding characterization at the test bench. Just the increasing complexity of modern engines, consequent the large number of variables that govern their operation, and the need to respond to higher and higher performance targets, justify the importance of appropriate methods of analysis able to describe the relevant phenomenology especially in the phase of design and prototype development [1].

The use of rigorous techniques of decision-making, such as optimization methods coupled with modern tools of numerical simulation, on the other hand, is today very effective to reduce costs, improve performance and reliability, and shorten the time to market of technical systems and components. In fact, numerical procedures may be used to generate a series of progressively improved solutions to a properly formulated optimization problem, starting from an initial one, until a given convergence criterion is satisfied [2]. In this perspective, automatic searching methods may strongly reduce the time needed for computational engine optimization effected through parametric analyses.

The state of the art of computational and optimization models for internal combustion engine development can be found in the book by Shi et al. [3]: engine optimization through parametric analysis is compared with optimizations realized by means of non-evolutionary methods or evolutionary methods. Several examples are provided.

In the present paper, the development of a simulation model able to predict both the mixture formation and the combustion processes occurring within the combustion chamber of a high-performance GDI engine working under a mixed mode boosting [4] is described.

Injecting the fuel directly into the combustion chamber of a spark ignition engine has major advantages of allowing a flexible control of the in-cylinder air-to-fuel ratio and of determining the reduction of the mixture temperature due to the subtraction of the fuel latent heat of vaporization from air, which limits the engine tendency to knocking [5, 6]. A further positive aspect is the possibility to realize the so-called mixed-mode GDI boosting, namely to make the engine working with either homogeneous (stoichiometric or rich) mixtures or with stratified charges. These last are characterized by a rich zone around the spark plug and leaner zones close to the walls, for overall lean engine operation and lower heat losses at the liner. As drawbacks of direct injection, one must consider emission problems such as excessive light load unburned hydrocarbons or particulate matter formation under late fuel injection conditions [7]. These aspects are strictly related to the impingement of the gasoline spray against the walls, which is often unavoidable or even intentional [8, 9]. In fact, in some engine configurations, as the wall-guided one, the piston exhibits a properly shaped nose adjacent to a cavity placed as opposite to the injector site, which has just the scope of redirecting the impinging spray droplets and vapor cloud toward the spark plug. The possibility to realize homogeneous or stratified charges depends on the choice of the time of injection, hence on the relative position of the piston with respect to the injector tip. Controlling the amount of impacting droplets, their trajectory and the evaporated gasoline mass is a challenging task. Not only one must consider, as in free sprays, the surface tension and inertia forces, but the outcome of the spray–wall interaction. The impact of droplets on hot surfaces may cause splashing or rebounding, and also sticking and formation of wallfilm. The heat transfer from the walls to the droplets or to the wallfilm determines the so-called secondary evaporation that is affected by pressure, temperature, and velocity of the surrounding gas (until local vapor saturation), as well as by the wall temperature. Injection time, during the compression or the intake phase, therefore, corresponds to different environmental conditions, which necessarily affect the whole mixture formation process and obviously determine a different evolution of combustion [9, 10].

In this scenario, the use of detailed methods of analysis, as CFD tools, possibly in conjunction with advanced optical diagnostics [11], may lead to a straightforward optimization of the mixture formation process over the whole engine working map [2, 12].

The 3D engine model presented in the following is just applied to two representative conditions of stoichiometric and lean operation. The main feature of the engine under study is the optical accessibility to the combustion chamber, which allows also collecting images of the in-cylinder spray evolution and flame development.

The occurrence of abnormal combustions is also analyzed under stoichiometric charge operation, in particular with reference to the phenomenon of spontaneous ignition arising in the so-called *end-gas* zone of the charge, not yet reached by the flame front initiated by the spark plug. The combustion process in a spark ignition engine regularly proceeds starting from the spark if the flame front propagates by progressively investing all the mixture within the combustion chamber. Various factors may affect the normal and complete ignition. Deposits of unburned hydrocarbons or local increase of wall temperature may originate, which cause spontaneous ignition of the mixture even before it is invested by the flame front triggered by the spark. Among abnormal combustions, the most relevant, without any doubt, is knocking, namely the self-ignition of the *end-gas* zones of the charge, not yet reached by the main flame front [13], which is here modeled through a reduced chemical kinetic scheme.

The developed simulation tool is finally applied to realize the best choice of the engine-governing parameters for the operation with the lowest fuel consumption and the avoidance of knocking.

2 Experimental Apparatus

The experimental apparatus employed for the collection of data to be used for the validation of the 3D engine model includes the following modules: the spark ignition engine, an electrical dynamometer, the fuel injection line, the data acquisition and control units, the emission measurement system, and the optical apparatus.

A GDI, inline four-cylinder, four-stroke, displacement of 1750 cm^3, turbocharged, high-performance engine is the object of the present study. The engine is equipped either with a Bosch seven-hole injector or with a Magneti Marelli six-hole injector, located between the intake valves and oriented at 70° with respect to the cylinder axis. Mixture formation is realized in the wall-guided mode, with the piston head properly shaped to redirect the spray and vapor cloud toward the top of the cylinder and the spark plug. The engine is equipped with a variable valve timing (VVT) system in order to optimize the intake and exhaust valves lift under each specific regime of operation. The engine is not equipped with after-treatment devices. Details are reported in Table 1.

An electrical dynamometer allows the operation under both motoring and firing conditions, hence detecting the in-cylinder pressure data and exploring the engine behavior under stationary and simple dynamic conditions. An optical shaft encoder is used to transmit the crank shaft position to the electronic control unit. The information is in digital pulses, the encoder has two outputs, the first is the top dead center (TDC) index signal with a resolution of 1 pulse/revolution, and the second is the crank angle degree marker (CDM) 1 pulse/0.2°. Since the engine is four-stroke, the encoder gives as output two TDC signals per engine cycle. In order to determine the right crank shaft position, one pulse is suppressed via the dedicated software.

Table 1 Characteristics of the engine under study

Unitary displacement [cm^3]	435.5
Bore [mm]	83
Stroke [mm]	80.5
Air supply	Exhaust gas turbocharger
Maximum boost pressure [bar]	2.5
Valve timing	Intake and exhaust VVT
Compression ratio	9.5:1
Max. power [kW]	147.1 @ 5000 rpm
Max. torque [Nm]	320.4 @ 1400 rpm

A quartz pressure transducer is installed into the spark plug to measure the in-cylinder pressure with a sensitivity of 19 pC/bar and a natural frequency of 130 kHz. Thanks to its characteristics, a good resolution at high engine speed is obtained. The in-cylinder pressure, the rate of heat release, and the related parameters are evaluated on an individual cycle basis and/or averaged over 300 cycles.

Tests presented in this paper are carried out at the engine speed of 1500 rpm. The absolute intake air pressure remains constant at 1300 mbar, the temperature is 323 K. Two operating conditions are considered for the 3D model validation, whose control parameters are summarized in Table 2. The first, hereafter indicated with the letter L, is characterized by an overall lean charge; the second one has a stoichiometric charge and is indicated with letter S. Start of injection (SOI) and start of spark (SOS) are expressed in crank angles before the top dead center (BTDC). The injection pressure is equal to 6 MPa in the lean case, to 15 MPA in the stoichiometric case. Commercial gasoline is delivered through the seven-hole Bosch injector. Intake valve opening (IVO), intake valve closing (IC), exhaust valve opening (EVO), and exhaust valve closing (EVC) are all reported in the table after (A) or before (B) the relevant dead center (TDC or bottom dead center, BDC).

The polar diagrams of Fig. 1 summarize the synchronization of injection, ignition, and valve timing for the cases L and S. The injection occurs entirely during the

Table 2 Operating conditions considered for the 3D model validation

	SOI [°BTDC]	SOS [°BTDC]	A/F	p_{inj} [MPa]
Overall lean charge (L)	70	13	21.5	6
Stoichiometric charge (S)	307	19	14.7	15

IVO@13°ATDC, IVC@52°ABDC, EVO@20°BBDC, EVC@15°BTDC

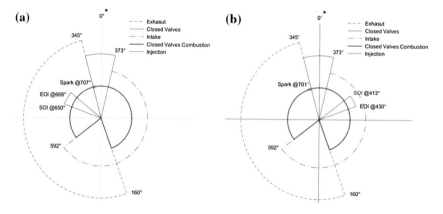

Fig. 1 Polar diagrams for the **a** lean burn and **b** stoichiometric case of validation

Table 3 Operating conditions considered for the study of the knocking occurrence

	SOI [°BTDC]	SOS [°BTDC]	A/F	p_{inj} [MPa]
No-knocking case	200	20	14.7	10
Incipient knocking case	200	25	14.7	10
Knocking case	200	30	14.7	10

IVO@22°BTDC, IVC@17°ABDC, EVO@5°BBDC, EVC@TDC

intake stroke in the S case, entirely in the compression stroke in the L case. Their length is comparable, though the delivered fuel amount is greater in the S case, due to the different injection pressure.

Three different operating conditions are instead employed with the aim of studying the knocking occurrence. The 92 RON fuel is delivered through the Magneti Marelli six-hole injector, with governing parameters as reported in Table 3. Figure 2 shows the in-cylinder pressure in the cases of Table 3. In particular, Fig. 2a reports the average over 300 consecutive cycles in a no-knocking situation, Fig. 2b, c represents the instantaneous pressure curves corresponding to the 150th cycle of 300 consecutive ones in an incipient knocking case and a knocking case,

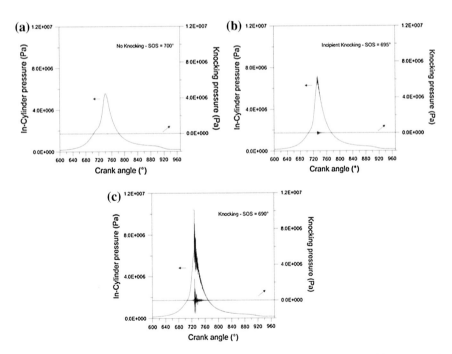

Fig. 2 In-cylinder pressure in the **a** no knocking, **b** incipient knocking, and **c** knocking cases and 5–30 kHz band-pass filter

respectively. In Fig. 2b, c the pressure traces show the typical ripples of knock, whose intensity increases as the spark advance is increased.

2.1 Optical Apparatus

The engine under study is optically accessible. Imaging measurements are performed by means of the optical experimental setup shown in Fig. 3. The optical accesses are realized in the engine head. A customized protective case for an endoscopic probe, equipped with an optical sapphire window (5 mm diameter), is installed in the engine head in the fourth cylinder. This system allows investigating an area including the spark and the gasoline spray through an endoscope exhibiting a viewing angle of 70°. The field of view is centered in the combustion chamber perpendicularly to the plane identified by the axes of the cylinder and the injector, hence perpendicular to the plane of tumble motion.

The endoscopic probe is coupled with two high spatial and temporal resolution CCD cameras. The first is an intensified cooled CCD camera (ICCD). It is equipped with a 78 mm focal length, f/3.8 UV Nikon objective. The ICCD has an array size of 512×512 pixels and a 16-bit dynamic range digitization at 100 kHz. The optical apparatus allows a spatial resolution of approximately 0.19 mm/pixel. Its spectral range spreads from UV (180 nm) to visible (700 nm).

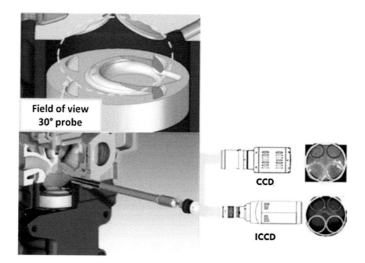

Fig. 3 Sketch of the experimental setup for optical investigation and detail of the combustion chamber: **a** CCD for 2D two-color technique, **b** ICCD for UV-visible acquisition

The ICCD operates at a digitizer offset of about 230 counts, but the dark noise fluctuation in the background is much smaller, less than 50 counts. Dark noise and photon statistical noise are both small compared with the measured intensity. The second camera is a digital CCD color camera equipped with a 50 mm focal length, f/3.8 Nikon lens. Its spectral range spreads from 400 to 700 nm and it allows performing a 2D flame visualization. The spatial resolution for this other optical assessment is of about 0.19 mm/pixel.

The gasoline injection phase is characterized through the ICCD camera and an intense strobe lamp, which is introduced in the spark location through fiber optics. For all the optical measurements, the synchronization between the cameras and the engine is made by the crank angle encoder signal through a unit delay.

3   Spray Dynamics Simulation

The complex phenomena of penetration, transport, and evaporation of gasoline have to be carefully controlled in order to allow the desired mixture preparation and the minimization of anomalies, such as the presence of liquid particles on the walls or any localized thickening of fuel deposits [14, 15].

The 3D submodel able to simulate the dynamics of the gasoline spray issuing from the considered injector is here developed in the context of the software AVL Fire^{TM} [16], in such a way to simulate preliminary experiments performed by mounting the injector and delivering sprays in an optically accessible vessel [12]. The followed approach is the classical coupling between the Eulerian description of the gaseous phase and the Lagrangian description of the liquid phase. The governing equations are here not reported for the sake of brevity; the interested reader may refer to the book by Ramos [17]. The train of droplets inserted in the computational domain in correspondence of the injector holes exit section suffers various concurring effects as it travels. Details of the model are given in the paper by Costa et al. [12]. Here it is only worth pointing out that the droplets breakup phenomenon is simulated according to the submodel of Huh–Gosman [18], whose constant C_1 (regulating the breakup time) is adjusted through a tuning procedure. The effect of the turbulent dispersion on the droplets dynamics is simulated through the submodel by O'Rouke [19], the coalescence through the submodel by Nordin [20], and the evaporation through the submodel by Dukowicz [21]. Initial size of droplets at the nozzle exit section, is considered as variable according to a probabilistic lognormal distribution, whose expected value is given by the following theoretical diameter, where τ_f is the gasoline surface tension, ρ_g the surrounding gas density, u_{rel} the relative velocity between the fuel and the gas, C_d a constant of the order of the unity (indeed taken equal to the unity), and the parameter λ^* deriving from the hydrodynamic stability analysis and indicating the dimensionless wavelength of the more unstable perturbation to the liquid–gas interface at the injector exit section:

$$D_{th} = C_d \left(\frac{2\pi\tau_f}{\rho_g u_{rel}^2} \right) \lambda^*. \tag{1}$$

The variance of the distribution, σ, is another submodel parameter to be properly tuned. The model tuning is effected through an automatic procedure, developed by authors, that solves a single objective optimization problem where the Nelder–Mead Simplex algorithm [22] is used to reduce the error between the results of the numerical computations and the experimental measurements relevant to the penetration length. The tuning procedure is hereafter described. Experimental tests are reproduced by simulating the spray dynamics within a domain reproducing in size and shape the used confined vessel. At each injection pressure, the lognormal distribution of the initial droplet size at the injector exit section is built starting from the value of σ chosen in the design of experiments (DOE), space and the expected value is computed according to Eq. (1). The distribution profile is transferred to the Fire^{TM} spray submodel, which also receives the value of C_1 from the DOE space. The model performs the spray computation and furnishes as an output the penetration length of the jets compounding the spray. The error between the numerically computed penetration length, as averaged over the six (or seven) jets and the experimentally measured one is minimized by the simplex algorithm. The objective function is defined as

$$Obj|_{\sigma, C_1} = \sum_{i=1}^{n} [l_{ex}(t_i) - l_{num}(t_i)]^2, \tag{2}$$

where n represents the number of discrete instants of time in which the injection interval of time is subdivided, t_i is the ith instant of time, and $l_{ex}(t_i)$ and $l_{num}(t_i)$ the values, respectively, of the experimentally measured and the numerically computed penetration length at t_i. The experimentally measured penetration length, indeed, is evaluated by means of a smoothing spline passing through the actual measurements points in the time-length plane.

Some results of the developed spray submodel for the seven-hole Bosch and the six-hole Magneti Marelli injectors are, respectively, given in Refs. [12, 23].

The 3D spray submodel also has to account for the spray–wall impingement, namely for the impact, rebound, deposition, and evaporation of gasoline droplets. Two different spray impingement models are here considered, the one proposed by Mundo and Sommerfeld [24] and the one proposed by Kunkhe [10]. The latter accounts for the dependence on the phenomenon on the wall temperature, namely it considers not only momentum, but also energy balance before and after the impact event.

In the model proposed in Ref. [24], authors distinguish between two regimes: deposition and splashing. In the deposition regime all of the liquid remains on the wall, while in the splashing regime part of droplets is deposited and another part reflected away. Transition from one regime to the other can be described by the so-called splashing parameter K, a function of the Reynolds and the Ohnesorge numbers

$$K = \mathrm{Re}^{1.25} Oh. \tag{3}$$

The dimensionless numbers Re and Oh are defined as follows:

$$\mathrm{Re} = \frac{\rho_d d_d u_{d,\perp}}{\sigma_d} \qquad Oh = \frac{\mu_d}{\sqrt{\rho_d \sigma_d d_d}}$$

being ρ_d the droplet density, d_d the diameter, $u_{d,\perp}$ the normal component of velocity, σ_d the surface tension, and μ_d the dynamic viscosity. For K < 57.7, droplets are deposited completely at the wall without bouncing or breaking up. The kinetic energy of droplet is dissipated, independently on wall roughness. In the splashing regime (K > 57.7), droplets are partially shattered to produce a different droplet size spectrum for the reflected droplets. With increasing K (higher momentum, less surface tension), the reflected droplets tend to be smaller and have a narrower bandwidth of size distribution. Empirical correlations, here not reported for the sake of brevity, are used to derive the droplets reflection angle (as a function of the incident angle), the reflected mass fraction, droplet size, and the tangential velocity component.

The model proposed by Kuhnke [10] distinguishes, instead, between four regimes. In the splashing regime, particles are atomized and smaller secondary droplets form after their impact on the wall. Under rebound conditions, a vapor layer between droplet and wall is formed, that prevents a direct contact and leads to a reflection of the impacting droplets, without formation of wallfilm. In the thermal regime, droplets also disintegrate into secondary ones, again without wallfilm formation. The Kuhnke's model is properly adjusted in order to keep into account the properties of the piston material. In particular, the heat penetration coefficient at the interface between the wall and the liquid is calculated with reference to the aluminum properties [25]. Within this work, a preliminary computation is performed reproducing an experimental campaign devoted to investigate the phenomenon of impingement. The injector is mounted orthogonally to a plate placed at a distance of 20 mm. Figure 4 shows the comparison between the experimentally collected and the numerically elaborated images of the spray impingement on a wall being at a temperature of 473 K, 700 and 1100 μs after the start of injection (ASOI). Injection pressure is of 55 bar.

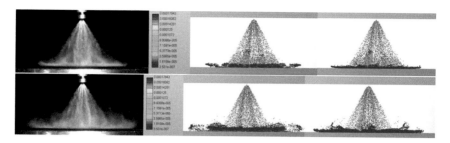

Fig. 4 Experimental images of the spray (*left*) impinging on a plate being at $T_w = 473$ K, 700 and 1100 µs ASOI as computed by the models of Mundo-Sommerfeld (*center*) and Kuhnke (*right*)

4 3D Engine Model

The formulation of the 3D engine model is carried out within the AVL Fire™ environment [16], though the choice of the software is not binding. The whole four-stroke cycle is simulated. Boundary conditions for the 3D model are obtained from test bench data. The discretization of the computational domain corresponding to the cylinder and the intake and exhaust ducts of the engine is made through the preprocessing module Fame Engine Plus (FEP) of the AVL Fire™ code, with part of the domain discretized "manually" to increase the mesh regularity and assure stability. Figure 5 shows a computational grid relevant to the closed valve period, where one may note the care devoted in the discretization of the zone surrounding the spark plug. The nose in the piston head is well visible on the bottom left of the figure.

The combustion process is simulated through the extended coherent flamelet model (ECFM) [26], NO formation follows the Zeldovich's mechanism [27]. The ECFM model is properly tuned to well catch the in-cylinder pressure curve by acting on the initial flame surface density and the flame stretch factor. The validation consisted in a preliminary verification of the results independency upon the grid size, as well as in the calculation of the motored cycle. For the sake of brevity, further details of the validation procedure of the 3D model are here not reported. Figure 6 shows the comparison between the measured pressure cycles (averaged over 300 consecutive cycles) and the cycles calculated numerically in the S and L cases of Table 2.

Fig. 5 Computational grid at TDC

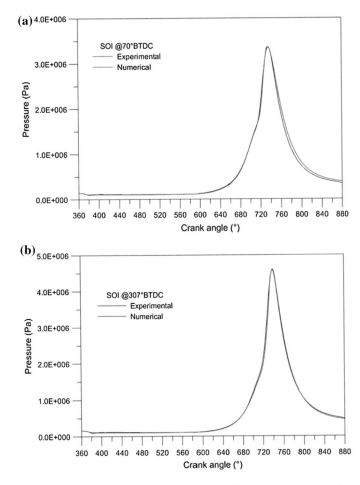

Fig. 6 Numerical-experimental comparison between in-cylinder pressure under **a** lean burn and **b** in the stoichiometric case

Information concerning the spray evolution within the combustion chamber can be derived from Fig. 7, where the numerically computed spray and the experimentally collected images are represented for two crank angles after SOI in the S case. The developed numerical model well reproduces the droplet dynamics and impingement on the piston wall, which is moving toward the TDC. Further confirmation of the good predictive capability of the numerical model is given by Fig. 8, which represents the propagation of the flame front in the S case in times immediately following the initiation of combustion. The experimental images represent the flame as averaged over the optical path, and the numerical ones represent the flame surface density distribution on a plane passing through the spark

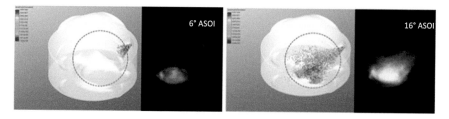

Fig. 7 Comparison between numerically computed (*left*) and experimentally collected images of the spray (*right*) at two crank angles ASOI, for the Case S

Fig. 8 Flame front in the S case: experimental (*left*) and numerical (*right*) at various times

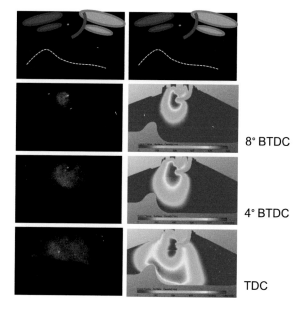

plug. The slight shift of the flame toward the exhaust valves, on the left side of the figure, is well reproduced numerically.

An example of the type of results suitable of being obtained through the developed numerical model is given in Fig. 9, where the trends of the mixture equivalence ratio are reported in the two conditions S and L.

The production of the main pollutants can instead be discussed with reference to Fig. 10. The amount of produced carbon monoxide is comparable, while a reduced NO amount is evident in the L case, due to the lower combustion temperature.

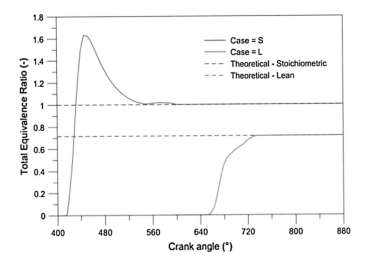

Fig. 9 In-cylinder equivalence ratio of the mixture in cases L and S

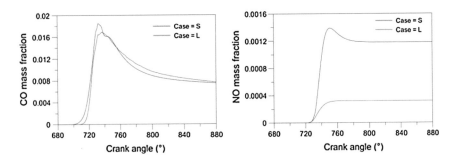

Fig. 10 In-cylinder CO (*left*) and NO (*right*) mass fraction in cases L and S

5 Knocking Prediction

In the knocking phenomenon, a part of the mixture, before being invested by the flame front ignited by the spark, reaches conditions that promote its spontaneous ignition. The self-ignition of a fuel–air mixture is the result of a series of preflame or low-temperature reactions, which leads to the start of the combustion process without the intervention of an external source of ignition, but through the formation of not stable products of partial oxidation (peroxides, aldehydes, hydroperoxides, etc.) and the release of thermal energy. When the energy of the exothermal chemical reactions exceeds the amount of heat transferred from the reagent system to the external environment, self-ignition takes place. As a result, the temperature of the mixture increases, rapidly accelerating the subsequent oxidation reactions.

The simulation of the process of self-ignition of an air–fuel mixture can be performed at different levels of approximation. A model that has proved being successful in predicting both spatially and temporally the occurrence of self-ignition, and that, at the same time, does not require excessive computational time, is the so-called Shell model, developed by Halstead et al. [28]. It comprehends the start of combustion, with the breakup of the carbon–hydrogen bonds and its development through the formation of oxygenated products. The main advantages deriving from the use of a reduced scheme with respect to what could be a detailed kinetic scheme, consist precisely in the identification of groups of radicals or radicals that lead to the branching of the reaction chains or to simple propagation of linear type, and in the possibility to follow the variation in time of the concentration of these radicals. Other kinetic mechanisms, developed subsequently to the Shell, have a much higher number of reactions and species involved, which makes of little interest their use within numerical simulations on multidimensional complex domains. The Shell model is here used in the so-called *end-gas* zone, namely in the volume of mixture not yet reached by the flame front. The combustion process resulting from the spark ignition is calculated using the ECFM model, as already pointed out in the previous paragraph.

Introducing the hydrocarbon RH, namely the fuel of composition C_xH_y, the Shell model is constituted by the following chemical reactions:

primary initialization

$$RH + O_2 \rightarrow 2R^*$$

preflame propagation

$$R^* \rightarrow R^* + P$$
$$R^* \rightarrow R^* + B$$
$$R^* \rightarrow R^* + Q$$
$$R^* + Q \rightarrow R^* + B$$

branching

$$B \rightarrow 2R^*$$

linear and quadratic termination

$$R^* \rightarrow NR$$
$$2R^* \rightarrow NR$$

where the letter P indicates the reaction products (CO_2, H_2O), and B and Q, respectively, represent branching agents and generic intermediate species. The term NR indicates inert compounds created at the end of the preflame reactions. Into detail, the model contemplates the start of combustion, with the breaking of the chains of carbon–hydrogen fuel and the formation of radicals R*, and its development through the formation of oxygenated products. As already mentioned, the species that have a similar role in the preflame kinetics are treated uniquely, as if they were a single entity.

The numerical 3D model is first adapted to reproduce the condition of Fig. 2a, then the spark advance is increased twice, each time of 5° crank angle, as in Table 3.

Figure 11 shows the in-chamber formation of Q for the three considered cases and well highlights that with more advanced ignition a more rapid formation of this group of species occurs. The location of the maximum value of the Q concentration at the crank angle of 729° (9° after top dead center, ATDC) is in excellent agreement with the experimental data of Fig. 2, in particular with the start of oscillations of the knocking case of Fig. 2c. The Shell model also allows drawing the distribution of the intermediate Q species in the combustion chamber, which may help to highlight the zone where knocking occurrence is most probable. Figure 12 represents, in the three situations of Table 3, the distribution of Q on a plane perpendicular to the cylinder axis (the one at which the maximum mass fraction of the species Q is attained) at the crank angle of knocking occurrence, namely 729°. Due to the symmetry assumption, only half plane is plotted. One may note the greater chemical reactivity in the area of the end gas located at the bottom of the figure, on the side of the injector. The shaped nose on the piston head, in fact, appears clearly in the figure. The species Q, therefore, can be used as an index of probability of knocking occurrence.

Based on the afore described calculations and on the analysis of experimental data, one may define criteria for knocking occurrence, either based on the evaluation of the in-cylinder amount of the Q species, or on the more traditional evaluation of the pressure gradient in the p-θ (pressure-crank angle) plane, where a threshold value can be established below which the engine operates regularly. By following this second route, and in agreement with the experimental data, the value of dp/dθ = 3.5 is fixed as threshold for the knocking occurrence.

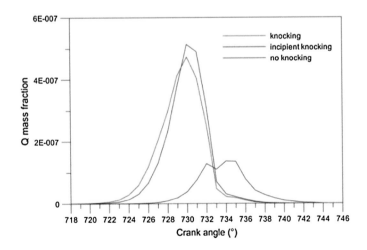

Fig. 11 Q formation in the combustion chamber in the three cases of Table 3

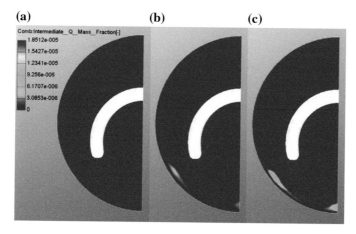

Fig. 12 Spatial distribution of the intermediate species of the preflame reactions on the plane orthogonal to the cylinder axis of maximum concentration 9° after TDC for the **a** no knocking, **b** incipient knocking, and **c** knocking cases

6   Engine Optimization

Although GDI engines are characterized by lower in-cylinder temperatures with respect to port fuel injection (PFI) engines, knocking occurrence remains an important issue, especially in approaching the design of a new prototype and with the aim of defining the control parameters leading to the best performance. A method is here proposed to explore the DOE space of the engine control variables, based on the coupling between the CFD engine model and an optimization algorithm able to point out the condition of maximum power output and simultaneously to avoid the occurrence of knocking. SOI and SOS are input parameters for the formulated optimization problem, whose flowchart is represented in Fig. 13. The engine speed, air-to-fuel ratio and valve timing are kept constant. The

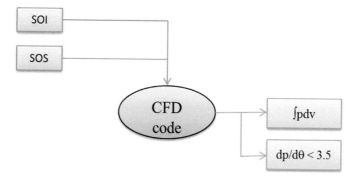

Fig. 13 Flowchart of the optimization problem

optimization algorithm, here chosen as the Simplex, runs the 3D engine model toward the optimal solution. Computed pressure cycles that do not satisfy the imposed constraint on the pressure derivative, defined at the end of paragraph 5, are discharged from the optimization results. This approach is preferred to the use of the previously presented Shell model in the 3D code, in order to avoid an excessive increase of computational time.

The obtained results exhibiting the highest value of the objective function are presented in Table 4 and in Fig. 14. Table 4 summarizes the value of the input variables SOI and SOS, the objective function and the maximum value of the pressure derivative in the p-θ plane. Figure 14 represents the in-cylinder pressure cycles in the closed valve period, for the IDs of Table 4. Case indicated with ID = 2 has a really high pressure gradient, as shown in Fig. 14. Therefore it must be discharged. The other cases have comparable power output, but also ID = 9 must not be considered due to the adopted constraint on the knocking occurrence. ID = 15 is, therefore, the optimal solution, having the same SOS of ID = 9 (knocking condition), but SOI occurring at 130° BTDC. This shows the favorable effect on knocking occurrence consequent an advanced injection.

Table 4 Cases computed with the optimization tool at different SOI and SOS and corresponding $dp/d\theta_{max}$ values

Id	SOI [°BTDC]	SOS [°BTDC]	$dp/d\theta_{max}$	$\int pdv$
2	126	40	8.08	272.40
9	122	12	5.38	277.59
15	130	12	3.31	274.44

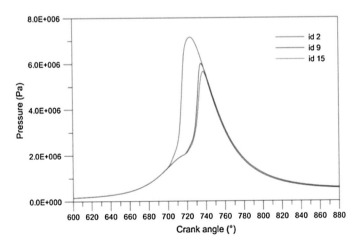

Fig. 14 Computed in-cylinder pressure cycles for the three cases of Table 4

7 Conclusions

Mixing control is fundamental in internal combustion engines. It assures flame stability, reduction of pollutants, improved combustion efficiency, reduced size, and greater lifetimes of combustors. Achievement of optimal charge conditions at all the engine loads and speeds in modern gasoline GDI spark ignition engines is undoubtedly a challenging task. This is the reason why fully automatic procedures to be used in the phase of definition of the engine-governing parameters are strongly demanded.

This work aims at presenting a 3D numerical model able to reproduce the in-cylinder processes of a high-performance GDI engine. The model includes a submodel for the spray dynamics tuned through an automatic procedure on the ground of an experimental campaign conducted in an optically accessible vessel, as well as a proper spray–wall impingement submodel that accounts for the heat transferred from the wall to the liquid deposits of fuel.

The considered engine has the head of a commercial 1750 cm^3 automotive power system, but is optically accessible due to properly made modifications to the piston and engine block. This allows capturing images of both the in-cylinder mixture formation and combustion processes.

The developed 3D engine model is shown to well reproduce the in-cylinder thermo-fluid-dynamics under both stoichiometric and lean charges. It allows determining with good accuracy the whole pressure cycle over the four-stroke period, as well as the flame propagation within the combustion chamber. A simplified model for the preflame reactions is also considered in the *end-gas* zone of the combustion chamber, not yet reached by the principal flame front, in order to detect the possible occurrence of knocking through numerical simulations. Computations show a good agreement with experiments as regards the knocking onset and its temporal location. The spatial position being the most probable for knocking is highlighted.

According to the performed analysis, a criterion is established to individuate the knocking occurrence from the pressure cycle trend.

The developed model is finally included within an optimization problem aimed at maximizing the engine power output by proper choice of the injection strategy and the time of spark ignition with the avoidance of knocking.

The described simulation approach may be employed in the phase of engine design to reduce the time to market of new technologies. If reliable experimental data available for engines of analogous configurations for validation are missing, the proposed approach may even furnish qualitative information useful for the development of control strategies.

Acknowledgments The work was supported by the government funding *PON 01_1517: Innovative methodologies for the development of automotive propulsion systems.*

References

1. Carling, R.W.: Predictive Simulation of Combustion Engine Performance in an Evolving Fuel Environment. Sandia National Laboratories (2010)
2. Thévenin, D., Janiga, G. (eds.): Optimization and Computational Fluid Dynamics. Springer, Berlin (2008)
3. Shi, Y., Ge, H.W., Reitz, R.D.: Computational Optimization of Internal Combustion Engines. Springer, London (2011)
4. Alkidas, A.C.: Combustion advancements in gasoline engines. Energy Convers. Manag. **48**, 2751–2761 (2007)
5. Taylor, A.M.K.P.: Science review of internal combustion engines. Energy Policy **36**(12), 4657–4667 (2008)
6. Zhao, H.: Advanced direct injection combustion engine technologies and development, Vol.1, Gasoline and Engines. Woodhead Publishing Limited (2010)
7. Shim, Y.S., Choi, G.M., Kim, D.J.: Numerical and experimental study on effect of wall geometry on wall impingement process of hollow cone fuel spray under various ambient conditions. Int. J. Multiph. Flow **35**, 885–895 (2009)
8. Wang, C., Xu, H., Herreros, J.M., Wang, J., Cracknell, R.: Impact of fuel and injection system on particle emissions from a GDI engine. Appl. Energy **132**, 178–191 (2014)
9. Ashgriz, N. (ed.): Handbook of Atomization and Sprays—Theory and Applications. Springer, New York (2011)
10. Kuhnke, D.: Spray wall interaction modeling by dimensionless data analysis. Ph.D. thesis, Technische Universität Darmstadt (2004)
11. Drake, M.C., Haworth, D.C.: Advanced gasoline engine development using optical diagnostic and numerical modeling. Proc. Combust. Inst. **31**, 99–124 (2007)
12. Costa, M., Sorge, U., Allocca, L.: CFD optimization for GDI spray model tuning and enhancement of engine performance. Adv. Eng. Softw. **49**, 43–53 (2012)
13. Zhen, X., Wang, Y., Xu, S., Zhu, Y., Tao, C., Xu, T., Song, M.: The engine knock analysis—An overview. Appl. Energy **92**, 628–636 (2012)
14. Stan, C.: Direct Injection Systems for Spark-Ignition and Compression-Ignition Engines. SAE Publication, Warrendale (2000)
15. Zhao, H., Ladommatos, N.: Engine Combustion Instrumentation and Diagnostics. SAE Int. Inc, Warrendale (2001)
16. https://www.avl.com/web/ast/fire
17. Ramos, J. I.: Internal Combustion Engine Modelling, CRC Press, New york, (1989)
18. Huh, K.Y., Gosman, A.D.: A phenomenological model of diesel spray atomization. In: International Conference on Multiphase Flows, Tsukuba, Japan (1991)
19. O'Rourke, P.J., Bracco, F.V.: Modeling of Drop Interactions in Thick Sprays and A Comparison With Experiments. IMECHE, London (1980)
20. Nordin, W. H.: Complex modeling of diesel spray combustion. PhD Thesis, Chalmers University of Technology (2001)
21. Dukowicz, J. K.: Quasi-steady droplet change in the presence of convection, informal report Los Alamos Scientific Laboratory. Los Alamos Report LA7997-MS (1979)
22. Nelder, J.A., Mead, R.: A Simplex method for function minimization. Computer Journal **7**, 308–313 (1965)
23. Costa, M., Marchitto, L., Merola, S.S., Sorge, U.: Study of mixture formation and early flame development in a research GDI engine through numerical simulation and UV-digital imaging. Energy **77**, 88–96 (2014)
24. Mundo, C., Sommerfeld, M., Tropea, C.: Droplet-Wall collisions: experimental studies of the deformation and breakup process. Int. J. Multiph. Flows **21**(2), 151–173 (1995)
25. Wruck, N.M., Renz, U., Transient phase-change of droplets impacting on a hot wall. Transient phenomena in multiphase and multicomponent systems: Research report, pp. 210–226 (2000)

26. Colin, O., Benkenida, A., Angelberger, C.: 3D Modeling of mixing, ignition and combustion phenomena in highly stratified gasoline engines. Oil. Gas. Sci. Technol.—Rev. IFP Energies Nouvelles, **58**, pp. 47–62 (2003)
27. Zeldovich, Y.B., Sadovnikov, P.Y., Frank-Kamenetskii, D.A.: Oxidation of nitrogen in combustion. Trans: by M. Shelef, Academy of sciences of USSR. Institute of Chemical Physics, Moscow-Leningrad, (1947)
28. Halstead, M.P., Kirsch, L.J., Quinn, C.P.: The auto-ignition of hydrocarbon fuel at high temperatures and pressures-fitting of a mathematical model. Combust. Flame **30**, 45–60 (1977)

Real-Time Radar, Target, and Environment Simulator

Halit Ergezer, M. Furkan Keskin and Osman Gunay

Abstract In this work, a real-time radar target and environment simulator (RTSim) is presented. RTSim provides a hardware-in-the-loop (HIL) test system for radar signal processing units (RSPU). RTSim provides repeatable experiments for radar developers in digitally controlled but complex environments. Moreover, it reduces the development costs by limiting expensive field tests. RTSim consists of three main components; a control computer that provides the user interface and scenario generation software, embedded processors for environment calculations, and field programmable gate arrays (FPGAs) for baseband radar signal generation. In hardware-in-the-loop operation scenario RTSim and RSPUs work in synchronization. RSPU sends the parameters of current pulse burst to RTSim and it generates baseband IQ signals using these parameters and user programmed environment parameters obtained from scenario generation software. RTSim can generate return signals for targets, jammers, clutter, and system noise. The generated baseband signals are sent to RSPU over fiberoptic lines.

Keywords Radar simulator · Hardware-in-the-loop · Target · Jammer · Clutter

1 Introduction

The need for real-time simulation tools to test radar systems is increasing in parallel with developments in radar technology. The algorithms used in RSPUs should be justified under real environment conditions prior to the integration with radar

H. Ergezer(✉) · M. Furkan Keskin · O. Gunay
MİKES, Microwave Electronic Systems Inc., Cankiri Yolu, 5.km, Akyurt, Ankara, Turkey
e-mail: halitergezer@gmail.com
URL: http://www.mikes.com.tr

M. Furkan Keskin
e-mail: mfurkankeskin@gmail.com

O. Gunay
e-mail: gunayosman@gmail.com

© Springer International Publishing Switzerland 2015
M.S. Obaidat et al. (eds.), *Simulation and Modeling Methodologies,*
Technologies and Applications, Advances in Intelligent Systems
and Computing 402, DOI 10.1007/978-3-319-26470-7_11

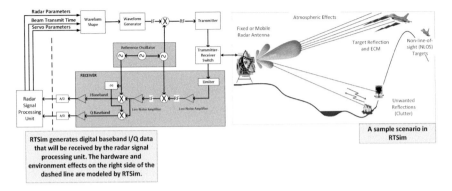

Fig. 1 RTSim and radar environment

hardware. The cost of real environment tests is very high and it is almost impossible to repeat the experiments under the same conditions. Therefore, this necessitates the development of a system that simulates the environment in which radar signals travel, and the targets that the radar is trying to detect.

Many of the radar target and environment simulators are designed as commercial products and their implementation details are not disclosed as academic publications [1–5]. All of these products are analog simulators that work in the RF (Radio Frequency) or IF (Intermediate Frequency) band and they are not closed-loop systems which means that they do not receive the radar parameters in real time. Another radar target generator that uses FPGAs (field programmable gate arrays) for baseband signal generation is proposed in [6]. Compared to these products, RTSim is a more flexible digital simulator that generates phase-coded baseband IQ signals. All parameters of a radar signal (RF frequency, waveform type, sampling rate, etc.) can be adjusted by radar for each pulse burst waveform.

The components of a typical radar, a test environment, and the role of RTSim is shown in Fig. 1.

The simulator generates signals whose properties are determined by the radar signal processing units, in the form of baseband IQ. The generated signal will contain all the effects described in the user defined scenario including; targets, jammers, chaff, decoys, environment (clutter and propagation effects), antennas, and radar hardware (amplifiers, mixers, etc.). The generated IQ signal is sent to the RSPUs over fiberoptic lines. In that sense RTSim provides a hardware-in-the-loop test environment that can account for all the effects that a radar signal encounters until it is received by the RSPUs.

2 Components of RTSim

RTSim consists of a control PC, embedded processors and FPGA (field programmable ate array) hardware as shown in Fig. 3. Control PC has the simulation engine and user interfaces. RTSim uses STAGE simulation engine [7] to calculate the nav-

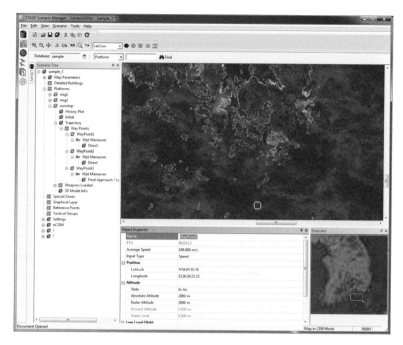

Fig. 2 RTSim user interface

igation and orientation parameters of radar and target platforms in real-time. It uses Digital Terrain Elevation Data (DTED) maps for terrain visualization. A screenshot of the main user interface is given in Fig. 2. All radar and target parameters can be adjusted from the user interfaces.

Target parameters include:

- Position, velocity, route,
- Scatter position and RCS (Radar cross section) parameters,
- RCS type (constant, orientation/frequency dependent),
- Multipath parameters,
- Scintillation parameters (swerling type and correlation time),
- Jammer parameters (noise jammer, deceptive jammer, activation time, etc.),
- Jet engine modulation, helicopter blade modulation.

Radar parameters include:

- Transmitted power, maximum range,
- Antenna scan pattern (circular or RSPU controlled),
- Antenna pattern tables,
- Waveform tables (for phase-coded signals),
- Pulse-repetition-interval (PRI) tables.

Fig. 3 RTSim and radar
environment

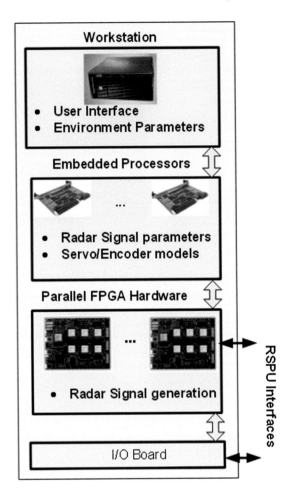

Environment parameters that can be programmed from the user interface are: statistical clutter model (distribution type, parameters), atmospheric attenuation, and rain attenuation.

Simulation parameters that are adjusted in the user interface are sent to the embedded processors and FPGAs before the simulation starts. Radar and target platforms' navigation information is sent to the embedded processors in real time. Embedded processors also receive radar beam parameters from RSPUs. The beam message defines the parameters of the radar signal that RTSim generates:

- RF frequency and sampling rate,
- Antenna pattern table ID,
- Waveform table ID,
- Pulse Width (PW) and Pulse-repetition-interval (PRI),
- PRI table ID,
- Antenna orientation.

When the beam message is received embedded processors calculate the parameters that FPGAs use for signal generation. Parameters defined in time units (PW, PRI, delay, etc.) are converted to FPGA clock units. Doppler calculations are performed for each platform. Channel attenuations for each platform, jammers and clutter are calculated and sent to the FPGAs. Servo and encoder model calculations are also performed by embedded processors to determine the orientation of the radar antenna. Encoder model controls the fixed speed circular motion of the antenna, whereas servo model is for the RSPU controlled motion of the antenna.

FPGAs generate the radar signals using the parameters sent by the embedded processors. Radar signal generator, receiver channels, clutter distribution generators, system noise generator, jammer noise generators are all implemented in the FPGAs.

3 RTSim Models

3.1 Target Modeling

Radar target generation approaches in the literature has focused on statistical modeling of target RCS. Jet engine modulation (JEM) and Helicopter Blade Modulation (HBM) effects are also studied for moving target identification [8–11]. In RTSim, targets are modeled as independent multiple scatters with different dynamic RCS tables. When the orientation of the target changes, the relative position of each scatter with respect to the radar changes (Fig. 4).

Dynamic RCS of targets is modeled by using the RCS value that corresponds to the orientation (yaw, pitch, roll) of the target. JEM and HBM are modeled using correlated complex Gaussian distributed signals. Target scintillation is modeled by four Swerling models [12]. Doppler effect is modeled by adding an additional doppler phase value to the phase codes of the radar signals at each clock cycle of the FPGA.

Doppler speed, which is the velocity component of the target in the direction of the radar, can be calculated as follows (see Fig. 5):

$$V_D = V_x \cos(\phi) \cos(\theta) + V_y \sin(\phi) \cos(\theta) + V_z \sin(\theta) \tag{1}$$

where ϕ is the azimuth and θ is the elevation angle between radar and the target. Doppler frequency is calculated using RF frequency (f) and speed of light (c) :

$$f_D = \frac{2f|V_D|}{c} \tag{2}$$

The doppler phase value should be incremented at specified intervals that depend on the doppler frequency, to give the doppler effect. The doppler phase is incremented by a specific amount at specific intervals. The interval and increment values and the sign of the velocity are sent to the FPGA as doppler parameters for each

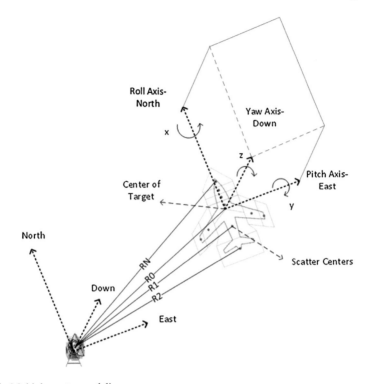

Fig. 4 Multiple scatter modeling

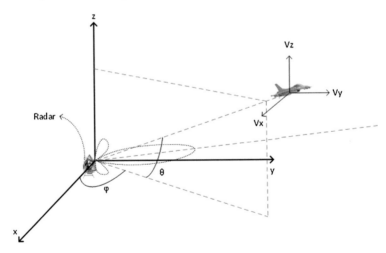

Fig. 5 Doppler modeling

beam message. The interval is calculated as the number of FPGA clock cycles:

$$P_c = \left| \frac{f_s \times k}{f_D(2^{nb} - 1)} + 0.5 \right| \tag{3}$$

where f_s is the sampling frequency, k is the increment value, nb is the number of bits of the phase codes. The value of k that gives the best frequency resolution is determined by the following optimization:

$$k = \arg \min_{x} \left| \frac{f_s \times x}{f_D(2^{nb} - 1)} - \left\lfloor \frac{f_s \times x}{f_D(2^{nb} - 1)} + \frac{1}{2} \right\rfloor \right| \tag{4}$$

where $x = 1, \ldots, 100$.

Multipath effects are also modeled in RTSim. Three multipaths can be used for each target platform. Multipaths are modeled in the same way as the targets but they have different complex attenuation constants that depend on reflection angle, radar frequency and the reflection surface [13].

For each receiver channel the received power from a target platform is calculated as follows:

$$P_r = \frac{P_t |K_1|^2 |K_2|^2 \lambda^2 \sigma}{(4\pi)^3 R_d^4 L_s} \tag{5}$$

where P_t is the transmitted power, K_1, K_2 are transmit and receive antenna gains, λ is the wavelength, σ is the RCS, R_d is the distance between radar and target, and L_s denotes the losses (polarization, atmospheric, rain). Using the received power the channel attenuation is calculated as follows (Fig. 6):

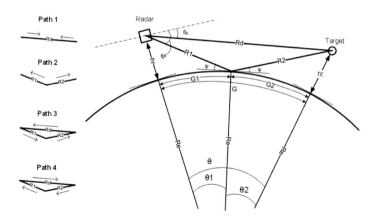

Fig. 6 Multipaths

$$C_A = e^{-j4\pi R_d/\lambda} \times e^{j(\angle K_1 + \angle K_2)} \times \sqrt{\frac{P_r}{D_{P_r}}} \times D_{SG} \qquad (6)$$

where D_{P_r} is the default received power that corresponds to the quantized signal amplitude D_{SG}. These two parameters are used to adjust the dynamic range of the 16-bit baseband IQ signal.

3.2 Jammer Modeling

The purpose of electronic warfare is to control the electromagnetic spectrum. RTSim employs different electronic attack (electronic counter measures) techniques to test the radar's performance under difficult scenarios. RTSim models spot, barrage, swept spot noise jammers, and Range/Velocity Gate Pull off/in (RVGPO/I) deception techniques [14–19]. Antenna gain, transmitted power, bandwidth and center frequency parameters can be adjusted for noise jammers. For swept spot noise different frequency patterns can be defined.

RGPO is implemented by adjusting the delays and PRIs of pulse burst radar waveforms. A sample RGPO scenario is given in Fig. 7. When VGPO is applied, velocity difference profile is defined instead of range difference. For each burst, range or velocity pull off/in amounts are calculated at 16 different points in the burst. It is observed that this resolution is satisfactory for 200 MHz sampling rate.

RGPO range difference from the beginning of a segment is calculated as follows:

$$R_F = R_0 + S_R \left(V_R \times t + \frac{1}{2} a_R \times t^2 + \frac{1}{6} J_R \times t^3 \right) \qquad (7)$$

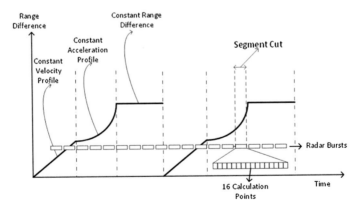

Fig. 7 RGPO scenario

where t is the time since the beginning of the segment, R_0 is the initial range difference, S_R is the sign of the range pull (off $= 1$, in $= -1$), $V_R(m/s)$ is the pull velocity, $a_r(m/s^2)$ is the pull acceleration, $J_R(m/s^3)$ is the pull jerk, and R_F is the final range difference. For each pulse burst the RGPO segment is determined and R_F values are calculated. Using these values PRI difference values that will be added to the radar PRI are calculated as follows:

$$\Delta P(m) = \left\lceil \frac{2(R_F(m) - R_F(1))}{c} \times F_s \right\rceil \quad m = 1, .., M \tag{8}$$

where c is the speed of light, F_s is the sampling rate. The delay for the generated RGPO/I signal is calculated as follows:

$$D = \left\lceil \frac{2R_D + 2R_F(1)}{c} \times F_s \right\rceil \tag{9}$$

where R_D is the distance between the radar and the jammer. These calculations are performed on the embedded processors and the results are sent to the FPGAs.

VGPO velocity difference from the beginning of a segment is calculated as follows:

$$V_F = V_0 + S_V \left(V_V \times t + \frac{1}{2} a_V \times t^2 \right) \tag{10}$$

where V_0 is the initial velocity difference, S_V is the sign of the velocity pull (off $= 1$, in $= -1$), $V_V(m/s^2)$ is the pull velocity, $a_V(m/s^3)$ is the pull acceleration, and V_F is the final velocity difference. Using these differences "phase counter (Φ_C)" and "phase increment (Φ_N)" values are calculated:

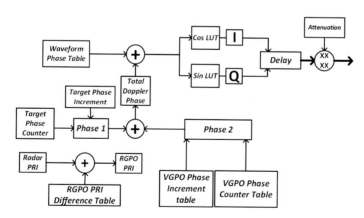

Fig. 8   RVGPO/I signal generator

$$f_D(m) = \frac{2f_{RF} \times |V_F(m)|}{c}$$

$$\Phi_C(m) = \left\lceil \frac{F_s \times \Phi_N}{f_D(m)(2^{nb} - 1)} \right\rceil \tag{11}$$

where f_D is the doppler frequency, f_{RF} is the radar RF frequency. VGPO is modeled by changing the phase of the baseband signal. After $\Phi_C(m)$ clock cycles phase difference is increased by Φ_N

Coordinated RVGPO/I signal generator is shown in Fig. 8. For coordinated implementation parameters should be set as $V_0 = V_R$, $V_V = a_R$, and $a_V = J_R$. The same signal generator can be used to generate target signals as well.

3.3 Clutter Modeling

RTSim implements statistical clutter models. Rayleigh, Weibull and K-distributions are supported. In Fig. 9 a typical clutter scenario is described. For each clutter patch a random number is generated corresponding to its range bin. The attenuation for clutter signal is determined by the patch area, grazing angle and distance from radar. The parameters of the random distributions depend on the grazing angle, polarization, radar frequency, surface's dielectric constant and conductivity.

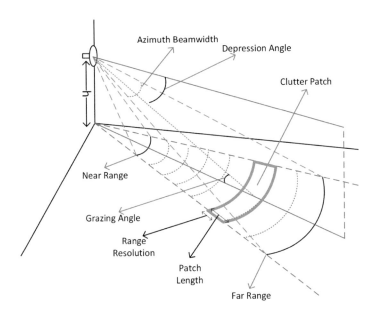

Fig. 9 Clutter scenario

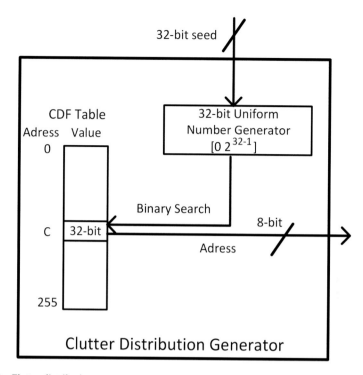

Fig. 10 Clutter distribution generator

Random clutter samples are generated using inverse CDF (Cumulative Distribution Function) method. In this method CDF of the distribution is generated as a table, then a uniform number is generated in the interval [0 1], the index of the CDF table that this number falls into is selected as the desired random number. The FPGA implementation of clutter distribution generator is shown in Fig. 10. CDF table contains 255 elements of 32-bit numbers which are obtained by quantizing the CDF function of the distribution. Uniform numbers are also 32-bits and the generated random numbers are 8-bits.

3.4 Atmospheric Attenuation Modeling

ITU recommendation P.676 [20] is used for atmospheric attenuation modeling. This model considers total attenuation caused by dry oxygen and water vapor and can be calculated as follows:

$$A = (\gamma_w(h,f) + \gamma_o(h,f)) \times R \tag{12}$$

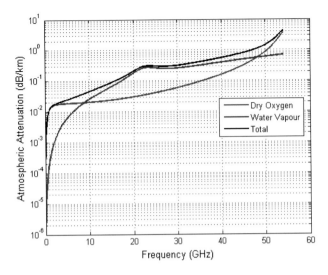

Fig. 11 Atmospheric attenuation versus frequency

where γ_w and γ_o denote attenuations caused by water vapor and dry oxygen respectively, h is height, f denotes radar frequency, and R is the distance between the radar and target platform. γ_w and γ_o are calculated and stored offline and in real-time simulation required values are obtained from the store tables. In Fig. 11 the plot of atmospheric attenuation versus frequency is displayed.

The method in [21] is used for rain attenuation modeling. In this method rain attenuation is modeled as follows:

$$\gamma_r(h,f) = a_r(f)r(h)^{b_r(h)} \tag{13}$$

where;

$$a_r(f) = \begin{cases} 6.39 \times 10^{-5} f^{2.03} & \text{if } f \le 2.9 \text{ GHz} \\ 4.21 \times 10^{-5} f^{2.42} & \text{if } 2.9 \text{ GHz} < f \le 54 \text{ GHz} \end{cases} \tag{14}$$

$$b_r(f) = \begin{cases} 0.851 \times f^{0.158} & \text{if } f \le 8.5 \text{ GHz} \\ 1.41 \times f^{-0.779} & \text{if } 8.5 \text{ GHz} < f \le 25 \text{ GHz} \end{cases} \tag{15}$$

4 Experimental Results

In the experiments the IQ signals generated by the FPGAs are compared to the theoretical calculations. In the first example there are four targets at distances 1343, 1460, 1656, and 1814 m and moving with speeds 29.83, −309.10, −347.19, −833.52 m/s.

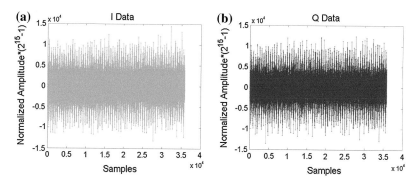

Fig. 12 I and Q samples for four targets, **a** I, **b** Q

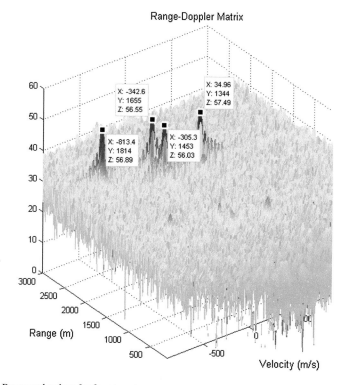

Fig. 13 Processed pulses for four targets

The radar parameters are, $PRI = 560$ samples, $PW = 100$ samples, number of pulses in each burst is 64. In Fig. 12 the I and Q signals generated by the simulator are displayed. These signals are analyzed using range doppler matrices. The ranges and velocities of detected targets are shown in Fig. 13. The results agree with the input parameters.

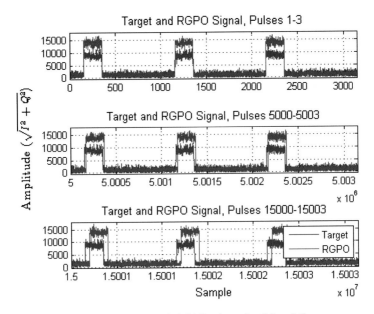

Fig. 14 Sample RGPO application. Scenario initialization, after 0.5 ve 1.5 s

In Fig. 14 sample RGPO and target signals are given. For this example range gate pull of is applied with $V_R = 500$ m/s. Radar signal parameters are; PRI = 100 μs, pulse width = 20 μs, chip width = 200 ns, number of pulses = 500 and sampling rate = 10 MHz. As seen in the figure in approximately 1.5 s 50 samples pull off is applied, and this corresponds to 750 m range difference at 5 μs.

In Fig. 15 spectrum estimations for target and VGPO signals are displayed. For this example a velocity gate pull off profile with $V_V = 150$ m/s^2 is applied. Radar signal parameters are; PRI = 30 μs, radar frequency = 10 GHz, number of pulses = 128, and sampling rate = 200 MHz. Using these parameters the maximum unambiguous velocity is 250 m/s, and velocity resolution is 7.8 m/s. As seen from the figure in 0.5 s the pull off rate is 71 m/s, and in 1 s it is 143 m/s, which are consistent with the VGPO profile.

In Fig. 16 the RGPO signal generated by RTSim is analyzed and the results are compared with the real constant acceleration profile. RGPO constant acceleration parameter is set to $a_R = 1000$ m/s^2. The generated signal has 10000 bursts each of which has a pulse of 100 μs. The sampling rate of the signal is 10 MHz. As can be seen from the figure, the generated signal's profile is consistent with the actual profile except for some quantization errors.

In Fig. 16 the VGPO signal generated by RTSim is analyzed and the results are compared with the real constant acceleration profile. VGPO constant acceleration parameter is set to $a_V = 1000$ m/s^3. The generated signal has 130 bursts each of which has 128 pulses of 30 μs. As can be seen from the figure, the generated signal's profile is consistent with the actual profile except for some quantization errors.

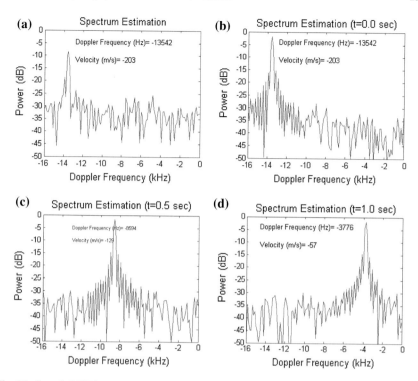

Fig. 15 Sample VGPO application, **a** target return, **b** VGPO return t = 0.0 s, **c** VGPO return t = 0.5 s, **d** VGPO return t = 1.0 s

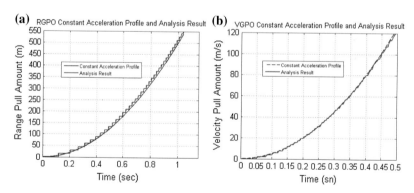

Fig. 16 Constant acceleration profile and analysis result for, **a** RGPO, **b** VGPO

The final experiment is clutter generation. K-distributed clutter with parameters shape = 4 and scale = 2.06 is generated. Radar signal has four pulses each with 10 k samples at 40 MHz sampling frequency. In Fig. 17a amplitude of the generated clutter samples is displayed. The histogram of the samples is analyzed and compared

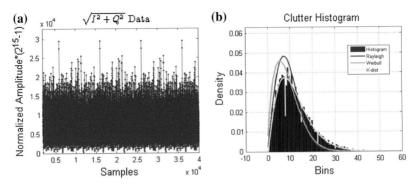

Fig. 17 Amplitude and histogram samples for clutter, **a** amplitude samples, **b** histogram

with the histograms of Rayleigh, Weibull and K-distributions and then the closest histogram is selected as the true histogram. The estimated parameters are shape $= 4$ and scale $= 2.11$ which are in close agreement with the true parameters. Histogram plots are shown in Fig. 17b.

5   Conclusions

In this work, a real-time hardware-in-the-loop radar target and environment simulator, RTSim, is described. The simulator can be used to test radar signal processing units even during the early stages of development. RTSim models moving and stationary targets, radars with multiple receiver channels, jammers, statistical clutter returns. RTSim is currently being used in some radar development projects in Turkish Defence Industry. Future work includes implementation of terrain-dependent more realistic clutter models and atmospheric propagation models.

Acknowledgments This work was supported in part by the Scientific and Technical Research Council of Turkey, TUBITAK, with grant no. TEYDEB-3120651.

Appendix: Hardware Components of RTSim

Hardware components of RTSim (white rack) and interface simulator for RSPUs (gray rack) is shown in Fig. 18. RSPU interface simulator is developed to test RTSim before integration with actual radar systems.

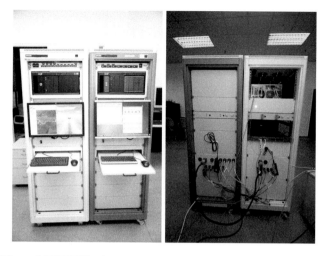

Fig. 18 RTSim and RSPU-IS hardware components

References

1. EW Simulation Technology: Radar Target & Environment Generator (2013). http://ewst.com.au/index.php?option=com_content&view=article&id=14&Itemid=28d
2. Intersoft Electronics: Radar Environment Simulator (2013). http://www.intersoft-electronics.com/HTML/RES.html
3. Saab Sensis: Radar Environmental Simulator (RES) (2013). http://www.saabsensis.com/products/radar-environmental-simulator-res/
4. Technology Service Corporation: Radar Environment Simulators (2013). http://www.tsc.com/Factsheets/RadarSimulatorsFactSheet.pdf
5. Utteridge, E.J.: Radar environment simulator. In: Radar—87; Proceedings of the International Conference, pp. 520–524 (1987)
6. Andraka, R., Phelps, R.: An FPGA based processor yields a real time high fidelity radar environment simulator. In: Military and Aerospace Applications of Programmable Devices and Technologies Conference, pp. 220–224 (1998)
7. Presagis: STAGE Scenario Generation Software (2013). http://www.presagis.com/products_services/products/modeling-simulation/simulation/stage/
8. Bell, M.R., Grubbs, R.A.: Jem modeling and measurement for radar target identification. IEEE Trans. Aerosp. Electron. Syst. **29**(1), 73–87 (1993)
9. Carriere, R., Moses, R.L.: Autoregressive moving average modeling of radar target signatures. In: IEEE National Radar Conference, pp. 225–229 (1988)
10. Phu, P., Adler, E., Innocenti, R., Paolella, A.: A test target generator for wideband pulsed doppler radars. In: IEEE MTT-S International Microwave Symposium Digest, vol. 2, pp. 973–975 (1995)
11. Sandhu, G.S., Saylor, A.V.: A real-time statistical radar target model. IEEE Trans. Aerosp. Electron. Syst. **AES-21**(4), 490–507 (1985)
12. Richards, M.: Fundamentals of radar signal processing. McGraw-Hill, New York (2005)
13. Skolnik, M.: Radar Handbook, 3rd edn. Electronics Electrical Engineering, McGraw-Hill Education, New York (2008)
14. Greco, M., Gini, F., Farina, A., Ravenni, V.: Effect of phase and range gate pull-off delay quantization on jammer signal. IEEE Proc. Radar Sonar Navig. **153**(5), 454–459 (2006)

15. Jing, Y., Mei-guo, G., Yun-jie, L.: Digital realization of pull-off jamming in R-V dimensions. In: 4th International Congress on Image and Signal Processing (CISP), vol. 4, pp. 2177–2181 (2011)
16. Kalata, P.R., Chmielewski, T.A.: Range gate pull off (RGPO): detection, observability and α-β target tracking. In: Proceedings of the Twenty-Ninth Southeastern Symposium on System Theory, pp. 505–508 (1997)
17. Neng-Jing, L., Yi-Ting, Z.: A survey of radar ecm and eccm. IEEE Trans. Aerosp. Electron. Syst. **31**(3), 1110–1120 (1995)
18. Schleher, D.C.: Electronic warfare in the information age. Artech House Radar Library, Artech House (1999)
19. Townsend, J.D.: Improvement of ECM techniques through implementation of a genetic algorithm. Master's thesis, Air Force Institute of Technology (2008)
20. International Telecommunication Union. P.676: Attenuation by atmospheric gases. Recommendation P.676, International Telecommunication Union, Geneva (2010)
21. Olsen, R., Rogers, D.V., Hodge, D.B.: The arbrelation in the calculation of rain attenuation. IEEE Trans. Antennas Propag. **26**(2), 318–329 (1978)

Computationally-Efficient EM-Simulation-Driven Multi-objective Design of Compact Microwave Structures

Slawomir Koziel, Adrian Bekasiewicz, Piotr Kurgan
and Leifur Leifsson

Abstract The size of microwave components has become an important design criterion for contemporary wireless communication engineering. Unfortunately, reduction of geometrical dimensions usually remain in conflict with electrical performance of the circuit, which makes it necessary to look for designs being a compromise between these two types of objectives. In this chapter, we discuss strategies for computationally-efficient multi-objective design optimization of miniaturized microwave structures. More specifically, we consider an optimization methodology based on point-by-point identification of a Pareto-optimal set of designs representing the best possible trade-offs between conflicting objectives, which include electrical performance parameters as well as the size of the structure of interest. Design speedup is achieved by performing most of the operations at the level of suitably corrected equivalent circuit model of the structure under design. Model correction is implemented using a space mapping technique involving, among others, frequency scaling. Operation and performance of our approach is demonstrated using a compact rat-race coupler designed with respect to the following objectives: bandwidth and the layout area. A representation of the Pareto set consisting of ten designs is obtained at the cost corresponding to less than thirty high-fidelity electromagnetic simulations of the structure.

S. Koziel (✉) · A. Bekasiewicz
Engineering Optimization & Modeling Center, School of Science and Engineering,
Reykjavik University, Menntavegur 1, Reykjavik IS-101, Iceland
e-mail: koziel@ru.is

A. Bekasiewicz
e-mail: bekasiewicz@ru.is

P. Kurgan
Faculty of Electronics, Telecommunications and Informatics,
Gdansk University of Technology, 80-233 Gdansk, Poland
e-mail: piotr.kurgan@eti.pg.gda.pl

L. Leifsson
Department of Aerospace Engineering, Iowa State University,
Ames 50011 IA, USA
e-mail: leifur@iastate.edu

© Springer International Publishing Switzerland 2015
M.S. Obaidat et al. (eds.), *Simulation and Modeling Methodologies,
Technologies and Applications*, Advances in Intelligent Systems
and Computing 402, DOI 10.1007/978-3-319-26470-7_12

Keywords Computer-aided design · Compact microwave structures · Electromagnetic simulation · Simulation-Driven design · Multi-objective optimization · Space mapping

1 Introduction

Miniaturization of microwave structures is an important design consideration for modern wireless communication systems. At the same time, it is a challenging task that requires satisfaction of several, often conflicting objectives such as size, bandwidth, phase response, etc. [1, 2]. Design closure (i.e., final tuning of the design) requires simultaneous adjustment of multiple (usually geometry) parameters of the structure under various (typically linear) constraints [3, 4] that most commonly originate from the necessity of maintaining highly compressed layouts, e.g., for folded or fractal-shaped couplers [5–8]. As a result of cross-coupling effects that take place between various parts of such circuits, simplified models (such as equivalent circuits) cannot be utilized for adequate evaluation of electrical performance. The latter can only be done by means of full-wave electromagnetic (EM) analysis.

Unfortunately, high-fidelity EM simulation is computationally expensive, which is a fundamental issue for simulation-driven design of compact microwave components [9, 10]. Conventional design strategies, such as repetitive parameter sweeps guided by engineering experience or direct EM-driven optimization—using, e.g., gradient-based or derivative free methods [11, 12]—require large number of EM analyses. Depending on the problem dimensionality, it might be dozens, hundreds, or even thousands of simulations [13, 14]. The computational cost of such process may be unacceptable from practical point of view or even prohibitive. On the other hand, alternative techniques for performance evaluation (e.g., exploiting transmission line theory) are highly inaccurate. This is particularly true for highly miniaturized circuits with closely fit building blocks (e.g., [15–18]).

The aforementioned difficulties of conventional methods can be alleviated, to some extent, by means of surrogate-based optimization (SBO) techniques [19–21], where direct handling of the expensive EM-simulated model is replaced by iterative construction and re-optimization of a fast replacement model also referred to as a surrogate. SBO techniques have proven their computational superiority over traditional optimization algorithms applied to the design of conventional microwave circuits [22–25]. SBO schemes benefit from low-cost surrogates that are aligned with high-fidelity EM models through adaptive corrections [26]. Because most of the operations are carried out on the corrected low-fidelity model, and the high-fidelity EM simulation is only launched occasionally (to verify the current design and update the surrogate model), the overall cost of the SBO process can be kept low. Probably, the most popular SBO technique in microwave engineering is space mapping (SM) [19, 23, 27]. One of the trademarks of space mapping is a parameter extraction process in which coefficients of the surrogate model (usually constructed by means of simple linear or affine transformations) are obtained by solving auxiliary nonlinear regression problems aiming at reducing the misalignment between the low- and high-fidelity models.

As opposed to conventional designs, compact structures are typically developed based on novel topologies and the influence of the structure size on its performance capabilities cannot be foreseen [28–31]. To eliminate the risk of design failure in case of excessively stringent specifications that cannot be met by a prototype circuit, multi-objective optimization becomes a necessity. The goal of multi-objective optimization, as opposed to a single-objective one, is to find the set of designs (a so-called Pareto set) representing the best possible trade-offs between non-commensurable objectives. The most popular solution approach is a population-based metaheuristics [1, 32–35]. While methods such as genetic algorithms or particle swarm optimizers are capable of identifying the entire Pareto set in one algorithm run, these methods are of limited use for compact circuit design due to a large number (from hundreds to tens of thousands) of objective function evaluations involved [33, 34].

In this chapter, we discuss a computationally efficient procedure for multi-objective simulation-driven design of compact microwave passives. Our methodology exploits surrogate-based optimization, an equivalent circuit representation of the structure, and space mapping correction techniques to perform a point-by-point Pareto set identification. In contrast to many other techniques, the algorithm considered here directly explores the Pareto front rather than attempts to identify it among all possible designs in the search space. Our approach is demonstrated using a compact rat-race coupler design.

2 Case Study: Compact Rat-Race Coupler

In this section, we provide a description of a specific miniaturized microwave circuit to be used for explaining and demonstrating the proposed multi-objective design optimization methodology. We also describe the design objectives that will be of interest.

2.1 Compact Rat-Race Coupler

Rat-race couplers (RRCs) are fundamental microwave components used in wireless communication systems to split an input signal between output ports with a 0°/180° phase shift or to combine two input signals [36–38]. A typical RRC is a ring of one-and-a-half wavelength circumference, and—for this reasons—occupies an excessively large estate area, especially for the lower parts of the frequency spectrum [7, 15, 39]. Here, we attempt to design a small-size RRC, simply by modifying

Fig. 1 Parameterized layout
of a folded rat-race coupler

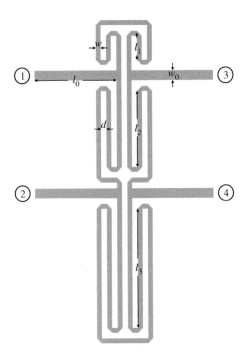

its topology. Thus, we consider a structure of an equal-split compact rat-race coupler (RRC) depicted in Fig. 1. Circuit size reduction is achieved by efficient routing of constitutive 70.7-ohm sections. The folded hybrid coupler has been designed on Taconic RF-35 dielectric substrate ($\varepsilon r = 3.5$, h $= 0.762$ mm, $\tan\delta = 0.0018$) and designated for 1-GHz operating frequency. System impedance is 50 Ω. Parameterized layout of Fig. 1 can be sufficiently described by vector of geometric dimensions, $x = [l_1\ l_2\ l_3\ d\ w]^T$, whereas $w_0 = 1.7$, $l_0 = 15$ remain fixed (all dimensions in mm). The low- and high-fidelity models of the circuit have been set up in Agilent ADS [40] and CST Microwave Studio Frequency Domain Solver [41] (~220,000 mesh cells; simulation time: ~15 min per design), respectively. Lower/upper bounds l/u of the solution space are given by the following vectors: $l = [2\ 10\ 17\ 0.2\ 0.5]^T$ and $u = [8\ 18\ 25\ 1.2\ 1.3]^T$. The center of the solution space has been chosen as the initial design $x = [5\ 14\ 21\ 0.7\ 0.9]^T$.

2.2 Design Objectives

In engineering practice, the design process of a compact RRC is a multifaceted task aimed at satisfying several, often conflicting objectives such as circuit layout area [7], operational bandwidth [15], insertion loss [42], coupling coefficient [3], phase response [38], etc. For the purpose of demonstrating the proposed multi-objective optimization scheme, we consider two objectives that are crucial from the practical standpoint: F_1—maximization of the bandwidth (defined as intersection of |S11| and |S41| below −20 dB) centered around the operating frequency of 1 GHz, and F_2—minimization of the RRC footprint (layout area). Typically, the chosen design criteria are non-commensurable, meaning that the improvement in terms of the circuit miniaturization rate remains in conflict with the enhancement of operational bandwidth [15]. The purpose of the proposed multi-objective design is to identify feasible trade-offs between the conflicting objectives. Information about these possible trade-offs is of utmost importance for the designer, especially when selecting a structure for a given application or testing the possibilities of an unconventional circuit.

3 Multi-Objective Design Methodology

In this section, we describe in detail a multi-objective design problem and methods for surrogate modeling by means of space mapping. We also discuss our approach for fast multi-objective optimization of compact microwave structures. The description of methodology is based on an exemplary rat-race coupler circuit of Sect. 2; however, they are applicable for any microwave structure with reduced footprint. The numerical results of rat-race coupler design and optimization are provided in Sect. 4.

3.1 Multi-Objective Design Problem Formulation

Consider $R_f(x)$ as a response of an accurate high-fidelity model of the microwave structure under consideration. $R_f(x)$ is normally obtained through EM simulations [3, 42, 46]. Vector x represents design parameters, i.e., dimensions of a circuit. Let $F_k(R_f(x))$, where $k = 1, ..., N_{obj}$, denote a kth design objective. The most common design objectives include reduction of a circuit footprint [7, 43] as well as requirements upon its electrical performance (e.g., reflection/isolation for a given operating frequency [3, 37], or frequency band [29, 44] below a defined level; desired in-band coupling [30, 42]; in-band transmission above defined level [9, 29], etc.). These requirements are critical for operation of a compact microwave structures. The necessity of handling several circuit specifications reveals multi-objective nature of a problem, where trade-offs between various (often conflicting) objectives must be sought [2, 34]. In a multi-objective optimization scheme, we seek for a Pareto-optimal set X_P, which is composed of non-dominated designs such that for any $x \in X_P$, there is no other design y for which the relation $y \prec x$ is satisfied [45]. In other words, $y \prec x$ (y dominates over x), if $F_k(R_f(y)) \leq F_k(R_f(x))$ for all $k = 1, ..., N_{obj}$, and $F_k(R_f(y)) < F_k(R_f(x))$ for at least one k) [45].

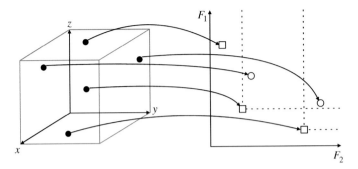

Fig. 2 Conceptual illustration of the mapping between the set of designs (●) within the solution space (here, with three independent design parameters) (*left*), and feature space with two conflicting objectives (*right*). The goal of the multi-objective optimization is to find a set of non-dominated solutions (□) that represents the Pareto-optimal set X_P. Solutions denoted as (○) are dominated

A set of design parameters considered in multi-objective optimization problem constitutes a solution space, whereas their responses reside in a so-called feature space [33, 45]. They are both bonded by a complex and unintuitive relations, which turns a multi-objective problem difficult to solve [33]. A conceptual illustration of the mappings between solution spaces for multi-objective optimization problem is illustrated in Fig. 2. For more detailed description of multi-objective optimization problem see, e.g. [2, 32, 33].

3.2 Low-Fidelity Models and Surrogate Modeling Using Space Mapping

The low-fidelity model R_c is a very fast, yet inaccurate counterpart of the high-fidelity one. Usually, R_c is an equivalent circuit [3, 9, 46] or a coarsely discretized EM realization of R_f; however, it can be constructed using interpolation models such as kriging [4, 47] or radial basis functions [48]. Discrepancies between R_c and R_f turns the former inappropriate for its direct utilization in optimization process of a compact microwave circuit. This problem is mitigated by utilization of surrogate modelling. A coarse model R_c is replaced during the optimization process by its corrected counterpart, i.e., surrogate model R_s. A specific type of model correction is determined based on the analysis of misalignment between the low- and high-fidelity model responses [4, 27, 46]. One should note that the equivalent circuit representations of R_s are particularly useful for the design of compact microwave circuits, especially due to their fast evaluation [9], capability of handling multiple design parameters [49] and variety of available correction techniques [46]. In case of miniaturized structures, implicit and frequency space mapping corrections [27, 46, 50] tend to be the most suitable for their surrogate-based modelling. The surrogate model may be formulated as

$$R_s(x) = R_{c.F}(x; f, p) \tag{1}$$

where $R_{c.F}$ is a frequency-scaled low-fidelity model, whereas f and p are frequency SM and implicit SM parameters, respectively.

We use the notation $R_c(x) = [R_c(x,\omega_1) \ R_c(x,\omega_2) \ ... \ R_c(x,\omega_m)]^T$, where $R_c(x,\omega_j)$ is a response of a circuit at frequency ω_j. Then, $R_{c.F}(x; f, p) = [R_c(x, f_0 + \omega_1 \cdot f_1, p) \ ... \ R_c(x, f_0 + \omega_m \cdot f_1, p)]^T$, with f_0 and f_1 being frequency scaling parameters [37]. Here, vector of implicit SM parameters p is composed of dielectric permittivity and thickness of the microstrip substrate of the circuit corresponding to selected groups of components as illustrated in Fig. 3. Discrepancy between R_s and R_f is minimized by solving the following parameter extraction problem

$$[f^*, p^*] = \arg \min_{f,p} ||R_f(x) - R_{c.F}(x; f, p)|| \tag{2}$$

Fig. 3 The equivalent circuit representation of a compact RRC of Fig. 1. The model is implemented in Agilent ADS circuit simulator [40]. Regions with different implicit SM parameters p are highlighted using different background shading

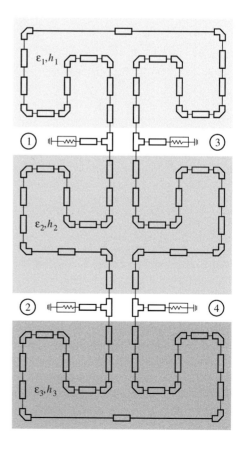

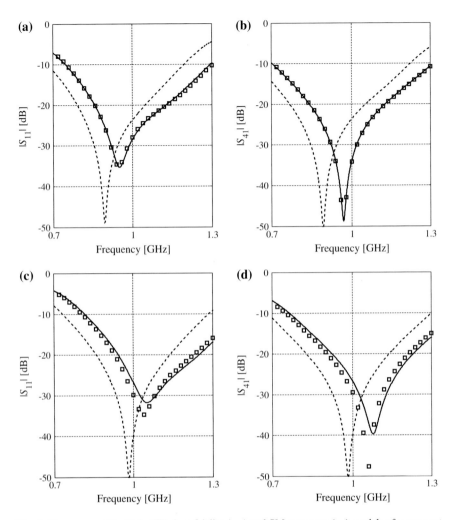

Fig. 4 Responses of the high- (□), low-fidelity (− −) and SM surrogate (—) models of a compact RRC: (a) $|S_{11}|$ at certain design $\boldsymbol{x}$ (at which the surrogate is extracted) and (b) $|S_{41}|$ at the same design as (a); (c) and (d) response of $|S_{11}|$ and $|S_{41}|$ at some design distinct from $\boldsymbol{x}$. Plots (c) and (d) indicates limited generalization capability of the surrogate model

The responses of the high- and low-fidelity models at certain design $\boldsymbol{x}$, as well as the response of the surrogate model $\boldsymbol{R}_s$ at the same design are shown in Fig. 4. The responses indicate that utilization of SM correction techniques considerably improves alignment between the models (see Fig. 4a, b). On the other hand, the generalization capability of a surrogate model $\boldsymbol{R}_s$ is limited (cf. Fig. 4c, d). Particularly, it is impossible to find a single set of SM parameters that would ensure accuracy of a surrogate model within entire solution space of interest. For that reason, iterative corrections of a surrogate model must be performed to provide satisfactory design solutions within predefined solution space frontiers.

3.3 Optimization Algorithm

Without a doubt, population-based metaheuristic algorithms [45] are the most common approaches for solving multi-objective optimization problems [1, 4, 33]. Particularly, evolutionary algorithms, e.g., genetic algorithms [34, 45] and particle swarm optimizers [35, 51] are of great interest for solving such problems, mostly due to their simple implementation and versatility [35, 45]. In contrary to conventional (e.g., gradient-based) algorithms, population-based metaheuristics can handle a set of solutions in a single iteration [2], which allows for identification of the entire Pareto set in a single algorithm run. On the other hand, evolutionary algorithms require thousands or even tens of thousands of objective function evaluations in order to identify a set of Pareto-optimal solutions [33, 45]. For that reason, their usefulness is limited to problems in which computational cost of objective evaluation is not of a major concern.

Unfortunately, high-fidelity EM models of compact microwave circuits are too expensive to be directly handled by population-based metaheuristics [34]. Instead, we utilize a faster low-fidelity model (cf. Fig. 2). The problem of limited generalization capability of a surrogate model (cf. Sect. 3.2), and large number of objective evaluations during multi-objective optimization is alleviated through point-by-point identification of the Pareto set. Initially, the solution space of interest is determined through optimization of the low-fidelity model of a miniaturized circuit with respect to single objective (in this case—electrical parameters). Subsequently, optimization toward second objective (circuits footprint miniaturization) is performed. The obtained solutions $F_1(\boldsymbol{R}_f(\boldsymbol{x}_p^{(1)}))$ and $F_2(\boldsymbol{R}_f(\boldsymbol{x}_p^{(2)}))$ define extreme designs of the Pareto-optimal set that are utilized for the determination of solution space frontiers. Next, we set the threshold values for the second objective $F_2^{(j)}$, and optimize the structure with respect to the first objective so that the above threshold value is preserved

$$x_p^{(j)} = \arg \min_{x,\ F_2(R_f(x)) \le F_2^{(j)}} F_1\left(R_f(x)\right) \tag{3}$$

Here, $\boldsymbol{x}_p^{(j)}$ is the jth element of the Pareto set. The optimization process is continued until $F_1(\boldsymbol{R}_f(\boldsymbol{x}_p^{(j)}))$ is still satisfactory from the point of view of given design specifications. Problem (3) is an iterative process solved using the SM surrogate model of Sect. 3.2

$$x_p^{(j,k)} = \arg \min_{x,\ F_2(R_s^{(j,k)}(x)) \le F_2^{(j)}} F_1\left(R_s^{(j,k)}(x)\right) \tag{4}$$

where

$$R_s^{(j,k)}(x) = R_{c.F}(x; f^{(j,k)}, p^{(j,k)}) \tag{5}$$

and

$$[f^{(j,k)}, p^{(j,k)}] = \arg\min_{f,p} ||R_f(x^{(j,k)}) - R_{c.F}(x^{(j,k)}; f, p)|| \tag{6}$$

The starting point for the algorithm (4) is $x_p^{(j-1)}$, which stands for the non-dominated design obtained during the previous algorithm iteration. Usually, only two iterations of (4) are sufficient to obtain $x_p^{(j)}$, which is because the starting point is already a good approximation of the optimum. In practice, the thresholds $F_2^{(j)}$ can be obtained as $F_2^{(j)} = \alpha \cdot F_2^{(j-1)}$ with $\alpha < 1$ (e.g., $\alpha = 0.95$), or $F_2^{(j)} = F_2^{(j-1)} - \beta$ with $\beta > 0$ (e.g., $\beta = 0.05 \cdot F_2^{(1)}$).

The computational cost of the proposed multi-objective optimization technique can be estimated (in terms of the number of EM simulations of the structure) as $N \cdot K$, where N is the number of designs in the Pareto-optimal set, and K is the average number of iterations (4) necessary to obtain the next design (in practice, $K \leq 3$). One should notice that the computational cost of algorithm grows linearly with density of Pareto-optimal solutions (i.e., a number of desired solutions). Moreover, the cost of algorithm execution is in the orders of magnitude lower in comparison to the conventional multi-objective optimization driven by population-based metaheuristics [2, 32, 34].

It should also be mentioned that another important design goal, i.e., 3 dB coupling of a structure ($|S_{21}| = |S_{31}|$) at the given operating frequency (here, 1 GHz), ensuring equal division of the signal power between ports 2 and 3 of the circuit, is enforced through an additional penalty function. More specifically, a penalty function to F_1 in (4) that restricts each design for which $||S_{21}| - |S_{31}|| > ds$ at the operating frequency is added (here, we use $ds = 0.1$ dB, as an acceptable inaccuracy level).

4 Numerical Results

The folded coupler of Sect. 2 has been designed based on the multi-objective optimization scheme introduced in Sect. 3. First, we used (4) without any area constraints to obtain a design solution with the widest operational bandwidth possible for this particular circuit topology. This resulted in bandwidth (defined as intersection of $|S_{11}|$ and $|S_{41}|$ below -20 dB) of 281 MHz and the layout area of 570 mm^2. Subsequently, by setting up $F_2^{(j)}$ to 540, 500, 475, 450, 425, 400, 375, 350, and 325 mm^2, respectively, nine other designs have been found. The acquired representation of the Pareto front is depicted in Fig. 4.

For $F_2^{(j)}$ set to 300 mm^2, the procedure failed to return a solution design with a positive value of the bandwidth. This value of the threshold can be considered as

Table 1 Multi-objective design optimization of the folded rat-race coupler: selected results

Design variables [mm]					Objectives	
l_1	l_2	l_3	d	W	Bandwidth [MHz]	Layout area [mm^2]
4.18	13.20	20.68	0.994	0.865	281	570
3.83	11.76	20.44	0.825	0.877	270	500
4.10	13.78	21.14	0.581	0.887	260	450
4.25	12.17	22.12	0.400	0.923	202	400
3.95	10.87	21.71	0.350	0.936	174	375
4.37	12.33	22.52	0.350	0.820	151	350

the lower limit (size-wise) for the usefulness of the designs obtained for the parameterization of Fig. 1. From the engineer's perspective, this is crucial information regarding limits of size reduction, which can be used, e.g., to discriminate between design solutions suitable for a given (in this example, space-limited) application.

Table 1 and Fig. 5 show the numerical data and frequency characteristics for the selected designs. The inspection of Fig. 4 indicates—when comparing the extreme points of the Pareto set—that the coupler size can be reduced by over 40 %, which is accompanied by over 45 % of bandwidth reduction. It should be reiterated that all designs from the Pareto set of Fig. 4 maintain acceptable performance.

Both Fig. 5 and Table 1 illustrate the conflicting nature of the objectives under consideration: minimization of the circuit layout area unavoidably leads to a degradation of its electrical performance (in this case, bandwidth).

It is noteworthy that all the design solutions along the Pareto front are nominally satisfying the fundamental design criteria (i.e., both $|S_{11}|$ and $|S_{41}|$ are below -20 dB and centered around the operating frequency of 1 GHz). Nonetheless, the designs with wider bandwidth (such as the one shown in Fig. 6a versus that in Fig. 6c) are electrically better because of higher chance of satisfying design specifications in case of unavoidable manufacturing tolerances (a consequence of which will be a

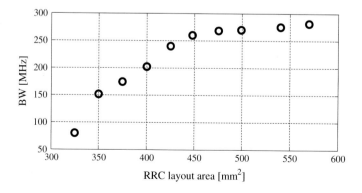

Fig. 5 Pareto set acquired by means of the proposed multi-objective design optimization scheme

Fig. 6 Transmission
characteristics of selected
RRCs, corresponding to the
layout area of 570 mm^2 (**a**),
448 mm^2 (**b**), and 375 mm^2
(**c**)

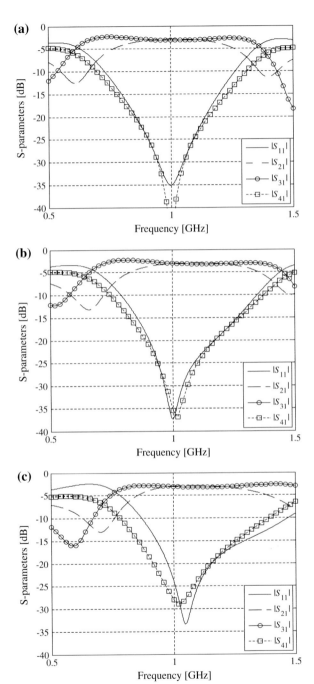

deviation of actual characteristics of the fabricated circuit with respect to the nominal ones).

The overall cost of the design procedure corresponds to less than 30 high-fidelity model evaluations (~7.5 h of CPU time), which includes the overhead associated with multiple evaluations of the circuit model Rc (the latter does not exceed 20 % of the overall EM simulation cost). It should be noted that direct multi-objective optimization of the high-fidelity EM RRC model Rf would not be possible within a reasonable time-frame (the expected cost of thousands or tens of thousands of model evaluations is practically prohibitive) (Fig. 6).

5 Conclusions

In this chapter, a cost-efficient approach to reliable multi-objective design optimization of compact microwave circuits has been presented. Numerical efficiency of the proposed technique has been achieved by using computationally cheap and suitably corrected equivalent circuit models representing the structure under consideration. In our approach, identification of the Pareto set is realized point-by-point through constrained single-objective optimizations. Consequently, the overall number of required evaluations of the high-fidelity EM model of the structure of interest has been dramatically reduced (to less than three EM simulations per one Pareto-optimal design). As shown in the provided example, a family of designs representing non-dominated solutions, i.e., the best possible trade-offs between conflicting objectives (here, operational bandwidth and RRC layout area) has been obtained at the cost corresponding to less than 30 EM evaluations of a high-fidelity model of the folded coupler. The overall cost of the proposed method is at least three orders of the magnitude smaller than the cost associated with the execution of population-based metaheuristic algorithms. According to our knowledge, this is the first successful attempt to solve the challenging problem of a reliable multi-objective design optimization of numerically demanding structure (here, a compact RRC).

Acknowledgments The authors thank Computer Simulation Technology AG, Darmstadt, Germany, for making CST Microwave Studio available. This work was supported in part by the Icelandic Centre for Research (RANNIS) Grant 13045051.

References

1. Yeung, S.H., Man, K.F.: Multiobjective optimization. IEEE Microwave Mag. **12**, 120–133 (2011)
2. Koziel, S., Bekasiewicz, A., Zieniutycz, W.: Expedite EM-Driven multi-objective antenna design in highly-dimensional parameter spaces. IEEE Antennas Wirel. Propag. Lett. **13**, 631–634 (2014)

3. Kurgan, P., Bekasiewicz, A.: A robust design of a numerically demanding compact rat-race coupler. Microw. Opt. Technol. Lett. **56**, 1259–1263 (2014)
4. Koziel, S., Bekasiewicz, A.: Fast multi-objective optimization of narrow-band antennas using RSA Models and design space reduction. IEEE Antennas Wirel. Propag. **14**, 450–453 (2014)
5. Tseng, C.-H., Chen, H.-J.: Compact rat-race coupler using shunt-stub-based artificial transmission lines. IEEE Microw. Wireless Comp. Lett. **18**, 734–736 (2008)
6. Ghali, H., Moselhy, T.A.: Miniaturized fractal rat-race, branch-line, and coupled-line hybrids. IEEE Trans. Microw. Theory Tech. **52**, 2513–2520 (2004)
7. Kurgan, P., Kitlinski, Z.: Doubly miniaturized rat-race hybrid coupler. Microw. Opt. Technol. Lett. **53**, 1242–1244 (2011)
8. Liao, S.-S., Sun, P.-T., Chin, N.-C., Peng, J.-T.: A novel compact-size branch-line coupler. IEEE Microw. Wireless Comp. Lett. **15**, 588–590 (2005)
9. Bekasiewicz, A., Kurgan, P., Kitlinski, Z.: A new approach to a fast and accurate design of microwave circuits with complex topologies. IET Microw. Antennas Propag. **6**, 1616–1622 (2012)
10. Koziel, S., Bekasiewicz, A., Kurgan, P.: Rapid EM-driven design of compact RF circuits by means of nested space mapping. IEEE Microw. Wireless Comp. Lett. **24**, 364–366 (2014)
11. Nocedal, J., Wright, S.: Numerical Optimization, 2nd edn. Springer, New York (2006)
12. Rios, L.M., Sahinidis, N.V.: Derivative-free optimization: a review of algorithms and comparison of software implementations. J. Global Opt. **56**, 1247–1293 (2013)
13. Kurgan, P., Bekasiewicz, A.: Atomistic surrogate-based optimization for simulation-driven design of computationally expensive microwave circuits with compact footprints. In: Koziel, S., Leifsson, L., Yang, X.S. (eds.) Solving Computationally Expensive Engineering Problems: Methods and Applications, pp. 195–218 (2014)
14. Kuwahara, Y.: Multiobjective optimization design of Yagi-Uda antenna. IEEE Trans. Antennas Propag. **53**, 1984–1992 (2005)
15. Bekasiewicz, A., Kurgan, P.: A compact microstrip rat-race coupler constituted by nonuniform transmission lines. Microw. Opt. Technol. Lett. **56**, 970–974 (2014)
16. Wincza, K., Gruszczynski, S.: Theoretical limits on miniaturization of directional couplers designed as a connection of tightly coupled and uncoupled lines. Microw. Opt. Technol. Lett. **55**, 223–230 (2013)
17. Kurgan, P., Filipcewicz, J., Kitlinski, Z.: Development of a compact microstrip resonant cell aimed at efficient microwave component size reduction. IET Microw. Antennas Propag. **6**, 1291–1298 (2012)
18. Kurgan, P., Filipcewicz, J., Kitlinski, Z.: Design considerations for compact microstrip resonant cells dedicated to efficient branch-line miniaturization. Microw. Opt. Technol. Lett. **54**, 1949–1954 (2012)
19. Koziel, S., Bandler, J.W., Madsen, K.: A space mapping framework for engineering optimization: theory and implementation. IEEE Trans. Microw. Theory Tech. **54**, 3721–3730 (2006)
20. Koziel, S., Bandler, J. W,. Madsen, K.: Towards a rigorous formulation of the space mapping technique for engineering design. Proc. Int. Symp. Circuits Syst. **1**, 5605–5608 (2005)
21. Koziel, S., Yang X.S. (eds.): Computational optimization, methods and algorithms. Series: Studies in Computational Intelligence, p. 356 (2011)
22. Cheng, Q.S., Rautio, J.C., Bandler, J.W., Koziel, S.: Progress in simulator-based tuning—the art of tuning space mapping. IEEE Microw. Mag. **11**, 96–110 (2010)
23. Bandler, J.W., Cheng, Q.S., Hailu, D.M., Nikolova, N.K.: A space-mapping design framework. IEEE Trans. Microw. Theory Tech. **52**, 2601–2610 (2004)
24. Bandler, J.W., Georgieva, N., Ismail, M.A., Rayas-Sanchez, J.E., Zhang, Q.-J.: A generalized space-mapping tableau approach to device modeling. IEEE Trans. Microw. Theory Tech. **49**, 67–79 (2001)
25. Koziel, S., Bandler, J.W.: Space mapping with multiple coarse models for optimization of microwave components. IEEE Microw. Wireless Comp. Lett. **18**, 1–3 (2008)

26. Koziel, S., Bandler, J.W., Madsen, K.: Quality assessment of coarse models and surrogates for space mapping optimization. Opt. Eng. **9**, 375–391 (2008)
27. Bandler, J.W., Cheng, Q.S., Dakroury, S.A., Mohamed, A.S., Bakr, M.H., Madsen, K., Søndergaard, J.: Space mapping: the state of the art. IEEE Trans. Microw. Theory Tech. **52**, 337–361 (2004)
28. Smierzchalski, M., Kurgan, P., Kitlinski, Z.: Improved selectivity compact band-stop filter with Gosper fractal-shaped defected ground structures. Microw. Opt. Technol. Lett. **52**, 227–232 (2010)
29. Kurgan, P., Bekasiewicz, A., Pietras, M., Kitlinski, M.: Novel topology of compact coplanar waveguide resonant cell low-pass filter. Microw. Opt. Technol. Lett. **54**, 732–735 (2012)
30. Kurgan, P., Kitlinski, Z.: Novel doubly perforated broadband microstrip branch-line couplers. Microw. Opt. Technol. Lett. **51**, 2149–2152 (2009)
31. Tsai, L.-T.: A compact dual-passband filter using stepped-impedance resonators. Microw. Opt. Technol. Lett. **55**, 2514–2517 (2013)
32. Koziel, S., Bekasiewicz, A., Couckuyt, I., Dhaene, T.: Efficient multi-objective simulation-driven antenna design using Co-Kriging. IEEE Trans. Ant. Prop. **62**, 5900–5905 (2014)
33. Bekasiewicz, A., Koziel, S., Zieniutycz, W.: Design space reduction for expedited multi-objective design optimization of antennas in highly-dimensional spaces. In: Koziel, S., Leifsson, L., Yang, X.S. (eds.) Solving Computationally Expensive Engineering Problems: Methods and Applications, pp. 113–147, Springer, Switzerland (2014)
34. Koziel, S., Ogurtsov, S.: Multi-Objective Design of Antennas Using Variable-Fidelity Simulations and Surrogate Models. IEEE Trans. Antennas Propag. **61**, 5931–5939 (2013)
35. Jin, N., Rahmat-Samii, Y.: Hybrid Real-Binary Particle Swarm Optimization (HPSO) in engineering electromagnetics. IEEE Trans. Antennas Propag. **58**, 3786–3794 (2010)
36. Opozda, S., Kurgan, P., Kitlinski, M.: A compact seven-section rat-race hybrid coupler incorporating PBG cells. Microw. Opt. Technol. Lett. **51**, 2910–2913 (2009)
37. Bekasiewicz, A., Koziel, S., Pankiewicz, B.: Accelerated simulation-driven design optimization of compact couplers by means of two-level space mapping. IET Microw. Antennas Propag. **9**, 618–626 (2014)
38. Koziel, S., Kurgan, P., Pankiewicz, B.: Cost-efficient design methodology for compact rat-race couplers. Int. J. RF Microwave Comput. Aided Eng. **25**(3):236–242 (2015)
39. Pozar, D.M.: Microwave Engineering, 4th edn. Wiley, Hoboken (2012)
40. Agilent A.D.S.: Agilent Technologies, 1400 Fountaingrove Parkway, Santa Rosa, CA 95403-1799, USA (2011)
41. CST Microwave Studio: Computer Simulation Technology AG, Bad Nauheimer Str. 19, D-64289 Darmstadt, Germany (2013)
42. Kurgan, P., Kitlinski, Z.: Slow-wave fractal-shaped compact microstrip resonant cell. Microw. Opt. Technol. Lett. **52**, 2613–2615 (2010)
43. Kurgan, P., Bekasiewicz, A., Kitlinski, M.: On the low-cost design of abbreviated multi-section planar matching transformer. Microw. Opt. Technol. Lett. **57**(3):521–525 (2015)
44. Bekasiewicz, A., Koziel, S.: Efficient multi-fidelity design optimization of microwave filters using adjoint sensitivity. Int. J. RF Microw. Comput. Aided Eng. **25**(2):178–183 (2015)
45. Deb., K.: Multi-Objective Optimization Using Evolutionary Algorithms. Wiley, New York (2001)
46. Koziel, S., Cheng, Q.S., Bandler, J.W.: Space mapping. IEEE Microw. Mag. **9**, 105–122 (2008)
47. Queipo, N.V., Haftka, R.T., Shyy, W., Goel, T., Vaidynathan, R., Tucker, P.K.: Surrogate-based analysis and optimization. Prog. Aerosp. Sci. **41**, 1–28 (2005)
48. El Zooghby, A.H., Christodoulou, C.G., Georgiopoulos, M.: A neural network-based smart antenna for multiple source tracking. IEEE Trans. Antennas Propag. **48**, 768–776 (2000)
49. Koziel, S., Bekasiewicz, A., Kurgan, P.: Nested space mapping technique for design and optimization of complex microwave structures with enhanced functionality. In: Koziel, S.,

Leifsson, L., Yang, X.S. (eds.) Solving Computationally Expensive Engineering Problems: Methods and Applications, pp. 53–86. Springer, Switzerland (2014)

50. Bandler, J.W., Cheng, Q.S., Nikolova, N.K., Ismail, M.A.: Implicit space mapping optimization exploiting preassigned parameters. IEEE Trans. Microw. Theory Tech. **52**, 378–385 (2004)

51. Chamaani, S., Mirtaheri, S.A., Abrishamian, M.S.: Improvement of time and frequency domain performance of antipodal vivaldi antenna using multi-objective particle swarm optimization. IEEE Trans. Antennas Propag. **59**, 1738–1742 (2011)

Simulation-Based Optimization in Design-Under-Uncertainty Problems Through Iterative Development of Metamodels in Augmented Design/Random Variable Space

Alexandros A. Taflanidis and Juan Camilo Medina

Abstract This contribution discusses design-under-uncertainty problems that employ a probabilistic performance as objective function and consider its estimation through stochastic (i.e., Monte Carlo) simulation. This simulation-based approach for estimating statistical quantities puts no constraints on the complexity of the numerical and probability models adopted to describe the system under consideration, but involves a high-computational cost, especially for applications utilizing complex, high-fidelity models. A framework relying on Kriging surrogates for approximation of the system response in an augmented input space is considered here to alleviate this cost. A subdomain of the design space is defined and a Kriging metamodel is built to approximate the system response (output) with respect to (augmented input) both the design variables and the random variables (i.e., uncertain model parameters). The established metamodel is then used within a stochastic simulation setting to approximate the system performance when estimating the objective function for specific values of the design variables as well as for obtaining derivative information. This information is used to search for a local optimum within the considered design subdomain. Only when the optimization algorithm drives the search outside this domain a new metamodel is generated. The process is iterated until convergence is achieved and an efficient sharing of information across these iterations is established to adaptively tune the characteristics of the metamodel as well as improve the accuracy of the Monte Carlo simulation through implementation of importance sampling. The explicit incorporation of the prediction error of the

A.A. Taflanidis (✉) · J.C. Medina
Department of Aerospace and Mechanical Engineering,
University of Notre Dame, Notre Dame, IN, USA
e-mail: a.taflanidis@nd.edu

J.C. Medina
e-mail: jmedina4@alumni.nd.edu

A.A. Taflanidis
Department of Civil and Environmental Engineering and Earth Sciences,
University of Notre Dame, Notre Dame, IN, USA

© Springer International Publishing Switzerland 2015
M.S. Obaidat et al. (eds.), *Simulation and Modeling Methodologies,*
Technologies and Applications, Advances in Intelligent Systems
and Computing 402, DOI 10.1007/978-3-319-26470-7_13

251

metamodel within the design problem formulation is also considered and is proven to greatly enhance the robustness of the optimization approach.

Keywords Optimization under uncertainty · Stochastic simulation · Kriging · Augmented metamodel input space · Target region design of experiments

1 Introduction

In any engineering design application, the performance predictions for the system under consideration involve some level of uncertainty, stemming from the incomplete knowledge about the system itself and its environment (representing excitation conditions) [2, 20]. Explicitly accounting for these uncertainties is exceptionally important for providing optimal configurations that exhibit robust performance, and a probability logic approach provides a consistent framework for this task [7]. This is established by assigning probability distributions to the characteristics of the model that are considered as uncertain. These distributions allow us to rationally incorporate our available, incomplete knowledge about the properties of the system into its modeling [1, 7]. In this setting, the objective function for a robust-to-uncertainties design may be expressed by the expected value (i.e., a probabilistic integral) of some chosen performance measure over the adopted probability distributions. For complex systems, this probabilistic integral can rarely be calculated or accurately approximated analytically, leaving stochastic simulation (i.e., Monte Carlo) techniques as the only practical alternative [18, 21, 22]. A challenge related to this approach is, though, the significant computational cost involved to estimate the objective function, especially for applications with complex numerical models.

A framework relying on surrogate modeling to approximate the system performance is discussed here to alleviate this burden. Kriging [15, 19] is utilized as surrogate model (metamodel) since it has been proven highly efficient for approximating complex response functions, while simultaneously providing gradient information. Metamodeling approaches for optimization under uncertainty problems are most commonly implemented either by (i) approximating the objective function in the design space [4] (metamodel is adopted to guide optimization with objective function directly evaluated through stochastic simulation utilizing the actual system model) or by (ii) approximating the system performance in the random variable space for specific design configurations [5] (metamodel is used to approximate the model response and subsequently estimate the objective function for each specific design configuration). Here a different implementation is investigated by considering an augmented input space composed of both the design variables and the random variables, an approach similar to the ideas discussed in [3] and [6]. The first of these studies focused only on reliability-based design optimization problems, whereas the second considered implementation within the entire design domain. By contrast, generalized optimization-under-uncertainty problems are examined here, whereas an iterative implementation is considered. The underlying assumption is that the system behavior within different regions in the design domain is radically different, something that can impose great challenges when trying to create a single augmented metamodel in this entire domain. This challenge stems from the fact that the regions of probabilistic importance for the random variables are radically different for the different design configurations examined. This leads to our preference for an iterative approach within sub-domains, and the focus is placed here on the adaptive sharing of information across the proposed iterations to improve the efficiency for the surrogate model development, as well as for the Monte Carlo simulation. The efficient incorporation of gradient information, obtained through the metamodel, is also considered to guide the optimization.

Within the subdomain of the design space defined in each iteration, high-fidelity model evaluations are obtained at properly selected support points (consisting of simultaneous variations with respect to both the design variables and the uncertain model parameters) and a Kriging metamodel is built to approximate the system response. This metamodel is then used within a stochastic simulation setting to approximate the system performance when estimating the objective function and its gradient for specific values of the design variables. This information is exploited to search for a local optimum within the current subdomain. When the optimization algorithm drives the search outside the sub-domain, a new metamodel is generated (over a new sub-domain), and the process is iterated until convergence is obtained. Additionally, an adaptive tuning of some characteristics of the Kriging metamodel is examined by sharing information across the iterations of the numerical optimization. For selecting the basis functions of the metamodel, a recently developed probabilistic global sensitivity analysis [9] is seamlessly integrated, quantifying the importance of each input variable toward the overall probabilistic performance. Higher order basis functions are assigned only to those with higher probabilistic sensitivity. Furthermore, an adaptive, sample-based approach is developed for selecting the support points, populating more densely those regions in the random variable space that have higher contribution to the integrand quantifying the probabilistic performance. The overall framework is demonstrated with an example considering the optimization of semi-active dampers for a half-car suspension model. In this example, the importance of explicitly including the Kriging prediction error in the evaluation of the performance function is investigated and shown to greatly enhance the robustness of the optimization approach.

2 Problem Formulation

Consider a system with design vector $\mathbf{x} = \begin{bmatrix} x_1 & x_2 & \dots & x_{n_x} \end{bmatrix} \in X \subset \mathbb{R}^{n_x}$, where X is the admissible design space, and uncertain model parameters (random variables) $\boldsymbol{\theta} = \begin{bmatrix} \theta_1 & \theta_2 & \dots & \theta_{n_\theta} \end{bmatrix} \in \Theta \subset \mathbb{R}^{n_\theta}$, where Θ denotes the set of their possible values. A probability density function (PDF) $p(\boldsymbol{\theta})$, which quantifies our available knowledge about the system is assigned to these parameters [22]. Let $\mathbf{z}(\mathbf{x}, \boldsymbol{\theta}) \subset \mathbb{R}^{n_z}$ be the response vector of the system model, assumed here to be obtained through a time-consuming call to a deterministic simulator, and let $h(\mathbf{x}, \boldsymbol{\theta}): \mathbb{R}^{n_x} \times \mathbb{R}^{n_\theta} \to \mathbb{R}$ be the performance function characterizing the favorability of that response, assumed here to represent a certain cost function (lower values for $h(\mathbf{x}, \boldsymbol{\theta})$ correspond to better performance). The probabilistic performance, $H(\mathbf{x})$, corresponding to the objective function for a robust-to-uncertainties design is given by

$$H(\mathbf{x}) = E_p[h(\mathbf{x}, \boldsymbol{\theta})] = \int_{\Theta} h(\mathbf{x}, \boldsymbol{\theta}) p(\boldsymbol{\theta}) d\boldsymbol{\theta}, \tag{1}$$

where $E_p[]$ denotes expectation under probability model $p(\boldsymbol{\theta})$. This leads to the following design problem

$$\mathbf{x}^* = \arg\min_{\mathbf{x} \in X} H(\mathbf{x}), \tag{2}$$

where any deterministic constraints can be incorporated in the definition of the admissible design space X.

As discussed in the introduction, estimation of the objective function through stochastic (i.e., Monte Carlo) simulation is examined, in particular though implementation of importance sampling (IS) [13, 16]. This is established by introducing a proposal density $q(\boldsymbol{\theta})$ (also referenced as IS density) to concentrate the computational effort within the stochastic simulation in regions of the random variables that have higher contribution to the integrand defining the probabilistic performance. The simplest choice is to set $q(\boldsymbol{\theta})$ equal to $p(\boldsymbol{\theta})$, representing direct Monte Carlo simulation (no IS implementation). An estimate of the objective function in Eq. (1) is then established by generating N samples from $q(\boldsymbol{\theta})$, with the sample set denoted as $\{\boldsymbol{\theta}^j\} = \{\boldsymbol{\theta}^j : j = 1, \ldots, N\}$, and approximating this function by the sample average

$$\hat{H}(\mathbf{x}|\{\boldsymbol{\theta}^j\}) = 1/N \sum_{j=1}^{N} h(\mathbf{x}, \boldsymbol{\theta}^j) p(\boldsymbol{\theta}^j)/q(\boldsymbol{\theta}^j), \tag{3}$$

where the dependence of the estimate on the sample set $\{\boldsymbol{\theta}^j\}$ is explicitly noted on the left-hand side of this equation. The accuracy of this approximation is quantified by the coefficient of variation of the estimator, which given by [16]

$$\delta = \frac{1}{\sqrt{N}} \sqrt{\frac{E_q\left[(h(\mathbf{x}, \boldsymbol{\theta}) p(\boldsymbol{\theta})/q(\boldsymbol{\theta}))^2\right]}{H^2} - 1}. \tag{4}$$

The accuracy improves inversely proportional to the computational effort $\sqrt{N}$, and it is also influenced by the exact selection of $q(\boldsymbol{\theta})$, impacting the expected value in the numerator of Eq. (4). The optimal IS density, maximizing this accuracy, is proportional to the absolute value of the integrand [13, 16] leading to PDF definition

$$\pi(\boldsymbol{\theta}|\mathbf{x}) = \frac{|h(\mathbf{x}, \boldsymbol{\theta})|p(\boldsymbol{\theta})}{H(\mathbf{x})} \propto |h(\mathbf{x}, \boldsymbol{\theta})|p(\boldsymbol{\theta}), \tag{5}$$

where $\propto$ denotes proportionality and the denominator in the first equation is simply a normalization constant. This density, representing the regions in Θ of bigger contributions toward the probabilistic performance, will play a key role within the adaptive Kriging formulation, beyond its aforementioned utility as a good IS density.

The optimization problem in Eq. (2) is ultimately solved by substituting the approximation of Eq. (3) as the objective function. This leads to a challenging simulation-based optimization problem because of (c.i) the existence of an unavoidable estimation error (i.e., noisy measurements for the objective function), (c.ii) the high computational cost associated with each objective function evaluation (requiring N evaluations of the system model response), and (c.iii) the challenges in obtaining gradient information for problem with complex black-box models. Various formulations have been proposed in the literature to directly address these challenges [13, 20–22]. In this contribution, an alternative indirect approach utilizing Kriging metamodeling for approximating the model response in an augmented input space is examined.

3 Design Optimization Through Kriging in Augmented Space

The augmented input space for development of the Kriging metamodel is defined as a tensor product between the design and uncertain spaces $X \otimes \Theta$. To improve accuracy of the metamodel a smaller set of the design domain is considered instead of the entire domain, leading to an iterative implementation. Let X_k denote the subdomain considered within X in the kth iteration (also known as trust region) and let G_{krig} represent the functional relationship for the Kriging-based optimization. The proposed iterative approach may be described by

$$\mathbf{x}_{k+1} = G_{krig}(\mathbf{x}_k \mid \{\boldsymbol{\theta}^j\}_k), \tag{6}$$

where the notation $\{\boldsymbol{\theta}^j\}_k$ is used to denote the sample set (for stochastic simulation) used within the kth iteration. Note that this sample set will ultimately change from iteration to iteration so that there is no dependence of the solution obtained on the sample set used, an approach corresponding to the concept of interior sampling [21].

Thus, the augmented input vector $\mathbf{y}$ for the Kriging metamodel is composed of the design and uncertain model parameter vectors $\mathbf{y} = [\mathbf{x} \ \boldsymbol{\theta}]$, whereas the output vector is chosen to correspond to the system response vector $\mathbf{z}(\mathbf{x},\boldsymbol{\theta})$. Note that the computational complexity of the performance evaluation model for estimating $h(\mathbf{x},\boldsymbol{\theta})$ based on $\mathbf{z}(\mathbf{x},\boldsymbol{\theta})$ is typically small for most practical engineering applications. Establishing an approximation for $\mathbf{z}(\mathbf{x},\boldsymbol{\theta})$, and then using the actual performance evaluation model to estimate $h(\mathbf{x},\boldsymbol{\theta})$ circumvents one level of approximations and can ultimately offer significant improvements in accuracy [10]. This approach further allows, as will be demonstrated in the illustrative example, the explicit consideration of the Kriging prediction error within the design problem formulation.

The Kriging model will provide an approximation for the response vector $\bar{\mathbf{z}}(\mathbf{y}) = \bar{\mathbf{z}}(\mathbf{x},\boldsymbol{\theta})$, resulting in an approximation to the performance function. The latter approximation is denoted by $\bar{h}(\mathbf{y}) = \bar{h}(\mathbf{x},\boldsymbol{\theta})$ and leads to the following expression for the Kriging-based objective function

$$H_{krig}(\mathbf{x}) = E_p[\bar{h}(\mathbf{x},\boldsymbol{\theta})] = \int_{\Theta} \bar{h}(\mathbf{x},\boldsymbol{\theta})p(\boldsymbol{\theta})d\boldsymbol{\theta}, \tag{7}$$

As will be demonstrated in the next section, gradient information for $\bar{\mathbf{z}}(\mathbf{x},\boldsymbol{\theta})$, and therefore $\bar{h}(\mathbf{x},\boldsymbol{\theta})$, can be easily obtained, leading to the following expression for the gradient of the objective function in Eq. (7)

$$\nabla H_{krig}(\mathbf{x}) = \nabla\left(\int_{\Theta} \bar{h}(\mathbf{x},\boldsymbol{\theta})p(\boldsymbol{\theta})d\boldsymbol{\theta}\right) = \int_{\Theta} \nabla(\bar{h}(\mathbf{x},\boldsymbol{\theta})p(\boldsymbol{\theta})d\boldsymbol{\theta}) = \int_{\Theta} p(\boldsymbol{\theta})\nabla\bar{h}(\mathbf{x},\boldsymbol{\theta})d\boldsymbol{\theta}, \tag{8}$$

under the additional assumption that the functions $\bar{h}(\mathbf{x},\boldsymbol{\theta})p(\boldsymbol{\theta})$ and $(\partial\bar{h}(\mathbf{x},\boldsymbol{\theta})/\partial x_i)p(\boldsymbol{\theta})$ are continuous in the domain $X \times \Theta$ and bounded, a necessary conditions in order for the differentiation and the expectation operators to commute [21].

These integrals can be evaluated through stochastic simulation, leading to

$$\hat{H}_{krig}(\mathbf{x}|\{\boldsymbol{\theta}^j\}_k) = 1/N \sum_{j=1}^{N} \bar{h}(\mathbf{x},\boldsymbol{\theta}^j)p(\boldsymbol{\theta}^j)/q_k(\boldsymbol{\theta}^j) \text{ and} \tag{9}$$

$$\nabla\hat{H}_{krig}(\mathbf{x}|\{\boldsymbol{\theta}^j\}_k) = 1/N \sum_{j=1}^{N} \nabla\bar{h}(\mathbf{x},\boldsymbol{\theta}^j)p(\boldsymbol{\theta}^j)/q_k(\boldsymbol{\theta}^j), \tag{10}$$

where sample set $\{\boldsymbol{\theta}^j\}_k$ is obtained from distribution $q_k(\boldsymbol{\theta})$, and an indexing notation for $q_k(\boldsymbol{\theta})$ is used to represent the ability to choose a different proposal density at each iteration of the numerical optimization.

Utilizing this information, especially the gradient approximation in Eq. (10), an appropriate gradient-based algorithm can be adopted to establish a local search within X_k [this local search defines G_{krig} in Eq. (6)]. The same metamodel is used within this entire local search, which is what contributes to the overall computational efficiency of the approach. Exploiting the numerical efficiency of the Kriging metamodel, the estimation of Eq. (10) [or Eq. (9)] for all examined design configurations $\mathbf{x}$ within X_k can be achieved at a significantly small computational cost even for larger N values, whereas the ability to obtain gradient information for the Kriging-based approximation of the objective function circumvents problems in obtaining derivative information for complex design problems. These characteristics directly address the challenges (c.ii) and (c.iii), discussed at the end of Sect. 3, that are associated with design optimization problems relying on stochastic simulation. Challenge (c.i) is partially addressed by the fact that derivative information is now readily available for the objective function, reducing the potential impact of noisy measurements. This impact can be further reduced by using the same sample set of common random numbers $\{\boldsymbol{\theta}^j\}_k$ within the entire kth iteration [21], an approach that corresponds to the concept of exterior sampling and that ultimately creates a consistent estimation error in the comparisons. This leads to selection of exterior sampling in the Kriging-based search within X_k (same random numbers) and interior sampling across the iterations of the proposed optimization formulation described by Eq. (6) (different sample set $\{\boldsymbol{\theta}^j\}_k$ in each iteration).

Considering now the local search within X_k, two are the possible outcomes: (i) converge to a local optimum within X_k or (ii) reach the boundary of the search domain, which means that the local search should stop to avoid extrapolations. The latter prompts the optimization algorithm to advance to the next iteration $\mathbf{x}_{k+1}$, and generate a new Kriging model if the overall optimization has not converged. This leads to the iterative Kriging implementation that is discussed in the next section, providing additional opportunity to adaptively select certain characteristics by sharing information across these iterations.

4 Iterative Kriging with Adaptive Characteristics

4.1 Review of Kriging Formulation

To form the Kriging metamodel, a database with n observations is utilized that provides information for the $\mathbf{y}$-$\mathbf{z}$ pair. For this purpose, n samples for $\{\mathbf{y}_l \; l = 1,\ldots, n\}$, also known as support points, are created and the model response $\mathbf{z}(\mathbf{y}_l)$ is evaluated for each of them. Using this dataset the metamodel is formulated providing the Kriging predictor which has a Gaussian nature with mean $\bar{z}_i(\mathbf{y})$ and standard deviation $\sigma_i(\mathbf{y})$. The response output can be approximated through this predictor as

$$z_i(\mathbf{y}) = \bar{z}_i(\mathbf{y}) + \bar{\varepsilon}_i(\mathbf{y}), \tag{11}$$

where $\bar{\varepsilon}_i$ is a Gaussian variable with zero mean and standard deviation $\sigma_i(\mathbf{y})$ [12]. The fundamental building blocks of Kriging are the n_p dimensional basis vector, $\mathbf{f}$ $(\mathbf{y})$, and the correlation function $R(\mathbf{y}^j, \mathbf{y}^k)$ defined through parameter vector

s. Selection of the former will be discussed later, whereas for the latter the popular generalized exponential correlation is used here

$$R(\mathbf{y}^j, \mathbf{y}^k) = \prod_{i=1}^{n_y} \exp[-s_i |y_i^j - y_i^k|^{s_{n_y}+1}]; \mathbf{s} = [s_1 \cdots s_{n_y+1}]. \tag{12}$$

Then for the set of n observations with input matrix $\mathbf{Y} = [\mathbf{y}^1 \ \dots \ \mathbf{y}^n]^T$ and corresponding output matrix $\mathbf{Z} = [\mathbf{z}^1 \ \dots \ \mathbf{z}^n]^T$, we define the basis matrix $\mathbf{F} = [\mathbf{f}(\mathbf{y}^1) \ \dots \ \mathbf{f}(\mathbf{y}^n)]^T$ and the correlation matrix $\mathbf{R}$ with the jk element defined as $R(\mathbf{y}^j, \mathbf{y}^k)$, j, $k = 1, \dots, n$. Also for every new input $\mathbf{y}$, we define the correlation vector $\mathbf{r}(\mathbf{y}) = [R(\mathbf{y}, \mathbf{y}^1) \ \dots \ R(\mathbf{y}, \mathbf{y}^n)]^T$ between the input and each of the elements of $\mathbf{Y}$. The Kriging mean prediction for vector $\mathbf{z}$ is given by [12]

$$\bar{\mathbf{z}}^T(\mathbf{y}) = \mathbf{f}(\mathbf{y})^T \boldsymbol{\alpha}^* + \mathbf{r}(\mathbf{y})^T \boldsymbol{\beta}^*$$
$$\text{where } \boldsymbol{\alpha}^* = (\mathbf{F}^T \mathbf{R}^{-1} \mathbf{F})^{-1} \mathbf{F}^T \mathbf{R}^{-1} \mathbf{Z}; \boldsymbol{\beta}^* = \mathbf{R}^{-1}(\mathbf{Z} - \mathbf{F}\boldsymbol{\alpha}^*). \tag{13}$$

Through the proper tuning of the parameters $\mathbf{s}$ of the correlation function, Kriging can efficiently approximate very complex functions. The optimal selection of $\mathbf{s}$ is based on the maximum likelihood estimation principle, where the likelihood is defined as the probability of the n observations, and maximizing this likelihood with respect to $\mathbf{s}$ ultimately corresponds to the optimization problem [12]

$$\mathbf{s}^* = \arg\min_{\mathbf{s}} \left[|\mathbf{R}|^{\frac{1}{n}} \sum_{i=1}^{n_z} \tilde{\sigma}_i^2 \right], \tag{14}$$

where |.| stands for determinant of a matrix and the process variance for each output $\tilde{\sigma}_i^2$, $i = 1,.., n_z$ is given by the diagonal elements of matrix $(\mathbf{Z} - \mathbf{F}\boldsymbol{\alpha}^*)^T \mathbf{R}^{-1}(\mathbf{Z} - \mathbf{F}\boldsymbol{\alpha}^*)$.

Beyond the mean Kriging predictions the error can be also explicitly considered in the optimization as will be illustrated in the example considered later. This requires estimation of the prediction error variance $\sigma_i^2(\mathbf{y})$ for z_i and input $\mathbf{y}$ which is given by

$$\sigma_i^2(\mathbf{y}) = \tilde{\sigma}_i^2[1 + \mathbf{u}(\mathbf{y})^T (\mathbf{F}^T \mathbf{R}^{-1} \mathbf{F})^{-1} \mathbf{u}(\mathbf{y}) - \mathbf{r}(\mathbf{y})^T \mathbf{R}^{-1} \mathbf{r}(\mathbf{y})] \text{ where } \mathbf{u}(\mathbf{y}) = \mathbf{F}^T \mathbf{R}^{-1} \mathbf{r}(\mathbf{y}) - \mathbf{f}(\mathbf{y}). \tag{15}$$

An additional benefit of the Kriging metamodel is that it facilitates a straight-forward estimation of gradient information for any functions that are dependent upon the system response. This is based on estimation of derivative information for $\mathbf{z}(\mathbf{y})$ and $\sigma_i^2(\mathbf{y})$ which are given, respectively, by

$$\nabla \bar{\mathbf{z}}(\mathbf{y})^T = \mathbf{J}_f(\mathbf{y})^T \boldsymbol{\alpha}^* + \mathbf{J}_r(\mathbf{y})^T \boldsymbol{\beta}^* \text{ and} \tag{16}$$

$$\nabla \sigma_i^2(\mathbf{y}) = 2\tilde{\sigma}_i^2[\mathbf{u}^T (\mathbf{F}^T \mathbf{R}^{-1} \mathbf{F})^{-1} (\mathbf{F}^T \mathbf{R}^{-1} \mathbf{J}_r(\mathbf{y}) - \mathbf{J}_f(\mathbf{y})) - \mathbf{r}(\mathbf{y})^T \mathbf{R}^{-1} \mathbf{J}_r(\mathbf{y})], \tag{17}$$

where $\mathbf{J}_f$ and $\mathbf{J}_r$ are the Jacobian matrices (with respect to $\mathbf{y}$) of $\mathbf{f}$ and $\mathbf{r}$, respectively.

4.2 Adaptive Kriging Implementation Within Iterative Approach

In the proposed framework, a Kriging approximation is developed in the augmented input space by sharing information across the iterations of the optimization algorithm described through Eq. (6). Within such a setting, the focus is on the adaptive design of experiments (DoE) to select the support points as well as the adaptive selection of the polynomial order of basis functions. To formalize these concepts, let $\mathbf{x}_k$ denote the design variable vector that has been identified at the end of the $k-1th$ iteration of the numerical optimization of Eq. (6). Evaluation of the approximation to the system performance will be also available for $\mathbf{x}_k$, $\{\bar{h}(\mathbf{x}_k, \boldsymbol{\theta}^j); j = 1, \ldots, N\}$ for the sample set $\{\boldsymbol{\theta}^j\}_{k-1}$ used to estimate the gradient through Eq. (10). A localized box-bounded design subdomain (trust region) is then defined X_k; this domain is centered on $\mathbf{x}_k$ and has an appropriate length for each design variable (defining the length vector $\mathbf{L}_k$) that ultimately prescribes the upper and lower bounds for the design vector $\mathbf{x}_k^l$ and $\mathbf{x}_k^u$, respectively. Any appropriate technique may be adopted for selecting the length vector $\mathbf{L}_k$. A relevant recommendation for this is that the length is gradually reduced as iterations progress to regions closer to a minimum, where one needs higher accuracy approximations [17]. A Kriging metamodel is then constructed within subdomain X_k for the augmented input vector $\mathbf{y}$ and then used in the local search described by the functional relationship G_{krig} in Eq. (6).

4.2.1 Design of Experiments

Space filling techniques or adaptive design of experiments [14, 25] are commonly preferred for the design of experiments in metamodeling applications. However, the former may not provide the necessary accuracy in targeted regions of importance, while the latter may significantly increase the computational cost for selecting the support points. Instead, an alternative DoE is advocated here. To establish this DoE we start by acknowledging that the two different components of the input vector $\mathbf{y}$ have different characteristics/demands related to their accuracy. In the case of $\mathbf{x}$, accurate approximations are needed within the entire domain X_k since the metamodel is ultimately used to compare different design choices within this domain to converge toward the optimal design configuration. This indicates that a space filling technique should be considered, for example, the popular Latin hypercube sampling (LHS) [25]. On the other hand, for $\boldsymbol{\theta}$ an accurate approximation is needed over the domain in the uncertain model parameters space Θ that provides higher contribution toward the integrand representing the objective function. Thus for $\boldsymbol{\theta}$ a target region DoE is needed, as advocated in [14]. An approach with small computational overhead is developed here for this purpose, exploiting the iterative optimization implementation.

The basis of the approach is the definition of the target region for $\boldsymbol{\theta}$ as the important region for the integrand of the objective function given by the density $\pi(\boldsymbol{\theta}|\mathbf{x})$ in Eq. (5) with the additional simplification of replacing the computationally complex $h(\mathbf{x},\boldsymbol{\theta})$ with the Kriging-based prediction $\bar{h}(\mathbf{x},\boldsymbol{\theta})$. This density, denoted $\bar{\pi}(\boldsymbol{\theta}|\mathbf{x})$ herein, may be efficiently approximated through samples for it that can be obtained utilizing the readily available evaluations of $\{\bar{h}(\mathbf{x}_k,\boldsymbol{\theta}^j); j=1,\ldots,N\}$. Recall that these evaluations correspond to sample set $\{\boldsymbol{\theta}^j\}_{k-1}$ following the proposal density $q_{k-1}(\boldsymbol{\theta})$. The sample set for $\bar{\pi}(\boldsymbol{\theta}|\mathbf{x})$, denoted $\{\boldsymbol{\theta}^a\}_k$, can be obtained [13] through implementation of rejection sampling by accepting the samples (within $\{\boldsymbol{\theta}^j\}_{k-1}$) for which

$$\frac{|\bar{h}(\mathbf{x}_k,\boldsymbol{\theta}^j)|p(\boldsymbol{\theta}^j)}{u^j q_k(\boldsymbol{\theta}^j)} > \max_j \left[\frac{|\bar{h}(\mathbf{x}_k,\boldsymbol{\theta}^j)|p(\boldsymbol{\theta}^j)}{q_k(\boldsymbol{\theta}^j)}\right] \tag{18}$$

holds, where $\{u^j; j=1,\ldots,N\}$ are independent uniformly distributed random samples in range [0 1]. The sample set $\{\boldsymbol{\theta}^a\}_k$ [samples within $\{\boldsymbol{\theta}^j\}_{k-1}$ satisfying Eq. (18)] represents the region in the Θ space that contributes more toward the probabilistic performance for $\mathbf{x}_k$ and, as such, provides a good approximation for the target region where higher accuracy is sought after for the Kriging metamodel (region of importance for probabilistic performance). A further approximation is established here by assuming that information for the behavior within X_k can be adequately represented through information for $\mathbf{x}_k$. If the system behavior within X_k is not radically different this is a reasonable assumption. For approximating the target region utilizing these samples any sample-based density approximation approach can be utilized [16]. This density will be denoted $f_k^s(\boldsymbol{\theta})$ herein. To guarantee that a sufficient number of samples are available for set $\{\boldsymbol{\theta}^a\}_k$ to inform the selection of $f_k^s(\boldsymbol{\theta})$ a further modification is introduced. Upon convergence to $\mathbf{x}_k$, an additional sample set, beyond the N samples in $\{\boldsymbol{\theta}^j\}_{k-1}$, is generated to obtain a large sample set consisting of N_p samples for which $\bar{h}(\mathbf{x}_k,\boldsymbol{\theta})$ is evaluated. The rejection sampling in Eq. (18) is then performed over this larger sample set. Given that evaluation of $\bar{h}(\mathbf{x}_k,\boldsymbol{\theta})$ involves a small computational effort this modification creates only a small additional burden.

Finally, it is important to consider that the Kriging metamodel needs to have sufficient accuracy even in regions beyond this specific target region, since erroneous approximations in such regions can still impact the estimation result. This consideration leads to the following two-stage DoE with the first stage aiming to obtain satisfactory global accuracy in the broader domain Θ and the second stage aiming to obtain higher accuracy in the target region. Initially (first stage) n_1^s samples are obtained adopting a space filling approach (LHS) within the domain of importance based on $p(\boldsymbol{\theta})$ (for example, 4–5 standard deviations away from the median values for each model parameters). Then, additional n_2^s are obtained from the density approximation $f_k^s(\boldsymbol{\theta})$. The total number of support points is, thus, $n = n_1^s + n_2^s$.

4.2.2 Selection of Basis Functions

Another important feature for the Kriging approximation is the selection of basis functions, i.e., the definition of $\mathbf{f}(\mathbf{y})$. Typically, polynomials of some lower order are used and then an important question is the exact polynomial order of the basis functions for each component of the input vector $\mathbf{y}$ [5]. Selecting the same higher order for all components might reduce the accuracy of the Kriging metamodel; ultimately components that exhibit higher sensitivity should have higher order associated with them, but the identification of the most appropriate basis function is in general a challenging task [5, 8]. This challenge is circumvented here by integrating a global sensitivity analysis [9], and selecting second-order polynomial functions only for the most important components and linear polynomial functions for the rest.

For vector $\mathbf{y}$ this sensitivity analysis is established by considering a density function similar to $\pi(\boldsymbol{\theta}|\mathbf{x})$ but defined in this case within the augmented input space

$$\pi(\mathbf{y}) = \pi(\mathbf{x}, \boldsymbol{\theta}) \propto |h(\mathbf{x}, \boldsymbol{\theta})| p(\boldsymbol{\theta}) p(\mathbf{x}), \tag{19}$$

where $p(\mathbf{x})$ corresponds to a uniform density in X_k. Comparison of this density to the prior joint distribution $p(\boldsymbol{\theta})p(\mathbf{x})$ for each component of $\mathbf{y}$ separately (comparison of the marginal distributions) reveals the importance toward the probabilistic performance with bigger differences corresponding to higher importance [9]. The comparison is efficiently performed utilizing samples for $\pi(\mathbf{y})$; such samples can be readily obtained utilizing the support points within X_k from the second DoE stage. Using rejection sampling again, these samples correspond to those [out of the larger set of samples from distribution $f_k^s(\boldsymbol{\theta})$] for which the following relationship holds:

$$\frac{|h(\mathbf{x}^j, \boldsymbol{\theta}^j)| p(\boldsymbol{\theta}^j)}{u^j f_k^s(\boldsymbol{\theta}^j)} > \max_{j=1,\ldots,n_2^s} \left[\frac{|h(\mathbf{x}^j, \boldsymbol{\theta}^j)| p(\boldsymbol{\theta}^j)}{f_k^s(\boldsymbol{\theta}^j)} \right]. \tag{20}$$

This approach leads to total of N_s samples, denoted $\{y_i^s\}$ for each component of y, and to the following approximation for the marginal distributions of interest, established utilizing kernel density approximation [9].

$$\tilde{\pi}(y_i) = \frac{1}{N_s} \sum_{s=1}^{N_s} \left(\frac{1}{t_i} \left[K\left(\frac{y_i - y_i^s}{t_i} \right) \right] \right); \quad K(y) = \frac{1}{\sqrt{2\pi}} e^{-\frac{y^2}{2}}, \tag{21}$$

where K is the Gaussian kernel and bandwidth t_i is given by $1.06 \cdot N_s^{-1/5} \sigma_i$ with σ_i corresponding to the standard deviation of the samples $\{y_i^s\}$. The importance of the different model parameters is quantified based on the relative entropy between the marginal distributions of interest, given by [9]

$$D(\pi(y_i)||p(y_i)) \approx \int_{b_{li}}^{b_{ui}} \tilde{\pi}(y_i) \log\left(\frac{\tilde{\pi}(y_i)}{p(y_i)}\right) dy_i, \qquad (22)$$

where b_{ui} and b_{li} are the upper and lower bounds, respectively, for the sample set $\{y_i^s\}$ and the integral in this equation can be readily obtained through one-dimensional numerical integration.

A threshold D_{min}^{re} can be set to determine the importance of the input vector components. Only if the value of relative entropy is larger than this threshold, then that particular parameter will be assigned a higher order basis function. This threshold is adaptively selected to correspond to a fraction of the highest relative entropy value. If the allowable percentage reduction of the maximum entropy among the entire input vector is s_e^{re}, then

$$D_{min}^{re} = s_e^{re} \max_i [D(\pi(y_i)||p(y_i))], \qquad (23)$$

and this formulation ultimately leads to consideration of higher order basis functions for parameters that correspond to relative entropy values higher or equal to s_e^{re} of the maximum entropy over the entire input vector.

4.3 Additional Considerations for Implementation Across Iterations

For the proposed implementation of Kriging across the iterations of the numerical optimization the following questions need to still be answered: (a) How can the IS density $q_k(\theta)$ for the estimations in Eqs. (9) and (10) be established? (b) How is convergence evaluated? (c) What are the recommendations for the selection of the length vector $\mathbf{L}_k$ defining the trust region?

Starting with (a) the IS selection, this density may be chosen based on the information from the already available sample set $\{\theta^a\}_k$ from distribution $\pi(\theta|\mathbf{x}_k)$ which as discussed earlier corresponds to the optimal IS density for design configuration $\mathbf{x}_k$. The density $\pi(\theta|\mathbf{x}_k)$ is expected to provide a satisfactory accuracy for the entire domain X_k if, as discussed previously, $\mathbf{x}_k$ provides an adequate representation about the behavior of the integrand for different design configurations within X_k. Exploiting the efficiency of the Kriging metamodel, a large number of samples N can be used in this case for the stochastic-simulation-based evaluation of the objective function and its gradient, described in Eqs. (9) and (10), respectively. As such, no special attention needs to be placed on a highly efficient IS formulation; improvement in accuracy is primarily sought after by adopting a larger number of N, though considerable advantages are also expected from the IS implementation.

For example, a simple parametric density approximation can be implemented, although more advanced approaches have been also recently proposed [13].

Moving now to (b) the convergence of the algorithm, this is established when the new identified optimum $\mathbf{x}_k^*$ is a local optimum of the trust region X_k. To further improve the quality of the obtained solution, a second optimization stage is proposed: upon convergence, the number of support points is increased to establish a higher accuracy Kriging metamodel and the optimization described by Eq. (6) is repeated. This allows the use of smaller number of support points $n = n_1^s + n_2^s$ in the initial iterations, until convergence is established. Ultimately, we are not concerned with obtaining high-accuracy estimates for the Kriging metamodel at the initial iterations; establishing an approximate search direction in the design domain toward the optimal design is sufficient (greedy optimization approach).

Finally, with respect to (c) the length vector selection $\mathbf{L}_k$, initially it can be considered as a specific fraction s_1^l of the design domain X, i.e., $\mathbf{L}_1 = s_1^l X$. At each iteration a specific reduction, s^r, of this proportionality can be implemented leading to selection $s_k^l = (s^r)^{k-1} s_1^l$ and $\mathbf{L}_k = s_k^l X$. Upon initial convergence, a further reduction by s_f^r can be established to localize the search around the candidate optimum. Figure 1 provides an example of how the algorithm progresses through the design space. The squares are the trust regions X_k for each iteration. The gray dots show the intermediate steps needed to find a local optimum within the trust region (only using evaluations of the Kriging model). The dash-dot line shows the second stage of the optimization that starts when the first stage has encountered an interior point local optimum. This stage has a reduced length and the number of support points within the domain is increased in order to improve the accuracy of the Kriging model near the optimum point.

Fig. 1 Evolution of trust regions within the Kriging-based optimization

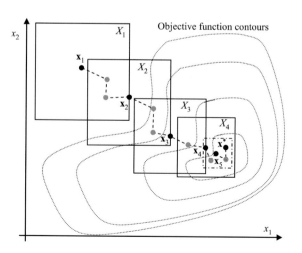

4.4   Algorithm for Adaptive, Iterative Kriging

When combining the previous ideas, one can formulate the following optimization algorithm utilizing adaptive Kriging. First, define the bounded design space X, the starting point of the algorithm $\mathbf{x}_1$, the number of support points for the two-stage DoE approach, n_1^s and n_2^s, respectively, as well as the respective numbers when the second optimization stage (convergence) has been reached, n_1^f and n_2^f. Select the number of samples N for the estimation of the objective function and its gradient utilizing stochastic simulation, the number of samples N_p for which $\bar{h}(\mathbf{x}_k, \boldsymbol{\theta})$ will be obtained, and the allowable percentage entropy reduction for the basis function formulation $s_e^{re} < 1$. Finally, choose the fraction parameter s_1^l defining the initial trust region as well as its reduction s^r per iteration and the final reduction upon convergence s_f^r.

At iteration k of the optimization algorithm [Eq. (6)] perform the following steps:

- Step 1 (trust region definition): Define box-bounded search domain X_k centered around $\mathbf{x}_k$ with length vector given by $\mathbf{L}_k = (s^r)^{k-1} s_1^l X$. If convergence has been established (last iteration) further reduce length vector by s_f^r.

- Step 2 (support points): Employing the two-stage DoE for $\boldsymbol{\theta}$, obtain n_1^s (n_1^f if convergence has been established) samples using a space filling approach (LHS) in the region of importance for $p(\boldsymbol{\theta})$, then obtain n_2^s (n_2^f if convergence has been established) samples from density $f_k^s(\boldsymbol{\theta})$ [$p(\boldsymbol{\theta})$ in first iteration]. For $\mathbf{x}$ obtain $n = n_1^s + n_2^s$ ($n = n_1^f + n_2^f$ if convergence has been established) samples using a space filling approach (LHS) in X_k. Reuse (as part of the n required support points) the support points from the previous iterations for which response has been already calculated. These correspond simply to support points whose $\mathbf{x}$ components lies within the current search space X_k.

- Step 3 (Evaluation of model response). For all the support points evaluate the model response $\{\mathbf{z}(\mathbf{x}^j, \boldsymbol{\theta}^j); j = 1, \ldots, n\}$ and ultimately the system performance function $\{h(\mathbf{x}^j, \boldsymbol{\theta}^j); j = 1, \ldots, n\}$.

- Step 4 (Selection of basis functions): Based on the evaluations on the performance function on the support points from the second stage $\{h(\mathbf{x}^j, \boldsymbol{\theta}^j); j = 1, \ldots, n_2^s\}$, obtain samples from $\pi(\mathbf{x}, \boldsymbol{\theta})$ through rejection sampling as in Eq. (20). Then calculate the entropy for each component of the output vector $D(\pi(y_i) \| p(y_i))$ using the approximation in Eq. (22). Consider higher order (quadratic) basis functions only for components of the input vector with relative entropy higher than the value given by Eq. (23) and lower order (linear) basis functions for the rest.

- Step 5 (Kriging model): employing the information in Steps 1–4, build the Kriging model in augmented input space through the approach discussed in Sect. 4.1.

- Step 6 (trust region local optimum): Simulate set of N samples from distribution $q_k(\boldsymbol{\theta})$ [$p(\boldsymbol{\theta})$ in first iteration] and perform local search within X_k utilizing the gradient information from Eq. (10) to identify local optimum $\mathbf{x}_k^*$.
- Step 7 (information for $\mathbf{x}_{k+1}$ and proposal density formulation for DoE); consider $\mathbf{x}_{k+1} = \mathbf{x}_k^*$ and evaluate the response and the performance function through the Kriging approximation for N_p samples. Obtain sample set $\{\boldsymbol{\theta}^a\}_{k+1}$ through Eq. (18) and formulate based on them $f_{k+1}^s(\boldsymbol{\theta})$.
- Step 8 (IS proposal density for iteration $k+1$): Utilizing the same sample set $\{\boldsymbol{\theta}^a\}_{k+1}$ formulate the IS proposal density $q_{k+1}(\boldsymbol{\theta})$.
- Step 9 (convergence check); if $\mathbf{x}_{k+1}$ is on the boundary of X_k, then convergence has not been achieved. Advance to iteration $k+1$ and start again from Step 1. If $\mathbf{x}_{k+1}$ is on the boundary of X_k, then convergence has been potentially attained and the second optimization stage needs to be implemented by repeating only Steps 1–6 with $n = n_1^f + n_2^f$ for support points and s_f^r for reduction of search space.

5 Illustrative Example

The framework is illustrated next with an example considering the optimization of semi-active, skyhook dampers for the suspension of a half-car nonlinear model driving on a rough road, modeled as a stochastic process [24]. The ride comfort and damper fatigue are considered as performance objectives, both characterized through their root mean square (RMS) statistics with these statistics calculated through time-domain simulation to address the various sources of nonlinearities included in the model.

5.1 Numerical and Probability Models

The schematic of the half-car model is shown in Fig. 2. The model is developed by using small angle assumption with primary state variables the vertical displacements of the chassis' center of mass, y_c, the front, y_{tf}, and rear, y_{tr}, tire displacements and the angular displacement (pitch) of the chassis, ψ_c. The vertical displacement of the front, y_{cf}, and rear suspensions, y_{cr}, are directly calculated based on y_c and ψ_c. The spring and dashpot forces are obtained based on these state variables as shown in Fig. 2, whereas for the suspension damper an idealized skyhook implementation is considered (also shown in that figure). All coefficients for the skyhook implementation are taken as design variables and separately selected for the front and rear dampers, leading to definition of design variable vector as $\mathbf{x} = \begin{bmatrix} C_f^t & C_f^s & C_r^t & C_r^s \end{bmatrix}^T$.

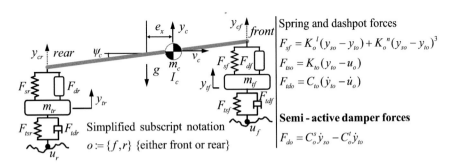

Spring and dashpot forces

$$F_{sf} = K_o'(y_{so} - y_{to}) + K_o''(y_{so} - y_{to})^3$$
$$F_{tso} = K_{to}(y_{to} - u_o)$$
$$F_{tdo} = C_{to}(\dot{y}_{to} - \dot{u}_o)$$

Semi - active damper forces

$$F_{do} = C_o^s \dot{y}_{so} - C_o^t \dot{y}_{to}$$

Fig. 2 Half-car model schematic and forces involved

The road surface input is modeled as a zero-mean Gaussian stationary stochastic process with the power spectral density proposed in [24], parameterized by the roughness coefficient, κ_i. A time-domain realization for this input is obtained [13] by the spectral representation method assuming that the car drives with a constant horizontal velocity v_c along a road with total length L_r.

Ultimately, the system of equations describing the half-car dynamical model are

$$m_c \ddot{y}_c + F_{sf} + F_{df} + F_{sr} + F_{dr} = m_c g$$
$$I_c \dot{\theta}_c + (F_{sf} + F_{df})(1 - e_x)L - (F_{sr} + F_{dr})(1 + e_x)L = 0 \qquad (24)$$
$$m_{to} \ddot{y}_{to} - F_{so} - F_{do} + F_{tso} + F_{tdo} = m_{tr} g; \; o = \{f, r\},$$

where g denotes gravity acceleration, e_x is the eccentricity between the geometric center of the chassis and its center of mass, L is the half-distance between the two suspensions, m_c, m_{tf}, m_{tr} are masses for the chassis, the front, and rear tires respectively, and I_c is the moment of inertia of the chassis. A numerical model for this dynamical system is developed [13] in SIMULINK [11] and finally the statistics (RMS_c) for any response c are obtained through the time-domain simulation results by

$$RMS_c = \sqrt{1/(L_r/v_c) \int_0^{L_r/v_c} c^2(t)dt}. \qquad (25)$$

The computational burden for one simulation, that is one evaluation of the RMS response, is on the average 3 s on a 3.GHz Xeon CPU (the numerical model is developed so that a compromise between accuracy and efficiency is established), meaning that an evaluation of the objective function within a stochastic simulation setting with $N = 600$ samples takes about half an hour. All model parameters apart from $L = 4$ m are considered as uncertain, leading to $n_\theta = 15$. Table 1 reviews the probability models.

Table 1 Probability models adopted for the different model parameters; μ corresponds to median, cv to coefficient of variation, and ρ to correlation coefficient

m_c	Lognormal $\mu = 580$ kg, $cv = 0.2$	m_{to}	Lognormal $\mu = 40$ kg, $cv = 0.2$	e_x	Uniform in [0.1 0.4]
I_c	Lognormal $\mu = 1180$ kg m^2, $cv = 0.2$	v_c	Lognormal $\mu = 60$ km/h, $cv = 0.2$	K_{to}	Lognormal $\mu = 190$ kN/m, $cv = 0.2$
C_{to}	Lognormal $\mu = 20$ N s/m, $cv = 0.2$	κ_i	Lognormal $\mu = $64e-6 m^2/cycle $cv = 0.1$	K_o^d K_o^n	Corr. Lognormal $cv = 0.2$, $\rho = 0.4$ $\mu = 23.5$ kN/m, for K_o^d $\mu = 435$ kN/m^3, for K_o^n

5.2 Performance Quantification and Adjustment for Kriging Error

The performance measure $h(\mathbf{x},\boldsymbol{\theta})$ is selected as the normalized linear combination of the fragilities related to the root mean square of the vertical acceleration at the center of mass RMS_{ac}, which in turn is a measure of passenger comfort, and of the root mean square of the suspension's damping forces at the rear and front of the car RMS_{df}, RMS_{dr}, respectively, which is a measure of suspension fatigue

$$h(\mathbf{x},\boldsymbol{\theta}) = 1/3 \sum_{i=ac,df,dr} \Phi[(\ln RMS_i - \ln b_i)/\sigma_{bi}], \tag{26}$$

where $\Phi[.]$ corresponds to the standard Gaussian Cumulative Distribution Function (CDF) and b_i is the threshold related to each response quantity of interest, taken here as 1 m/s^2 for the acceleration, and 160 N for the damper forces, and σ_{bi} is the coefficient of variation for the fragilities, assumed as 5 % for all of them. The introduction of the fragilities through the CDF can be also viewed as addressing unmodeled uncertainties [23]; rather than having a binary distinction of the performance, i.e., perform acceptably when the response is smaller than threshold b_i and unacceptably when not. In this context each quantity within the sum in Eq. (26) corresponds equivalently to probability $P[RMS_i > b_i\varepsilon_i]$ with ε_i having a lognormal distribution with median equal to one and logarithmic standard deviation σ_{bi}. Analytical integration of the influence of ε_i leads ultimately to the CDF fragility expressions in Eq. (26). The objective function $H(\mathbf{x})$ is the average failure probability over the three different RMS response quantities and is constrained within the [0 1] range.

The Kriging approximation is formulated directly for the log values of the RMS response, since these are the ones appearing in the performance function, thus $\mathbf{z}(\mathbf{x},\boldsymbol{\theta}) = [\ln(RMS_{ac})\ \ln(RMS_{df})\ \ln(RMS_{dr})]^T$. Furthermore, the prediction error stemming from the Kriging metamodeling is directly incorporated into the performance function definition, exploiting the equivalent representation discussed above. This is established by considering the following transformation of the probability $P[RMS_i > b_i \varepsilon_i]$

$$P[RMS_i > b_i \varepsilon_i] = P[\ln(\varepsilon_i) \leq \ln(RMS_i) - \ln(b_i)] = P[\ln(\varepsilon_i) \leq \bar{z}_i + \bar{\varepsilon}_i - \ln(b_i)]$$

$$= P[\ln(\varepsilon_i) - \bar{\varepsilon}_i \leq \bar{z}_i - \ln(b_i)] = \Phi\left[(\bar{z}_i - \ln(b_i))/\sqrt{\sigma_i^2 + \sigma_{bi}^2}\right], \quad (27)$$

where $\bar{z}_i$ corresponds to the Kriging approximation for $\ln(RMS_i)$ and the last equality is based on the fact that since $\ln(\varepsilon_i)$ and $\bar{\varepsilon}_i$ are zero mean independent Gaussian variables with variances, σ_i^2 and σ_{bi}^2, respectively, their sum (or difference in this case) is also a Gaussian variable with zero mean and variance $\sigma_i^2 + \sigma_{bi}^2$. This leads to the following approximation to the performance function:

$$\bar{h}(\mathbf{x}, \boldsymbol{\theta}) = 1/3 \sum_{i=ac,df,dr} \Phi\left[(\bar{z}_i - \ln(b_i))/\sqrt{\sigma_i^2 + \sigma_{bi}^2}\right]. \quad (28)$$

The gradient of this expression, needed in the optimization, is

$$\nabla\bar{h}(\mathbf{x}, \boldsymbol{\theta}) = 1/3 \sum_{i=ac,df,dr} \phi\left[\frac{\bar{z}_i - \ln(b_i)}{\sqrt{\sigma_i^2 + \sigma_{bi}^2}}\right]\left(\frac{\nabla\bar{z}_i}{\sqrt{\sigma_i^2 + \sigma_{bi}^2}} - \frac{(\bar{z}_i - \ln(b_i))}{2(\sigma_i^2 + \sigma_{bi}^2)^{3/2}}\nabla\sigma_i^2\right) \quad (29)$$

where $\phi[.]$ corresponds to the Gaussian PDF, and evaluation of all required gradients in the last equation was discussed in Sect. 4.1.

5.3   Numerical Details for Optimization

The design domain X has upper bound $[400 \quad 4000 \quad 400 \quad 4000]$ N s/m, and lower bound $[0 \quad 0 \quad 0 \quad 0]$ N s/m. For the trust region definition, the length of the initial region $\mathbf{L}_1$ is initially selected as 20 % of the design domain X, i.e., $s_1^l = 0.2$ with a reduction in size of 5 % with every iteration, i.e., $s^r = 0.95$. When the optimization has reached the last stage, the reduction in the trust region is set to 50 % ($s_f^r = 0.5$). For the local search within X_k a trust-region-reflective algorithm is adopted utilizing exterior sampling whereas, as discussed earlier, the overall implementation is formulated as interior sampling. The number of support points for the two stage DoE approach is selected as $n_1^s = 200$ and $n_2^s = 700$, whereas the number of support points for the second optimization stage, for increasing the accuracy of the Kriging approximation, is taken to be double these values. The number of samples for the estimates in Eqs. (9) and (10) for the local search is taken as $N = 2000$. The total number of simulations for $\bar{h}(\mathbf{x}_k, \boldsymbol{\theta})$ to inform the selection of IS densities and the sampling density for the second stage of the DoE approach is set to $N_p = 10000$. For the basis functions selection, the percentage reduction (defining the cut-off entropy with respect to the maximum entropy over the entire input vector) is set to $s_e^{re} = 0.4$.

Apart from the fully adaptive Kriging implementation that additionally incorporates the Kriging error in the objective function formulation, two additional cases are examined, leading to a total of three different optimization approaches. The first, denoted AK (adaptive Kriging), and second, denoted AKE (adaptive Kriging with error), correspond to the proposed algorithm that adaptively employs the probabilistic sensitivity analysis to select the order of the basis function for the Kriging model, and also adopts the proposed DoE. The only difference is that the first approach does not include the Kriging error in the objective function while the second one does. The former is established simply by taking $\nabla \sigma_i^2 = \sigma_i^2 = 0$ in Eqs. (28) and (29). The third approach, denoted LK (Latin Hypercube Kriging), employs the traditional sampling technique of LHC for both $\mathbf{x}$ and $\boldsymbol{\theta}$ (it still incorporates the Kriging prediction error).

To judge the quality of the obtained solutions, the optimization problem was additionally solved using the simultaneous perturbation stochastic approximation algorithm [21] coupled with the highly efficient adaptive IS formulation presented recently in [13]. The optimal benchmark solution was found to be $\mathbf{x}^* = [175.2\ 1645.3\ 190.1\ 1495.4]$ N s/m with respective performance $H(\mathbf{x}^*) = 0.066\,\%$, and the total number of model evaluations needed to converge to this solution was close to 265,000. This large computational effort should be attributed to the fact that the performance close to the optimum corresponds to a rare event (small failure probability) requiring a large number of samples for accurate approximation.

5.4 Results and Discussion

Results are reported in Table 2 for four different trials corresponding to different initial conditions $\mathbf{x}_1$ for the algorithm. In particular, the following quantities are reported: the optimal solution $\mathbf{x}^*$, the total number of simulations of the system high-fidelity model till convergence is established, N^{tot}, the objective function value obtained through the use of the Kriging metamodel, $\hat{H}_{krig}(\mathbf{x}^*)$, which is provided directly by the optimization algorithm, as well as the objective function value obtained through the use of the actual system model, $\hat{H}(\mathbf{x}^*|\{\boldsymbol{\theta}^c\})$. For the latter the same sample set $\{\boldsymbol{\theta}^c\}$ (common random numbers) is used across all comparisons for each design problem to enable a consistent comparison. $N = 10000$ samples are used in this comparison which facilitates a small coefficient of variation for $\hat{H}(\mathbf{x}^*|\{\boldsymbol{\theta}^c\})$, close to 4 %. A threefold comparison can be established based on these results: (i) Comparison between $\hat{H}_{krig}(\mathbf{x}^*)$ and $\hat{H}(\mathbf{x}^*|\{\boldsymbol{\theta}^c\})$ shows the accuracy of the Kriging implementation, (ii) comparison of the N^{tot} for different approaches shows the computational efficiency for convergence of the algorithm, (iii) comparison between $\hat{H}(\mathbf{x}^*|\{\boldsymbol{\theta}^c\})$ and the benchmark optimal solution of 0.0066 % shows

Table 2 Optimization results for different cases considered and different trials

Case	Trial	$\mathbf{x}^*$ (N s/m)	N^{tot}	$\hat{H}_{krig}(\mathbf{x}^*)(\%)$	$\hat{H}(\mathbf{x}^*|\{\boldsymbol{\theta}^c\})$ (%)
AK (adapt. kriging)	1	[122.5 2121.1 142.8 1884.5]	8817	0.025	0.079
	2	[161.3 1484.9 145.7 1725.3]	6607	0.023	0.091
	3	[149.4 1780.3 168.6 1568.5]	12,046	0.020	0.073
	4	[159.1 1541.0 160.2 1532.8]	8698	0.022	0.089
AKE (adapt. kriging with error)	1	[136.9 2091.9 154.5 1814.3]	5721	0.054	0.066
	2	[159.4 1819.0 189.8 1521.3]	4751	0.053	0.063
	3	[158.6 1886.3 153.6 1762.9]	5512	0.059	0.063
	4	[171.3 1847.5 170.1 1708.0]	4703	0.059	0.063
LK (Latin hyperc. kriging)	1	[163.2 1630.6 152.3 1963.1]	19,850	0.22	0.086
	2	[193.7 1786.5 156.0 2189.1]	21,757	0.165	0.139
	3	[386.4 936.1 373.3 779.5]	3821	2.601	2.121
	4	[136.3 1937.0 306.3 859.2]	19,591	0.119	0.191

the robustness of the approach in converging to an optimal (as opposed to a sub-optimal) solution.

The results for AKE demonstrate a remarkable *computational efficiency* and *robustness*. The identified solution $\mathbf{x}^*$ is always in the vicinity of the benchmark optimal solution and, more importantly, the attained performance is always comparable or even better than the benchmark performance. This is accomplished with a small number of model evaluations, not exceeding 6000. Note that the differences between these trials are well expected since there is a strong dependence on the initial conditions. Overall the reported efficiency corresponds to tremendous computational savings (265,000 simulations of the expensive numerical model needed before), something that is accomplished primarily through the augmented metamodel approach.

Moving now to the accuracy of the Kriging implementation (still for AKE), assessed through the comparison between $\hat{H}_{krig}(\mathbf{x}^*)$ and $\hat{H}(\mathbf{x}^*|\{\boldsymbol{\theta}^c\})$, there is an overall good agreement. When compared against the accuracy of the Kriging implementation when the prediction error is not included in the performance function estimate [compare the values of $\hat{H}_{krig}(\mathbf{x}^*)$ and $\hat{H}(\mathbf{x}^*|\{\boldsymbol{\theta}^c\})$ for AK] it is evident that the explicit consideration of that error provides significantly improved estimates, i.e., closer values of $\hat{H}_{krig}(\mathbf{x}^*)$ and $\hat{H}(\mathbf{x}^*|\{\boldsymbol{\theta}^c\})$.

The more interesting comparison is, however, between AKE and the alternative approaches (AK or LK) in terms of computational efficiency [comparison of N^{tot} for same trial] and more importantly robustness [comparison of $\hat{H}(\mathbf{x}^*|\{\boldsymbol{\theta}^c\})$ for same trial]. In all instances it is shown that the other two approaches do not share the robustness of the proposed AKE implementation, as they converge for some trials to a significantly suboptimal performance. The differences are perhaps more evident for LK and secondary for AK. This is an important result; it shows that a space-filling DoE, even though it might provide a good global accuracy, leads to

significant errors in regions of the model parameters that are of importance for the
probabilistic performance and ultimately to erroneous identified optimal designs.
Similarly ignoring the prediction error, not only decreases the accuracy of the
estimated performance as argued in the previous paragraph, but, and perhaps more
importantly, can provide erroneous optimal solutions. Even though calculation of
this error does involve a higher computational burden compared to using only the
mean Kriging approximation [8], it is evident that its explicit consideration provides
significant enhancements that counteract this burden.

A final remark is warranted for the benefits of the iterative Kriging implementation. The half-car model exhibits drastically different behavior for different design
configurations within X, with objective function values ranging from 30 % to
0.063 %. This means that the ranges within Θ contributing to the integrand for the
objective function significantly change leading to great differences in the distri-
bution $\pi(\theta|x)$ for different x. The benefits from getting support points for θ based on
the targeted region that $\pi(\theta|x)$ represents have been already illustrated above
(comparison between AKE and LK cases). Therefore, development of a single
Kriging metamodel within the entire domain X would encounter significant challenges since the target region would be fundamentally different for different design
configurations, reducing the overall accuracy of the metamodel. Implementation of
an iterative formulation even simply with larger subdomain definition (s_1^l of close to
0.35) posed significant challenges and led to frequent convergence to suboptimal
solutions. This validates the claim that for such type of applications the iterative
Kriging formulation may offer significant advantages, especially when coupled with
an appropriate trust region definition.

6   Conclusions

An adaptive, iterative implementation of Kriging metamodeling was considered to
reduce the computational burden associated with optimization-under-uncertainty
problems adopting a simulation-driven (stochastic simulation) approach for evaluation of the objective function. The formulation of the Kriging metamodel was
established in the augmented model parameter and design variable space, whereas
the local prediction error associated with the Kriging approximation was explicitly
considered in the objective function estimation. Two important aspects for tuning of
the Kriging metamodel were adaptively addressed within this implementation by
seamlessly sharing information across the iterations of the numerical optimization:
(i) design of experiments (DoE) for selecting support points aimed at improving the
accuracy over targeted regions, the ones contributing most toward the probabilistic
performance, and (ii) selection of the order of basis functions for the different inputs
of the metamodel. The illustrative example showed the computational efficiency

(convergence with small number of evaluations of the high-fidelity system model) as well as robustness (convergence to solutions that are close to the true optimum) established through the proposed iterative, adaptive Kriging implementation in the augmented input space. The proposed DoE for a targeted region was shown to greatly enhance the accuracy of the Kriging approximation and its ability to avoid convergence to suboptimal solutions. Finally, the explicit incorporation of the prediction error in the objective function estimation improved not only the accuracy of the objective function estimates but additionally supported a more robust optimization.

References

1. Beck, J.L., Taflanidis, A.: Prior and posterior robust stochastic predictions for dynamical systems using probability logic. J Uncertainty Quantification **3**(4), 271–288 (2013)
2. Beyer, H.-G., Sendhoff, B.: Robust optimization—a comprehensive survey. Comput. Method Appl. Mech. **196**, 3190–3218 (2007)
3. Dubourg, V., Sudret, B., Bourinet, J.-M.: Reliability-based design optimization using kriging surrogates and subset simulation. Struct. Multidiscip. Optim. **44**(5), 673–690 (2011)
4. Gasser, M., Schueller, G.I.: Reliability-based optimization of structural systems. Math. Method Oper. Res. **46**, 287–307 (1997)
5. Gavin, H.P., Yau, S.C.: High-order limit state functions in the response surface method for structural reliability analysis. Struct. Saf. **30**(2), 162–179 (2007)
6. Janusevskis, J., Le Riche, R.: Simultaneous kriging-based estimation and optimization of mean response. J. Global Optim. **55**, 313–336 (2013)
7. Jaynes, E.T.: Probability Theory: The logic of science. Cambridge University Press, Cambridge (2003)
8. Jia, G., Taflanidis, A.A.: Kriging metamodeling for approximation of high-dimensional wave and surge responses in real-time storm/hurricane risk assessment. Comput. Method Appl. Mech. **261–262**, 24–38 (2013)
9. Jia, G., Taflanidis, A.A.: Sample-based evaluation of global probabilistic sensitivity measures. Comp. Struct. **144**, 103–118 (2014)
10. Jin, R., Chen, W., Simpson, T.W.: Comparative studies of metamodelling techniques under multiple modelling criteria. Struct. Multidiscip. Optim. **23**(1), 1–13 (2001)
11. Klee, H., Allen, R.: Simulation of Dynamic Systems with MATLAB and SIMULINK. CRC Press, Boca Raton (2007)
12. Lophaven, S.N., Nielsen, H.B., Sondergaard, J.: DACE-A MATLAB Kriging Toolbox. Technical University of Denmark (2002)
13. Medina, J.C., Taflanidis, A.: Adaptive importance sampling for optimization under uncertainty problems. Comput. Method Appl. Mech. **279**, 133–162 (2014)
14. Picheny, V., Ginsbourger, D., Roustant, O., Haftka, R.T., Kim, N.H.: Adaptive designs of experiments for accurate approximation of a target region. J. Mech. Des. **132**(7) (2010)
15. Rasmussen, C.E., Williams, C.K.I.: Gaussian Processes for Machine Learning (Adaptive Computation and Machine Learning). The MIT Press, Cambridge (2005)
16. Robert, C.P., Casella, G.: Monte Carlo statistical methods, 2nd edn. Springer, New York (2004)
17. Rodríguez, J.F., Renaud, J.E., Wujek, B.A., Tappeta, R.V.: Trust region model management in multidisciplinary design optimization. J. Comp. App. Math. **124**(1), 139–154 (2000)
18. Royset, J.O., Polak, E.: Reliability-based optimal design using sample average approximations. Probabilist Eng. Mech. **19**, 331–343 (2004)

19. Sacks, J., Welch, W.J., Mitchell, T.J., Wynn, H.P.: Design and analysis of computer experiments. Stat. Sci. **4**(4), 409–435 (1989)
20. Schuëller, G.I., Jensen, H.A.: Computational methods in optimization considering uncertainties—an overview. Comput. Method Appl. Mech. **198**(1), 2–13 (2008)
21. Spall, J.C.: Introduction to stochastic search and optimization. Wiley-Interscience, New York (2003)
22. Taflanidis, A.A., Beck, J.L.: An efficient framework for optimal robust stochastic system design using stochastic simulation. Comput. Method Appl. Mech. **198**(1), 88–101 (2008)
23. Taflanidis, A.A., Beck, J.L.: Reliability-based design using two-stage stochastic optimization with a treatment of model prediction errors. J. Eng. Mech. **136**(12), 1460–1473 (2010)
24. Verros, C., Natsiavas, S., Papadimitriou, C.: Design optimization of quarter-car models with passive and semi-active suspensions under random road excitation. J. Vib. Control **11**(5), 581–606 (2005)
25. Wang, G.G., Shan, S.: Review of metamodeling techniques in support of engineering design optimization. J. Mech. Des. **129**(4), 370–380 (2007)

Social Aggravation Estimation to Seismic Hazard Using Classical Fuzzy Methods

J. Rubén G. Cárdenas, Àngela Nebot, Francisco Mugica,
Martha-Liliana Carreño and Alex H. Barbat

Abstract In the last years, from a disasters perspective, risk has been dimensioned to allow a better management. However, this conceptualization turns out to be limited or constrained, by the generalized use of a fragmented risk scheme, which always consider first, the approach and applicability of each discipline involved. To be congruent with risk definition, it is necessary to consider an integral frame, and social factors must be included. Even those indicators that could tell something about the organizational and institutional capacity to withstand natural hazards, should be invited to the table. In this article, we analyze one of the most important elements in risk formation: the social aggravation, which can be regarded as the convolution of the resilience capacity and social fragility of an urban center. We performed a social aggravation estimation over Barcelona, Spain and Bogota, Colombia considering a particular hazard in the form of seismic activity. The Aggravation coefficient was achieved through a Mamdami fuzzy approach, supported by well-established fuzzy theory, which is characterized by a high expressive power and an intuitive human-like manner.

J.R.G. Cárdenas(✉)
IUSS UME School, Via Ferrata 45, Pavia, Italy
e-mail: ruben.gonzalez@umeschool.it

À. Nebot · F. Mugica
Soft Computing Group, Technical University of Catalonia, Jordi Girona
Salgado 1-3, Barcelona, Spain
e-mail: angela@lsi.upc.edu

F. Mugica
e-mail: fmugica@lsi.upc.edu

M.-L. Carreño · A.H. Barbat
Centre Internacional de Mètodes Numèrics En Enginyeria (CIMNE), Universitat
Politècnica de Catalunya, Jordi Girona Salgado 1-3, Barcelona, Spain
e-mail: alex.barbat@upc.edu

A.H. Barbat
e-mail: liliana@cimne.upc.edu

© Springer International Publishing Switzerland 2015
M.S. Obaidat et al. (eds.), *Simulation and Modeling Methodologies,
Technologies and Applications*, Advances in Intelligent Systems
and Computing 402, DOI 10.1007/978-3-319-26470-7_14

Keywords Fuzzy sets · Risk management · Natural hazards · Vulnerability index ·
Social vulnerability · Seismic vulnerability · Fuzzy inference system

1 Introduction

Social vulnerability is one of the key factors to assembly risk in space and time, however, such important element is largely ignored over ex-ante, ex-post, and cost/lost
estimation reports, in part because the measurement of social vulnerability is not
quite understood, and in part because the presence of epistemology oriented-based
discrepancies along vulnerability definition, which binds a particular methodology
with the orientation where such definition has been used, i.e., ecology, human, physical, etc. Therefore, there is a concept discrepancy when a social vulnerability model
is about to be built. Diverse models have been used to obtain social vulnerability
estimations. For example, Cutter et al. [1] used a hazard-of-place model to exam-
ine the components of social vulnerability to natural hazards among US counties
through the development of a vulnerability index based on the reduction of variables
by a factor analysis plus an additive model. Kumpulainen [9] using ESPON Hazards
integrative model, created a vulnerability index map for all Europe regions based on
an aggregated model, considering that regional vulnerability is measured as a combination of damage potential (anything concrete that can be damage) and the coping
capacity. The principal difference between these models rely on one basic definition:
while in Cutter's model the hazard potential is dependent on risk and mitigation, in
ESPON model risk is a combination of the same hazard potential and the regional
vulnerability.

Carreño et al. [6] proposed an seismic aggravation risk model based on Cardona's
conceptual framework of a risk model analysis for a city considering a holistic perspective, thus describing seismic risk by means of indices [2] and assessing risk
with the expression known as Mocho's equation in the field of in the field of disaster
risk indicators. They propose that seismic risk is the result of physical risk (those
elements susceptible to be damage or destroyed) and an aggravation coefficient that
includes both: the resilience and the fragility of a society.

In this article, we propose a complete Mamdani fuzzy social aggravation model
starting from the aggravation descriptors described in Carreño et al. [6]. The aggravation model synthesizes the social aggravation characteristics of a city struck by an
earthquake that could conduct to social vulnerability enhancement or moderation. A
main advantage of the proposed model is its white box nature that results in a high
level understandability model. Moreover, the fuzzy approximation used in this paper
is well established and with a solid background.

2 Previous Models

Cardona [2] proposed a holistic model of seismic risk at urban level which considers a structuralist and figurative vision by using representations of the interaction between human settlements and their surroundings. One of the main points in Cardona's risk model is the assumption that vulnerability has identifiable components, whom can be regarded as a reflection of two main components: fragility or physical susceptibility (exposition) and social fragility and lack of resilience. By means of an index characterization, the model branches among different indicators running through these two previous risk components, where each indicator is a representative value of a defined descriptors set.

Carreño et al. [6] made a slight modification of the Cardona original model, following the consideration that holistic risk could be regarded as it were hazard-function (considering the hazard intensities) and social and physical vulnerability on a period of time, but considering that risk might be viewed as a function of the potential damage on asset plus the socioeconomic aggravation onto the urban system produced by the lack of resilience and fragility reported at site. Therefore in Carreño model, for seismic risk modeling, the formulation of the index is based, in one hand; on seismic damage scenarios (or the hazard and physical vulnerability convolution) and in the other, on the estimation using a set of descriptors of social vulnerability based on fragility and resilience indicators, but grouped into a single module called: aggravation.

A conceptualization of Cardona's modified seismic risk model can be seen in the Fig. 1.

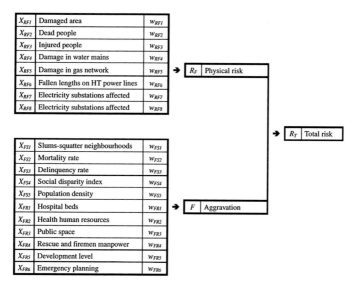

Fig. 1 Carreño et al. [5, 6] Holistic seismic risk model

Table 1 Descriptors used for aggravation estimation [6]

Aggravation descriptors
Marginal slums
Population density
Mortality rate
Delinquency rate
Social disparity
Hospital beds
Human health resources
Emergency and rescue personnel
Development level
Emergency operability

Many times the strength of a vulnerability model becomes weakened not because the type or resolution of the models themselves but because the lack of information and accurate data, in such a way that the results achieved are misleading in many ways.[1] Furthermore, the lack on understanding about how accurately measure vulnerability is one of the major uncertainty sources among social models. In most of the cases, social vulnerability is described using the individual characteristics of people (age, race, health, income, type of dwelling unit, employment, gross domestic product (GDP), income, etc.) Just in recent time, vulnerability models started to include place inequalities, such as level of urbanization, growth rates, and economic vitality [6].

Although there is a general consensus about some of the major factors that influence social vulnerability, disagreement arise in the selection of specific variables to represent these boarder concepts [1].

The descriptors used by [6] for aggravation estimation can be seen in the Table 1.

2.1 Index Method

Carreño et al. [6] obtained a seismic risk evaluation at urban level by means of indicators that leads to the calculation of a total risk index. This is obtained by direct application of Moncho's equation described in 1:

$$R_T = R_{Ph}(1 + F) \qquad (1)$$

where R_T is the total risk, R_{Ph} is the physical risk, and F is a aggravation coefficient.

---

[1]Sometimes redirecting toward a definition staying that vulnerability is a characteristic and not a condition, leading toward the assumption that without damage, or a specific hazard, vulnerability places could stand forever.

Thus, considering seismic risk as produced for physical and an aggravation coefficient; the risk index provides an approximate vision of the state of the social capital infrastructure.

The physical risk is evaluated by using the Eq. 2

$$R_{Ph} = \sum_{i=1}^{p} w_{R_{Ph}k} F_{R_{Ph}k} \qquad (2)$$

where $F_{R_{Ph}k}$ are the physical risk descriptors, and $w_{R_{Ph}k}$ are their weights and p the total number of considered descriptors in the estimation. As we have said, the physical risk descriptors values can be obtained from previous physical risk evaluation (damage scenarios) already made at the studied location.

The F coefficient depends on a weighted sum of an aggravation factors set associated to socioeconomic fragility of the community (F_{SFi}) and lack of resilience of exposed context (F_{LRj}), according to Eq. 2.

$$F = \sum_{i=1}^{m} w_{SFi} F_{SFi} + \sum_{i=1}^{n} w_{LRj} F_{LRj} \qquad (3)$$

where w_{SFi} and w_{LRj} are the assessed weights on each factors calculated by an analytic hierarchy process [6, 10], and m and n the total number of descriptors, of fragility and lack of resilience, respectively. The descriptors of the socioeconomic fragility and lack of resilience of exposed context are obtained from existent databases and statistical data for the studied area.

When using Moncho's equation for estimate total risk, came to arise the consideration that F can be up to much twice the value of P_R, which is not always accomplished, because some times the indirect effects are much larger than the direct effects, leading a mislead in risk estimation.

2.2 Carreño's Fuzzy Method

Taking the objective to build a more flexible risk management tool when information is incomplete or is not available, Carreño et al. proposed the use of fuzzy logic tools and expert opinion to replace indexes by fuzzy sets. The same descriptors are used and the sequences of calculations are similar to those made in the conventional index method, however, the aggravation's descriptors values which were originally obtained by demographic data bases are replaced by local expert opinions. Using linguistic qualifiers, instead of using numerical values, the aggravation value can be evaluated. Distinct linguistic descriptors qualifiers where proposed, which range in five levels of aggravation description: *very low, low, medium, high, very high*. Using local expert opinion, a membership function was defined for each linguistic level used to link the reported demographic or expert opinion value to one level of aggravation.

With the positive link between a reported data and its suitable linguistic level, the level is then grouped into another set of membership functions, (based on expert opinion or strictly arbitrary) which plays as a homogenizer since it blends the original qualifier level into a new single fuzzy set.

They calculated the fuzzy union between social fragility and lack of resilience descriptors, $\mu_f\left(x_{SF}, x_{LR}\right)$, and applied on each of these new membership functions, μ, the weights, w, corresponding to the level of aggravation, L_F, of each descriptor x_{SFi} and x_{LRj}, as defined in Eq. 3.

$$\mu_f\left(x_{SF}, x_{LR}\right) = max\left(w_{SF1}\mu_{FL1}\left(L_{F1}\right)...w_{LR1I}\mu_{FLI}\left(L_{F1}\right)\right) \tag{4}$$

The proposed weighted and union methods between social fragility and lack of resilience descriptors can be seen in Fig. 1.

In the same way of index's method, weights are assigned to each fuzzy set by using an analytic hierarchy process. The aggravation coefficient F is calculated as the centroid abscise of the area beneath the curve obtained with Eq. 3.

However, we think that the Carreño's fuzzy model is not entirely appropriate because it is a nonconventional fuzzy approach, which may be questionable due to the fact that fuzzy mathematical raised in the inference process is not well established and accurately validated.

3 Classical Fuzzy Method

Behind the holistic risk proposal is the consideration of an urban center as it behaves as a complex dynamic system; in which a collection of various structural and non-structural elements are connected and organized in such a way as to achieve some specific objective through the control and distribution of material resources, energy, and information, [2]. The hypothesis considers then, that there are some system elements (or a collection of them) not necessarily structural or geological (but social) that can be identified in terms of their true affectation or affectation predisposition of the complex system state. In this way, the complex dynamic systems theory considers that risk is in fact, a state characterization of the complex system which is, at all time, in a *potentially at crisis situation* or, in a instability state. Methodologically, this can be seen as:

$$P_C = T_a I_c \tag{5}$$

where P_C is potential crisis, T_a is a trigger agent capable to produce such crisis, and I_c the instability conditions of the system [3, 4].

The system elements identify as related with the creation of the instability conditions when considering seismic risk, are assumed to be the social fragility and the resilience capacity of a urban center, along with the physical infrastructures that could be damaged. At the other hand, the trigger event, in this particular case, is the earthquake itself. In this way, an urban center which is meant to last, must find the ways to decrease the reachable factors that leads toward the crisis state. This is obviously done trough risk management processes and, at the end, with a sustainability development scheme.

The model proposed in this research pretend to build an aggravation coefficient by re-defining Carreño et al. descriptors into three different Fuzzy Inference Systems (FIS), called: resilience, fragility, and aggravation. Each subsystem is defined by a set of rules directly over the aggravation descriptors. A conceptualization of the different steps along the proposed model can be seen in Fig. 3. The variables involved in each subsystem are presented in the left-hand side of Fig. 2. FIS #1, corresponds to the Social fragility model and has as input variables the Marginal Slums (MS), the social disparity index (SDI) and the population density (PD). The output of FIS #1 is the level of Fragility. On the other hand, FIS #2 corresponds to the Resilience model and has as input variables the human health resources (HHR),the emergency operability (EO), and the development level (DL). The output of FIS #2 is the resilience level. The aggravation model (FIS #3) takes as inputs the fragility and resilience

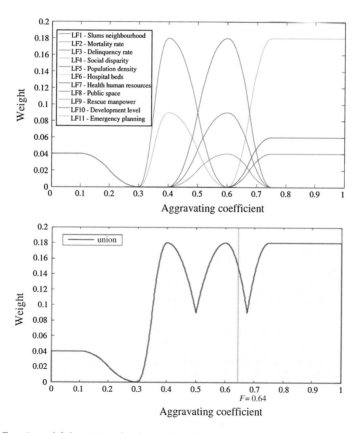

Fig. 2 Carreño weighting (*up*) and union method (*low*) for San Martí District, Barcelona Spain (taken from Carreño et al. [6])

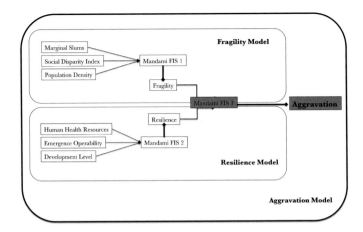

Fig. 3 Conceptualization of fuzzy classical model to estimate aggravation coefficient

levels that are the output of FIS #1 and #2, respectively, and infers the aggravation coefficient. All the fuzzy inference systems proposed in this research are based on the Mamdani approach [7], since it is the one that better represents the uncertainty associated to the inputs (antecedents) and the outputs (consequents) and allows to describe the expertise in an intuitive and human-like manner. Our main objective is to develop a fuzzy aggravation model as much interpretable as possible and with high expressive power. In our approach, the original 10 variables presented in Table 1 are reduced to six variables. Population density, slum area or marginal slums, human health resources, and development level remain the same, and social disparity index and Emergence operability are redefined in such a way that subsume the other variables.

The reduction or simplification of the original variables was made by taking advantage of certain descriptors that are linked and could englobe various descriptors in one single class considering its social nature, for example: the descriptors called: mortality rate and delinquency rate, are related between them and are reflecting social consequences produced by a social structure failure (could be lack of access) to certain social advantages, such as having an efficient public health program, or no marginalization dynamics, or access to education and effective justice and law policies. Therefore, we consider these descriptors could be enclosed within the descriptor called social disparity index, which is a fragility descriptor as well. In the case of resilience descriptor, we merge descriptors called: public space, hospital beds, and emergency Personnel, into the descriptor called emergence operability, because the former descriptors acts when the emergency is being or has recently occurred, and therefore are related with the capacity of the city to face an emergence situation, and the assets that a city has to confront it. We modify fuzzy classes by reducing the number of linguistic levels defined for each descriptor up to 3 (*low, medium, high*) along their respective universe of discourse, but we kept the same five levels for the final output (resilience, fragility, and aggravation). We think that three classes are enough to represent accurately the input variables of the resilience and fragility models. Moreover, a reduction of the number of classes implies also a more compacted and reduced set of fuzzy rules. In the same way, to improve model's sensibility, we adjust membership functions forcing them to be more *data-based* kind of type, and thus considering the reported aggravation data as embedded along membership functions limits definition. With these new membership functions we build a set of fuzzy logic rules that could infer the behavior of the aggravation coefficient components using the three Mamdani Fuzzy Inferences Systems mentioned before (see Fig. 3).

The developing of the fuzzy rules was established for consider all possible combinations between the input descriptor's linguistic levels, giving a total of 27 rules for calculating fragility and resilience values, respectively. The rules were intended to follow risk management literature which could suggest possible outcomes when three of these elements interact to form resilience or fragility. The Mamdani aggravation model, that has as input variables the resilience and the fragility, discretized into five classes each, is composed of 25 fuzzy rules.

In Table 2 the rules of the Mamdani resilience model are presented as an example. As mentioned before, the use of classical fuzzy systems, with well established fuzzy inference theory, allow a high level understandability model and easily manageable by experts which in turn leads toward a deepest discussion in the topic of social vulnerability description and casual interrelation.

Let's describe the inference process by following the example of the proposed Resilience FIS. The fuzzy inference engine combines the fuzzy *if-then* rules (see Table 2) into a mapping from fuzzy sets in the input space $U \subset R^n$ to fuzzy sets in the output space $V \subset R$, based on fuzzy logic principles. Let's $U = U_1 \ x \ U_2 \ x \ U_3 \subset R^n$ and $V \subset R$, where U_1, U_2 and U_3 represents the universes of discurse of human health resources, development level, and emergency operability input variables, respectively, and V the universe of discourse of Resilience. In our case each input variable contains three fuzzy sets and the output variable is discretized into five fuzzy sets. Then, the fuzzy rule-based shown in Table 2 can be expressed in a canonical form as shown in Eq. 6.

$$R^{(l)} \ : \ IF x_1 is A_1^l and...and x_n is A_n^l THEN y is B^l \tag{6}$$

where A_1^l and B^l are fuzzy sets in U_i and V, respectively, $x = (x_1, x_2, x_3) \in U$ are human health resources, development level, and emergency operability linguistic variables, $y \in V$ is the resilience linguistic variable and $l = 1, 2, ..., 27$ is the rule number. Consider now the fuzzy facts: x_1 is A_1', x_2 is A_2', x_3 is A_3', being A_1', A_2' and A_3' fuzzy sets.

The generalized modus ponens allows the deduction of the fuzzy fact y is B' by using the compositional rule of inference (CRI), defined trough the fuzzy relation between x and y, as defined in Eq. 7.

$$B' = A' \circ R \tag{7}$$

Table 2 Logic rules used for resilience estimation

1.	If (HHR is L) and (DL is L) and (EO is L) then (R is VL)
2.	If (HHR is M) and (DL is M) and (EO is M) then (R is M)
3.	If (HHR is H) and (DL is H) and (EO is H) then (R is VH)
4.	If (HHR is M) and (DL is L) and (EO is L) then (R is L)
5.	If (HHR is H) and (DL is H) and (EO is L) then (R is M)
6.	If (HHR is L) and (DL is M) and (EO is L) then (R is L)
7.	If (HHR is M) and (DL is M) and (EO is L) then (R is M)
8.	If (HHR is H) and (DL is M) and (EO is L) then (R is H)
9.	If (HHR is L) and (DL is H) and (EO is L) then (R is M)
10.	If (HHR is M) and (DL is H) and (EO is L) then (R is M)
11.	If (HHR is H) and (DL is H) and (EO is L) then (R is H)
12.	If (HHR is L) and (DL is L) and (EO is M) then (R is L)
13.	If (HHR is M) and (DL is L) and (EO is M) then (R is M)
14.	If (HHR is H) and (DL is L) and (EO is M) then (R is H)
15.	If (HHR is L) and (DL is M) and (EO is M) then (R is M)
16.	If (HHR is H) and (DL is M) and (EO is M) then (R is H)
17.	If (HHR is L) and (DL is H) and (EO is M) then (R is M)
18.	If (HHR is M) and (DL is H) and (EO is M) then (R is H)
19.	If (HHR is H) and (DL is H) and (EO is M) then (R is H)
20.	If (HHR is L) and (DL is L) and (EO is H) then (R is M)
21.	If (HHR is M) and (DL is L) and (EO is H) then (R is H)
22.	If (HHR is H) and (DL is L) and (EO is L) then (R is H)
23.	If (HHR is L) and (DL is M) and (EO is H) then (R is H)
24.	If (HHR is M) and (DL is M) and (EO is H) then (R is VH)
25.	If (HHR is H) and (DL is M) and (EO is H) then ((R is VH)
26.	If (HHR is L) and (DL is H) and (EO is H) then (R is H)
27.	If (HHR is M) and (DL is H) and (EO is H) then (R is VH)

HHR Human health resources, *DL* Development level, *EO* Emergency operability, *R* Resilience, *VH* Very high, *H* High, *M* Medium, *L* Low, *VL* Very low

where $A' = (A'_1, A'_2, A'_3)$. The simplest expression of the compositional rule of inference can be written as Eq. 8.

$$\mu_{B'^i}(y) = I\left(\mu_{A^i}(x_0), \mu_{B^i}(y)\right) \tag{8}$$

when applied to the ith rule; where:

$$\mu_{A^i}(x_o) = T\left(\mu_{A^i_1}(x_1), \mu_{A^i_2}(x_2), \mu_{A^i_3}(x_3)\right)$$

where $x_0 = (x_1, x_2, x_3)$. Here, T is a fuzzy conjuctive operator and I is a fuzzy implicator operator.

Once the inference is performed by means of the compositional rule of inference scheme, the resulting individual (one for each rule) output fuzzy sets are aggregated into an overall fuzzy set by means of a fuzzy aggregation operator and then a defuzzification method is employed to transform the fuzzy set into a crisp output value, i.e., the resilience level following the example.

The defuzzification method used in this work is the center of gravity (COG), which slices the overall fuzzy set obtained in the inference process into two equal masses. The center of gravity can be expressed as Eq. 9.

$$COG = \frac{\int_a^b x\mu_B(x)dx}{\int_a^b \mu_B(x)dx}$$

(9)

where B is fuzzy set on the interval $[a, b]$.

4 Results and Comparison

To obtain a final social aggravation inference value, we used the aggravations linguistic levels that can be viewed in Fig. 4. In the case of index method, we used the levels of aggravation that can been seen in Table 3. Both: linguistic classes and levels of aggravation were reported by Carreño et al. [6].

4.1 Barcelona

Figure 5a shows the estimated spatial distribution of the aggravation coefficient and its correspondent level for the 10 administrative districts, of the city of Barcelona,

Fig. 4 Membership functions for levels of aggravation. Carreño et al. [6]

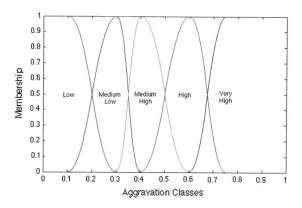

Table 3 Levels of aggravation used in index method, Carreño et al. [6]

Level	Aggravation
Low	[0–0.19]
Medium low	[0.20–0.39]
Medium high	[0.40–0.54]
High	[0.55–0.64]
Very high	[0.65–1.00]

Fig. 5 Aggravation coefficient: **a** Proposed fuzzy model. **b** Carreño fuzzy method. **c** Index method. Districts: (1) Ciutat Vella, (2) Eixample, (3) Sants-Montjuic, (4) Les Corts, (5) Sarrià-Sant Gervasi, (6) Gràcia, (7) Horta-Guinardó, (8) Nou Barris, (9) Sant Andreu, (10) Sant Martí

achieved through the proposed model, (b) and (c) shows the aggravation coefficient calculated by Carreño et al. using fuzzy methods, and the aggravation coefficient estimated using index method respectively.

The proposed model, as well as the other two alternative methods, estimates that highest aggravation is spread mostly over the northeast part of the city. But only in the proposed model and index method levels of *very high* are reached over Sant Martí district. In our model, the level of *high* is reached over San Andreu, while in the index and Carreño method is for Nou Barris. *Medium-high* values for L'Eixample, Horta Guinardo, and Ciutat Vella are estimated by the proposed model while the rest of the city presents values of *medium-low* aggravation level. The index method estimate that only Ciutat Vella have a *Medium-high* value and the rest of the city ranges between *low* and *medium-low* aggravation values, while Carreño method gives a level of aggravation of *medium-high* for almost all the city, except in Sarria-Saint Gervasi, where it gives a value of *medium-low*. The first thing that we noted is that the proposed fuzzy model resembles more the index method rather than Carreño's method. Even if the spatial distribution is not the same (which was not the aim of our model), we observe that the aggravation classes distribution in both models has a similar spread. As expected, the actual distribution of the level of aggravation is not the same, thus index method could be regarded as if giving lowest aggravation values, but it is necessary to remember that the limit levels to define the aggravation classes are not the same, making the two models (FIS and Index) impossible to coincide in this part. Although, we do observe a under and overestimation on the actual aggravation values estimated by our proposed model, as we will discuss next.

Figure 6a–c shows the aggravation coefficient numerical value obtained by the proposed fuzzy model, Carreño fuzzy method, and index method, respectively. Districts are ordered from lower to highest aggravation level. In these figures, we can see that even there is no correct total match among the two methods, all of them preserve quite the same order in terms on higher and lower aggravation levels.

When comparing the numerical aggravation value obtained from the proposed model to a robust method like index models [8], it suffer by a slight under and overestimation of the aggravation values by district. In the proposed method this issue could be addressed with the inclusion of weights to each descriptor, as the other methods do. Nevertheless, we consider that even with these small numerical dissimilarities, the proposed fuzzy model limits the different aggravation levels in a suitable way, allowing the identification of more potentially problematic zones with a good resolution and reduced computation time.

Figure 6d shows the same as (a), (b), and (c) but without ordering the districts by aggravation value, showing how the aggravation values behaves along the different districts. As it can bee seen, even if the explicit aggravation coefficient value is not the same for each district, a similar trend shape come to appears (with the inherent over and underestimation aggravation level), which leads to the conclusion that the general behavior of the proposed model is coherent with the result achieved by Index Method.

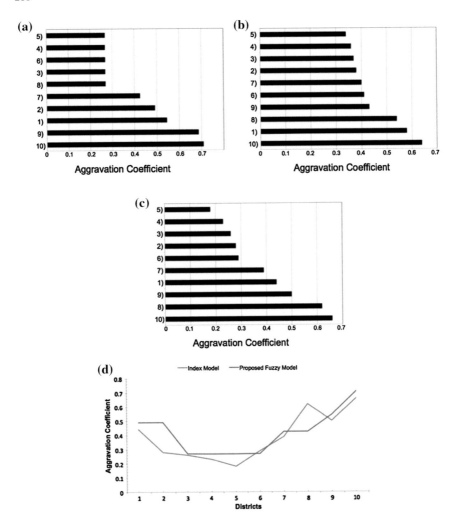

Fig. 6 Aggravation coefficient values by district, sorted from lower to higher: **a** Proposed fuzzy model. **b** Carreño fuzzy method. **c** Index method. **d** Aggravation coefficient comparison over the 10 Barcelona Districts (numeration as Fig. 5)

4.2 Bogota, Colombia

Colombia's Capital is divided since 1992 into 20 administrative districts. However in our study, we took into account only 19 on these because the district called Sumapaz correspond basically to the rural area of the city. For the Social Aggravation Coefficient estimation on each district we used statistical and demographic data from 2001 [6].

In Fig. 7a–c we can see the Aggravation coefficient value obtained by the fuzzy proposed model, Carreño fuzzy method and Index method respectively. The general

Aggravation level seems to be underestimated by the FIS model, however, the FIS spatial pattern distribute the highest values of aggravation at the South West part of the city as reported by index method, this corresponding to the districts of: Ciudad Bolivar, Bosa, Usme, and San Cristobal. The East part of the city remains with *medium low*, and the North West part of the city presents *medium-high* aggravation value. The index method reach a *very high* value at South West part of the city

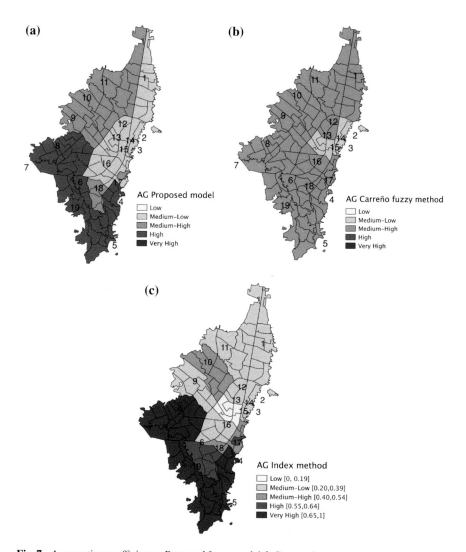

Fig. 7 Aggravation coefficient: **a** Proposed fuzzy model. **b** Carreño fuzzy method. **c** Index method. Localities: (1) Usaquén, (2) Chapinero, (3) Santa Fe, (4) San Cristóbal, (5) Usme, (6) Tunjuelito, (7) Bosa, (8) Ciudad Kennedy, (9) Fontibón, (10) Engativá, (11) Suba, (12) Barrios Unidos, (13) Teusaquillo, (14) Mártires, (15) Antonio Nariño, (16) Puente Aranda, (17) Candelaria, (18) Rafael Uribe, (19) Ciudad Bolvar

while the northern part presents mostly a *medium-low* aggravation value. Carreño fuzzy method presents an almost homogeneous level of aggravation, with values of *medium low* for Teusaquillo and Chapierno districts.

Figure 8a–c show the Aggravation Coefficient numerical value using the fuzzy proposed model, the Carreño fuzzy method and index method, respectively. Districts are ordered from lower to highest aggravation level. As in the case of Barcelona, we can note that even there is no correct total match among the three methods, all of them preserve quite the same order in terms on higher and lower aggravation levels.

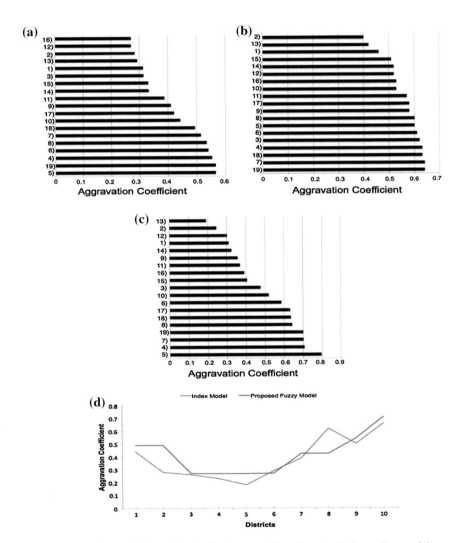

Fig. 8 Aggravation coefficient values by district, sorted from lower to higher: **a** Proposed fuzzy model. **b** Carreño fuzzy method. **c** Index method. **d** Aggravation coefficient comparison over the 10 Barcelona Districts (numeration as Fig. 5)

Figure 8d shows the trend line of the Aggravation Coefficient over the 19 administrative districts of Bogota Colombia, obtained by the three methods announced previously; where it is noted that the underestimation referred on previous lines. Similar to Barcelona case, the trend is quite alike with the one estimated using index Method.

4.3 Discussion

According to the previous analysis, with the use of classical fuzzy inference system methodology it is plausible to reproduce the results obtained from a more analytical method such as indexes, for example: in terms of district aggravation classification, or in reproducing similar spatial pattern of aggravation. In first term, the proposed inference model allows a useful simplification for the large quantity of variables required for social aggravation analysis, in the spirit of reduce the subjectivity associated with aggravation descriptors suitability designation by using a more flexible and small descriptors set in which the underlying links between them can be more easily observed, enabling a more understandable analysis scheme for social aggravation inference estimation. Building rules directly over the aggravation descriptors allows to assemble a compositional rule of inference over the very same descriptors that are assumed to create aggravation itself, and therefore, the inference process can be made using rules designed to follow risk management knowledge, allowing the model to represent, with a certain degree of freedom, the actual understanding of aggravation formation, and at the same time, it allows a real discussion of the rule's structure strength; which can be absolutely improved with a deepest debate.

Fuzzy logic inference capabilities can be exploited in a more suitable way because the outputs from each FIS used in the model are always fuzzy sets, giving the chance to connect them trough a new FIS without loosing consistency, allowing model completeness.

At the other hand, the proposed model slightly over and underestimated aggravation values for some districts when comparing with index model, as it is also de case of Carreño's fuzzy model. However, if necessary, the proposed fuzzy model can be further tuned if descriptors are weighted.

4.4 Future Work

The flexibility of the model enables its adaptation to several conditions which could be used in more general studies of social vulnerability and that can also help to fill some gaps among analytic methods. For example, the same procedure can be applied to a more general social vulnerability model that considers not only physical, and aggravation inputs, but environmental, economic, and even completely subjective descriptors can be add as well, such as solidarity or brotherhood.[2] All of these can then be embedded into one single inference model. One of the main problems of risk ex-ante and ex-post models is that they do not necessarily consider the interconnectivity of social characters (sectors) in a real scenario, for example, the lack of hospitals in one geographic area does not necessarily mean that human health resources is zero at that place. It will be like assuming that the fire department can only help those who are in close proximity. Assuming interconnectivity, the potential damage to the social network-connections in case of disaster is the real issue that must be addressed, and we consider it plausible to be approach using fuzzy methods. Although the proposed model was intended to be applied to assess the risk over an urban environment when it's strike by an earthquake, the structure of the social vulnerability module of the model, (the one who deals with resilience and fragility), can easily be transformed in a non-disaster dependent analytic framework. Therefore adapting itself to other types of disasters, or even to the study and analysis of social vulnerability by itself.

5 Conclusions

We obtain a inference fuzzy model to make an estimation of social aggravation over the cities of Barcelona and Colombia using the descriptors proposed in [6]. Building inference compositional rules over the selected descriptors, we were able to obtain a robust method that resembles the identification of relevant aspects and characteristics of seismic risk at urban level already achieved by two other consolidated methods. The proposed model displays more simplicity, flexibility, and resolution capacities and can be rapidly transformed into a non-disaster event model type with the inclusion of new type of variables, englobing a more detailed social vulnerability scheme and interconnectivity issues.

References

1. Cutter, S.L., Boruff, B.J., Shirley, W.L.: Social vulnerability to environmental hazards. Soc. Sci. Q. **84**, 242 (2003)
2. Cardona, O.D.: Holistic evaluation of the seismic risk using complex dynamic systems (in Spanish). PhD Thesis. Technical University of Catalonia, Barcelona, Spain (2001)
3. Cardona, O.D.: Environmental management and disaster prevention: two related topics? a holistic risk assessment and management approach. In: Igleton, J. (ed.) Natural Disaster Management. Tudor Rose, IDNDR, London, UK (1999)
4. Cardona, O.D.: Prevencion de desastres y preparativos para emergencias: Aspectos tecnico-cientificos, sociales, culturales e institucionales Centro de Estudios sobre Desastres y Riesgos Naturales. CEDERI) Universidad de Los Andes, Bogota, Colombia (1995)

---

[2]Loosing in this way its event-base model characterization.

5. Carreño, M.L., Cardona, O.D., Barbat, A.H.: Disaster risk management performance index. Nat. Hazards **40**, 1–20 (2007)
6. Carreño, M.L., Cardona, O.D., Barbat, A.H.: New methodology for urban seismic risk assessment from a holistic perspective. Bull Earthq. Eng. **10**, 547–565 (2012)
7. Mamdani, E.H., Assilian, S.: An experiment in linguistic synthesis with a fuzzy logic controller. Int. J. Man-Mach. Stud. **7**(1), 1–13 (1975)
8. Marulanda, M.C., Cardona, O.D., Barbat, A.H.: Robustness of the holistic seismic risk evaluation in urban centers using the USRi. Nat. Hazards **49**, 501–516 (2009)
9. Kumpulainen, K.: Vulnerability Concepts in Hazard and Risk Assessment Geological Survey of Findand. vol. 42, pp. 65–74 (2006)
10. Saaty, T.L., Vargas, L.G.: Prediction, Projection, and Forecasting: Applications of the Analytical Hierarchy Process in Economics, Finance, Politics, Games, and Sports. Kluwer Academic Publishers, Boston (1991)

Fuzzy Cognitive Mapping and Nonlinear Hebbian Learning for the Qualitative Simulation of the Climate System, from a Planetary Boundaries Perspective

Iván Paz-Ortiz and Carlos Gay-García

Abstract In the present work, we constructed a collective fuzzy cognitive map for the qualitative simulation of the Earth climate system. The map was developed by considering the subsystems on which the climate equilibrium depends, and by aggregating different experts opinions over this framework. The resulting network was characterized by graph indexes and used for the simulation and analysis of hidden patterns and model sensitivity. Then, linguistic variables were used to fuzzify the edges and aggregated to produce an overall linguistic weight for each one. The resulting linguistic weights were defuzzified using the center of gravity technique, and the current state of the Earth climate system was simulated and discussed. Finally, a nonlinear Hebbian learning algorithm was used for updating the edges of the map until a desired state was reached, defined by target values for the concepts. The results are discussed to explore possible policy implementation, as well as environmental decision making.

Keywords Fuzzy cognitive maps · Qualitative simulation · Nonlinear hebbian learning · Climate system

1 Introduction

Nowadays, the widespread concern over the climate stability and the resilience of natural ecosystems under pressure, have pointed out the necessity of developing more flexible tools to simulate the present conditions, to asses future scenarios, and

Currently PhD student at the Soft Computing Research Group, BarcelonaTech.

I. Paz-Ortiz (✉) · C. Gay-García
Programa de Investigación en Cambio Climático,Universidad Nacional
Autónoma de México Edificio de Programas Universitarios, Circuito de la
Investigación Científica s/n, Ciudad Universitaria, Mexico City, Mexico
e-mail: ivnpaz@gmail.com; ivanpaz@cs.upc.edu
URL:http://www.pincc.unam.mx

C. Gay-García
e-mail: cgay@unam.mx

© Springer International Publishing Switzerland 2015
M.S. Obaidat et al. (eds.), *Simulation and Modeling Methodologies,
Technologies and Applications*, Advances in Intelligent Systems
and Computing 402, DOI 10.1007/978-3-319-26470-7_15

to explore possible policy implementations oriented to environmental decision making and management.

Recent publications show a strong emerging tendency to explore the use of fuzzy cognitive maps (FCMs) for the development of qualitative models in environmental sciences. Examples include FCMs as a methodological framework for environmental decision making and management [12]. Cases of study, like "Visions for the Future of Water in the Seyhan Basin in Turkey" [1], and the description of current system dynamics together with the development of land cover scenarios in the Brazilian Amazon [20].

FCMs have the capability for including quantifiable and nonquantifiable concepts in a model [11], and do not require neither large capacity of computation nor to have numeric equations of the analyzed phenomena. By its flexibility, they also allow us to integrate information scattered in various places, as is the case in environmental systems, where the information must be gather from several disciplines (e.g., atmospheric sciences, biology, and geophysics). Moreover, the latest models based on cognitive mapping seek not only to simulate systems, but also to control them. This approach allows us to create recommendations and to explore possible policy implementations in environmental decision making and management. The usual techniques for the training and tuning of FCMs are performed by using Hebian learning algorithms [15]. These algorithms aim to find appropriate weights between the concepts of the map so the model equilibrates into a desired state.

In the present work, we used FCMs to analyze the dynamic of the climatic system based on a planetary boundaries framework, proposed by Rockström et al. [18], by considering the Earth subsystems identified as those on which the Earth's climate equilibrium depends. The paper is structured as follows: In Sect. 2, we present the basis of the FCMs theory. Section 3 describes the planetary boundaries framework. Section 4 presents the FCM simulation and a comparison of the results versus the current system state, Sect. 4.3 describes the implementation of a nonlinear Hebian learning algorithm to adjust the weights of the map until the map reaches a desired state. And finally, Sect. 5 includes the conclusions and further work.

2 Fuzzy Cognitive Maps

Fuzzy cognitive maps are directed graph structures for representing causal reasoning between variable concepts [7]. The concepts are represented as nodes ($C_1, C_2, ..., C_n$) in an interconnected network. Each node C_i represents a variable concept, and the edges e_{ij} which connect C_i and C_j (denoted as $C_i \rightarrow C_j$) are causal connections and express how much C_i causes C_j. Edges can have either a numerical or a linguistic value. $w_{ij} > 0$ indicates a positive causality between concepts C_i and Cj, $w_{ij} < 0$ indicates a inverse (or negative) causality, and $w_{ij} = 0$ indicates no causality. Linguistic quantifiers (such as low, medium, or high) are also used to represent the value of the weights; they indicate the qualitative relation among concepts. Figure 1 shows

Fig. 1 A four concepts
FCM (numbered I, II, III,
IV). The table shows the
weights for the edges
associated with the
relationships between
concepts

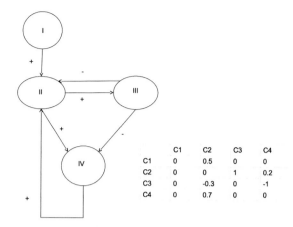

	C1	C2	C3	C4
C1	0	0.5	0	0
C2	0	0	1	0.2
C3	0	-0.3	0	-1
C4	0	0.7	0	0

a FCM containing four concepts (or nodes) numbered I, II, III, and IV. The table
shows the assigned weights for the relationships between concepts.

FCMs are built using expert's opinion. They define, considering their expertise,
the main components of the system as the concepts of the network. Once these are
defined, the causal relations among them, as well as the weights for each relation,
are established.

2.1 The FCM Inference Process

The dynamic of a fuzzy cognitive map can be simulated analytically through a specific inference process [11]. For this process, we considered FCM concepts having
values in a specific numerical interval (e.g., [0,1]). The value of each C_i at time t is
called the "activation level," and is interpreted as the state of activation (or quantity)
of this concept.

The values of the concepts $C_i^{(t)}$ at each time t are represented in a row state vector
$A^{(t)} = [C_1^{(t)}, ..., C_n^{(t)}]$ which describes the state of the system at each time step or iteration t [8]. Given, a particular input $A^{(0)}$, the state of the system during the iterations
can converge into an equilibrium point, into a limit cycle or diverge. The inference
process is calculated using the following equation:

$$A_i^{t+1} = f\left(\sum_{i=1,i\neq j}^{N} W_{ij}A_j^{(t)}\right) \tag{1}$$

where f is a threshold function, W is the matrix of weights, and A is the state vector.

2.2   FCM Graph Indexes

The structural properties of cognitive maps can be analyzed by using graph theory indexes. Through this analysis we gain intuition of the general structure and quality of the network. This is important since cognitive maps are a subjective representation of the real system, hardly dependent on the expert's opinion. Graph indices allow us to analyze how experts are regarding and structuring the system. Those include, among others, the number of concepts, connections, the ratio concepts/connections, and the density, defined as the ratio of the number of connections in the map and the square value of concepts.

The complexity of the concepts is analyzed through structural measures like the conceptual centrality index for a particular node C_i defined by Kosko [7] as:

$$CEN(C_i) = IN(C_i) + OUT(C_i) \qquad (2)$$

where:

$$IN(C_i) = \sum_{k=1}^{n} \overline{w}_{ik} \qquad (3)$$

$$OUT(C_i) = \sum_{k=1}^{n} \overline{w}_{ki} \qquad (4)$$

$IN(C_i)$ represents the number of concepts that causally act on concept C_i. Similarly, the row sum $OUT(C_i)$ is the number of concepts in which C_i causally acts [7]. Then, the conceptual centrality represents the importance of C_i to the causal flow on the map. When fuzzy weights values are used, the causal amount counts, i.e., a node can be connected to fewer nodes than another and still have greater conceptual centrality [7].

The density index (D) shows how connected the network is. It is defined as the number of existing connections divided by the maximum number of possible connections among all nodes (given by N^2). A map with high density will have a large number of causal relations.

The structure analysis of a cognitive map also includes the transmitter, receiver, and ordinary variables. These are defined by their associate values of $IN(C_i)$ and $OUT(C_i)$, called indegree and outdegree, respectively. Nodes whose outdegree is positive and their indegree is 0 are defined as transmitter variables. Receiver variables, on the contrary, are those nodes whose outdegree is 0 and their indegree is positive. Variables with nonzero outdegree and indegree are called ordinary variables [2, 10]. Receiver and transmitter variables show the structure of the map. For example, transmitter variables can be seen as forcing functions, which influence the system but cannot be controlled by other system's variables. The complexity of the system, as well as the degree of elaboration, can be analyzed by considering the number of receiver and transmitter variables. A great number of receiver or transmitter variables could be interpreted as the map is not being well elaborated or the relationships among its variables are not well known, which means that the causal relations among components are not clear for the experts [2, 11].

Collective maps are used to integrate different perspectives of a particular system. When collective cognitive maps are developed each expert creates a map, and then, the different versions are condensed either by grouping subgraphs in a single node, or by maintaining the nodes when several experts coincide. The centrality is used to decide which concepts will be represented in the collective map.

3 Subsystems of the Climate System: A Planetary Boundaries Framework

The stability analysis of the Earth climate system is based on the planetary boundaries framework proposed by a group of scientist heading by Rockstöm [18], and later discussed by Foley [4]. They established a planetary boundaries frame by identifying and quantifying the boundaries associated with the planet's biophysical subsystems that must not be transgressed in order to prevent an unacceptable environmental change. Based on the review of those reports we created an expert's opinion knowledge base for the construction of the fuzzy cognitive map. To do this, we grouped in a single map the concepts and relations identified in each article by its authors. They mentioned nine processes suggested as those in which is necessary to define planetary boundaries: climate change; rate of biodiversity loss (terrestrial and marine); interference with the nitrogen and phosphorus cycles; stratospheric ozone depletion; ocean acidification; global freshwater use; change in land use; chemical pollution; and atmospheric aerosol loading. These concepts and their relationships are briefly described below.

Climate Change
It refers to the increase in the mean temperature of the Earth. More precisely, to change in climate variability in terms of the extreme and mean values [5, 6]. Specifically, we refer to antropogenic climate change, which is a consequence of the human activity. Climate change causes changes in vegetation distribution, and so threats the ecological live-support systems as well as human activities. The climate change is described in terms of two variables having critical thresholds that qualitatively separate different climate system states [18], these are the atmospheric CO_2 concentration and the radiative forcing.

Changes in Atmospheric CO_2 Concentration
Defined as the increase in the parts per million of CO_2 molecules in the atmosphere [5, 6]. Most models suggest that, as atmospheric CO_2 increases also does global temperature. For example, doubling atmospheric CO_2 will lead to a rise about 3 °C (with a probable uncertainty of $2 - 4.5$ °C).

Changes in Radiative Forcing
The radiative forcing is the rate of energy change per unit area of the globe as measured at the top of the atmosphere.

Rate of Biodiversity Loss
Refers to the extinction rate, the number of species loss per million per year. Mace and collaborators [9], define biodiversity as the variability of living organisms, included terrestrial and marine ecosystems, other aquatic ecosystems and the ecological systems in which they reside. It comprises the diversity within species, among species, and within ecosystems. Mace emphasizes three levels of biodiversity: genes, species, and ecosystems. Biodiversity loss during the industrial period has grown notably. The species extinction rate is estimated against the fossil record. The extinction rates per million per year varies for marine life between 0.1 and 1 and for mammals between 0.2 and 0.5. Today, the rate of extinction of species is estimated to be 100–1,000 times more than what could be considered natural [18].

Ocean Acidification
Defined as the ocean pH increase, mainly in the surface layer. The acidification process is closely related with the CO_2 emission level. When the atmosphere CO_2 concentration increases, the amount of carbon dioxide dissolved in water as carbonic acid increases, which in turn, modifies the surface pH. Normally, the ocean surface is basic with a pH of approximately 8.2. Nevertheless, the observations show a decline in pH to around a value of 8. These estimations are made using the levels of aragonite (a form of calcium carbonate) that is created in the surface layer. This concept has an important relation with biodiversity loss as many organisms (like corals and phytoplankton), basic for the food chain, use aragonite to produce their skeletons or shells. As the aragonite value decreases, the ocean ecosystems weaken [4].

Phosphorus and Nitrogen Cycles
The human activity at the planetary scale is perturbing the global cycles of phosphorus (P) and nitrogen (N). The agriculture activity convert around 120 million tones of N_2 from the atmosphere per year into reactive forms [18]. The pressure that this changes exerted over the environment threats the balance of natural equilibrium. For example, the nitrous oxide is one of the most important greenhouse gases and its grown directly increases the radiative forcing. In the case of phosphorus around 20 million tones are mined every year and around 8.5–9.5 million tonnes flow into the oceans perturbing the marine ecosystems. To establish boundaries, the changes in P and N cycles are estimated with the quantity of P going to the oceans, measured in million tones per year, and with the amount of N_2 removed from the atmosphere for human use, also in million tones per year.

Change in Land Use (Urban Growth and Agriculture Use)
The IPCC [5] defines the change in land use as the percentage of global land converted into cropland. A general definition of land use change includes any type of human use. This transformation, either to cropland or urban, increases the biodiversity loss, which is associated with the destruction of ecosystems. In order to establish

the difference in land use between urban growth and agriculture use, and their different consequences, we include both concepts as nodes in the map.

Chemical Pollution

It refers to the emitted quantity, persistence, or concentration of organics pollutants, plastics, heavy metals, chemical, and nuclear residues, etc., which affect the dynamic of ecosystems.

Global Freshwater Use

Defined as the increase in its current use. Today, the annual use of freshwater from rivers, lakes, and groundwater aquifers is of $2,600\,km^3$. From that, 70 % is destined for irrigation, 20 % for industry, and 10 % for domestic use. This extraction causes the drying and reduction of body waters [5, 6].

Atmospheric Aerosol Loading

Referred as the concentration of particles in the atmosphere. These can be lead, copper, magnesium, iron, traces of fire, ashes, etc.

Stratospheric Ozone Depletion

O_3 depletion is estimated according to the ozone concentration in the atmosphere in Dobson Units.[1]

With the previous subsystems and the highlighted relations among them we built the cognitive map. Even though FCM are a subjective representation of the reality, this representation is not arbitrarily because its constructions is reviewed and processed carefully to extract the system knowledge of the experts.

4   A Cognitive Map of the Earth Climate System

The FMC constructed was based on four different cognitive maps. Two of them based on the analysis of Rockström et al. [18] and Foley [4], and others proposed by Paz-Ortiz [16], and Gay-García [17]. Then, we analyzed the maps separately and a collective cognitive map was created. For maps based on [4, 18] we extracted the concepts mentioned by the authors as those who described the Earth climate stability system, as well as the relations among the concepts. For each relation, we assigned a value of 1 or -1 (according to the positive or negative causality described by the author) to be the weight of the arc representing it. If an author did not mentioned a concept, it was included in its map's matrix without relation (writing 0 in the adjacent matrix) with other concepts. In that way, all the matrices had the same dimensions. The adjacent matrix of the collective map was defined as the sum of all adjacent matrices of the individual maps. Fig. 2 and 3 show the cognitive maps derived from [4, 18]. Cognitive maps of [16, 17] are not shown. Figure 4 shows the collective cognitive

---

[1]Dobson unit is a measure of the ozone layer thickness, equal to 0,01 mm of thickness in normal conditions of pressure and temperature (1 atm and 0 C respectively), expressed as the molecule number. DU represents the existence of 2.69×10^{16} molecules per square centimeter.

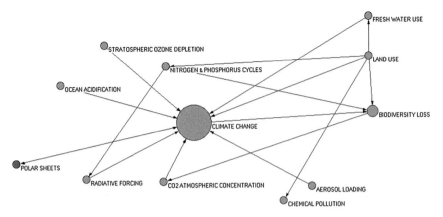

Fig. 2 Cognitive map of the Earth climate system based on Rockström et al. [18]

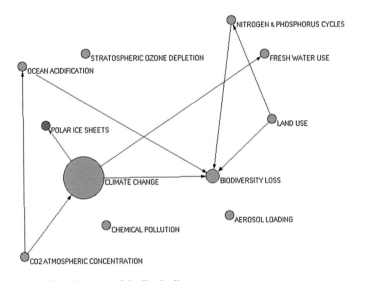

Fig. 3 Foley [4] Cognitive map of the Earth climate system

map created considering the four cognitive maps. Figures were developed in pajek software [http://vlado.fmf.uni-lj.si/pub/networks/pajek/]). Table 1 shows the average (±SD) graph theoretical indices [11] of the individual FCMs and the indices of the collective FCM. The indices allowed us to evaluate the descriptive strength of the model.

In Table 1 the index for the number of connections (W) shows values of 40 for the collective map and 18.5 for the individual ones. Also the number of variables (N) is 11.5 for the individual and 14 for the collective. This illustrate how experts have different perspectives respecting to nodes and relations. The integration of those different perspectives in one model is another powerful characteristic of FCMs.

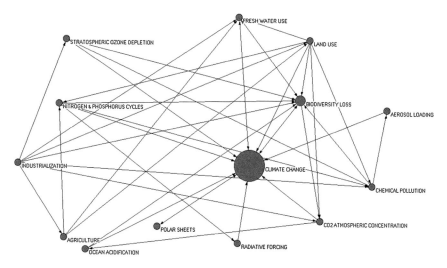

Fig. 4 Collective cognitive map

Table 1 Cognitive map indexes

Indices	Individual maps	Collective CM
Maps	4	1
Variables (N)	11.5 ± 0.5	14
Number of connections (W)	18.5 ± 6.42	40
No. of transmitter variables (T)	2 ± 1.22	1
No. of receiver variables (R)	1.25 ± 0.43	0
Connection/Variable (W/N)	1.84 ± 0.26	2.85
Density $(D = W/N^2)$	0.13 ± 0.04	0.20

As said, a measure of how a cognitive map is connected or sparse, is the *density*, expressed as the number of connections divided by the maximum number of possible connections (N^2). The collective cognitive map is highly connected compared with individual maps.

The transmitter and receiver variables indicate how the experts structured causal relations in a cognitive map. As referred by Papageorgiou [11], cognitive maps containing a larger number of receiver variables consider many outcomes that are a result of the system, while maps containing many transmitter nodes show the "flatness" of a cognitive map where causal arguments are not well elaborated. The number of transmitter and receiver variables diminishs in the collective map in comparison with individual maps. The indices ratio connection/variable (W/N) and Density also show that the collective map provides a strongest model in comparison with individual ones.

4.1 Simulation

To simulate the dynamic of the map, the inference process described in Eq. 1 was used. For the simulation we represented the earth process in the state vector ordering the nodes as follows: 1 Climate Change, 2 Ocean acidification, 3 Stratospheric ozone depletion, 4 Nitrogen and Phosphorus cycles, 5 Fresh water use, 6 Land use, 7 Biodiversity loss, 8 Aerosil loading, 9 Chemical pollution, 10 CO_2 atmospheric concentration, 11 Radiative forcing, 12 Polar sheets, 13 Agriculture, 14 Industrialization. For the basic simulation process, in order to analyze the general behavior of the network, we considered three cases. These will give us information about hidden patterns (obtained when the network is forced), and the net sensitivity (when we use random values). Knowing the hidden patterns (or feedback process) allows policy makers to pay attention in the processes that could be irreversibles. While network sensitivity shows how much the output of the network depends on the initial values, and therefore, how much our approximations for the edges values will influence the results. This allow the stakeholders to consider the model but taking this perspective into account.

Case 1: Initializing the network with random values in the interval [0,1] for nodes and in the interval (0,1] for edges. This consideration is made since we assumed that the links established by the experts cannot have a value of zero, in which case there would be no such link. In the process, all randomly generated values were generated independently.

Case 2: Initializing the network with the same random values used in Case 1 for nodes and edges. And forcing Industrialization (Node 14) by giving a value of 1 to this node after each iteration.

Case 3: Forcing the Industrialization (Node 14), by inducing a recurrent value of 1 in this node after each iteration, and considering an initial value of 0 for all nodes.

Table 2, summarizes these three cases. It contains the Initial values of the nodes, the initial weights used, the resulting vector, and the number of the iterations where the convergence is reached.

Table 2 Basic simulation of the cognitive map

Initial values of nodes	Initial weights	Resulting vector	Number of iterations
Case 1 random values $\in$ [0,1]	random values $\in$ (0,1]	1, 1, 0, 0, 1, 0, 1, 0, 0, 1, 0, 0.3, 0, 0	6
Case 2 Same random values of Case 1 $\in$ [0,1] Industrialization =1 for all t	random values (Case 1) $\in$ (0,1]	1, 1, 1, 0.83, 1, 0.81, 1, 0.6, 1, 1, 0.08, 0.3, 0.3, 1	5
Case 3 Industrialization = 1 for all t	1 for all edges	1, 1, 1, 1, 1, 1, 1, 1, 1, 1, 1, 1, 1, 1	3

Case 1: Cycles and feedback processes.

The resulting vector in the first simulation shows the feedback processes of the network. Nodes inside feedbacks appear with nonzero values in the resulting vector. In our case, we can see that climate change (Node 1), ocean acidification (2), fresh water use (5), biodiversity loss (7), and CO_2 atmospheric concentration (10) converge into a value of 1, and that polar sheets (12, only activated in the map by node climate change and that got an edge random value of 0.3), remain "ON" (with these values). All these nodes are inside feedback processes in the network, for this reason they remain "ON" even when the driver has gone out. As said, these nodes are key to the design of policies, since disturbances in these systems could trigger irreversible processes. In this type of simulation, i.e., initializing some nodes without forcing the network, the resulting values will depend on the values of the nodes and weights [17]. In some cases, where the weights are small enough (compared with concept's values), the value of the concepts is damped to zero as the system is iterated.

Case 2: Dependency of initial conditions (values of nodes and edges).

The resulting vector in Case 2 shows the effect of the weights of the edges in the resulting vector. Even though the vector converges, the values to which each node converges will depend on the strength of the causal connections between them. This hypothesis is supported by Case 3. In which we have forced Industrialization (14) but considering a value of 1 for all connections. Cases 2 and 3 show the sensitivity of the network.

4.2 Fuzzy Weights

In order to have more information form the model, we considered fuzzy weights for the edges. These are fuzzy linguistic variables associated with the relationship between two concepts that determine the grade of causality between them. We created and associated these variables as proposed in [11]. We used 13 fuzzy quantities taken from the strength suggested for each relation by each one of the expert in its original map. The quantities used were T{influence}= {negatively very very strong, negatively very strong, negatively strong, negatively medium, negatively weak, negatively very weak, zero, positively very weak, positively weak, positively medium, positively strong, positively very strong, positively very very strong}. Then, the results were aggregated by using the SUM method [12] to produce an overall linguistic weight for each edge. The resulting linguistic weights were defuzzified using the center of gravity (CoG) [22] that takes all the linguistic weights and creates numerical values within the interval [−1, 1]. The overall linguistic values are shown in Table 4.

To obtain initial values for the edges we considered the current state of the earth-system processes referred by [18] (shown in Table 3). We can see that subsystems: Climate change, Rate of biodiversity loss and Nitrogen cycle (i.e., three of nine interlinked planetary boundaries), have already transgressed their proposed limits. Stratospheric ozone depletion and Ocean acidification are just above. Phosphorus

Table 3 Current state of the planetary boundaries

Earth-system processes	Parameters	Proposed boundary	Current status
Climate change	CO_2 concentration, changes in radiative forcing	350 ppm, 1 W/m^2	280 ppm, 1.5 W/m^2
Rate of biodiversity loss	Extinction rate	10	> 100
Nitrogen (with P cycle)	N_2 removed for human use	35	121
Phosphorus (with N cycle)	Quantity of P flowing into the oceans	11	8.5–9.5
Stratospheric ozone depletion	O_3 concentration	276	283
Ocean acidification	Global saturation of aragonite in surface sea	2.75	2.90
Global fresh water use	Consumption (km^3/year)	4000	2600
Change in land use	Percentage of land cover converted to cropland	15	11.7
Atmospheric aerosol loading	Particulate concentration, on a regional basis	to be determined	–
Chemical pollution	Amount emitted to, or concentration	to be determined	–

cycle, global fresh water use, and Change in land use are close to the boundary. While Atmospheric aerosol loading and chemical pollution systems have not yet an established boundary. We assigned for each concept a fuzzy weight by considering its current position with respect to the proposed boundary. For the industrialization, agriculture, and polar sheets processes, which are not considered in this frame, we used the description given by [17, 18] to set the initial values. Again, the overall fuzzy quantifiers were defuzzified by using the CoG method into a numerical values for the system's simulation. The initial defuzzified numerical values for the concepts are shown in the initial vector at the top of Table 4. For this simulation the map was iterated while keeping the forcing in node Industrialization (14) with its initial value of 0.4. The resulting vector converged after 10 iterations to:

0.69, 0.21, 0.25, 0.27, 0.62, 0.32, 0.90, 0.17, 0.56, 0.43, 0.08, 0.42, 0.24, 0.4 (button of the Table 4).

It can be seen that nodes climate change (1) and biodiversity loss (7) referred as concepts who have already crossed the boundary appeared with the highest values of 0.69 and 0.90, respectively. However, the node representing nitrogen and phospho-

Table 4 Comparison between initial and final vector (with initial and final weights). Nodes-relation and linguistic (initial and final) quantifiers and its respective associated numerical value

Initial Vector	0.6, 0.1, 0.3, 0.8, 0.1, 0.1, 0.7, 0.2, 0.3, 0.7, 0.8, 0.2, 0.3, 0.4 Iterations: 0
Nodes—relation	Linguistic value—initial weight—final weight
1 - 5 Climate change - Fresh water use	positively very weak, 0.3 -
1 - 12 Climate change - Polar sheets	positively strong, 0.6, -
1 - 7 Climate change - Biodiversity loss	positively weak, 0.3, -
2 -1 Ocean acidification - Climate change	positively very weak, 0.1, -
2 - 7 Ocean acidification Biodiversity loss	positively very weak, 0.1, -
3 - 1 Stratospheric ozone depletion - Climate change	positively weak, 0.2, -
3 - 7 Stratospheric ozone depletion - Biodiversity loss	positively very weak, 0.1, -
4 - 9 N P cycles - Chemical pollution	positively medium, 0.5, -
4 - 7 N P cycles - Biodiversity loss	positively very weak, 0.1, -
4 - 1 N P cycles - Climate change	positively very weak, 0.1, -
4 - 11 N P cycles - Radiative forcing	positively weak, 0.3, -
5 - 7 Fresh water use - Biodiversity loss	positively very weak, 0.1, -
5 - 1 Fresh water use - Climate change	positively very weak, 0.1, -
6 - 9 Land use - Chemical pollution	positively strong, 0.6, -
6 - 4 Land use - N P cycles	positively medium, 0.4, -
6 - 7 Land use - Biodiversity loss	positively strong, 0.6, -
6 - 1 Land use - Climate change	positively weak, 0.2, -
6 - 5 Land use - Fresh water use	positively weak, 0.3, -
6 - 10 Land use - CO_2 concentration	positively weak, 0.2, -
7 - 10 Biodiversity loss - CO_2 concentration	positively very weak, 0.1, -
7 - 1 Biodiversity loss - Climate change	positively very weak, 0.05, -
8 - 1 Aerosol loading - Climate change	positively very weak, 0.05, -
9 - 8 Chemical pollution - Aerosol loading	positively weak, 0.3, -
9 - 7 Chemical pollution - Biodiversity loss	positively medium, 0.5, -
9 - 3 Chemical pollution - Stratospheric ozone depletion	positively weak, 0.3, -
10 - 2 CO_2 concentration - Ocean acidification	positively medium, 0.5, -
10 - 1 CO_2 concentration - Climate change	positively very strong, 0.7, -
11 - 1 Radiative forcing - Climate change	positively medium, 0.4, -
12 - 1 Polar sheets - Climate change	positively weak, 0.2, -
13 - 4 Agriculture - NP cycles	positively strong, 0.6, -
13 - 6 Agriculture - Land use	positively medium, 0.5, -
13- 5 Agriculture - Fresh water use	positively medium, 0.5, -

(continued)

Table 4 (continued)

Initial Vector	0.6, 0.1, 0.3, 0.8, 0.1, 0.1, 0.7, 0.2, 0.3, 0.7, 0.8, 0.2, 0.3, 0.4 Iterations: 0
14 - 13 Industrialization - agriculture	positively strong, 0.6 , 0.5
14 - 6 Industrialization - Land use	positively medium, 0.5, 0.3
14 - 5 Industrialization - Fresh water use	positively medium, 0.5, 0.45
14 - 10 Industrialization - CO_2 concentration	positively very strong, 0.7, 0.45
14 - 9 Industrialization - Chemical pollution	positively strong, 0.6, 0.55
14 - 7 Industrialization - Biodiversity loss	positively weak, 0.2, 0.15
14- 3 Industrialization - Stratospheric ozone depletion	positively weak, 0.2, 0.1
FINAL VECTOR WITH INITIAL WEIGHTS	0.69, 0.21, 0.25, 0.27, 0.62, 0.32, 0.90, 0.17, 0.56, 0.43, 0.08, 0.42, 0.24, 0.4 Iterations: 10
FINAL VECTOR WITH FINAL WEIGHTS	0.49, 0.14, 0.17, 0.20, 0.49, 0.22, 0.67, 0.13, 0.45, 0.29, 0.06, 0.29, 0.2, 0.4 Iterations: 24

rous cycles (4) exhibited a value of 0.27. This can be a consequence that although the threshold is established considering both processes, the distance of each process in respect to its proposed boundary is slightly different, so when the concepts were defuzzified the model could have lost accuracy. Also nodes fresh water use (5), and chemical pollution (9) exhibited high values of 0.62 and 0.56, respectively. In the first case, it is possible that the established edges, as well as the relations, are over-estimating the state of the system. In the second case, since there are no data on the system state, this simulation is interesting since it allows us to estimate the state of this node.

4.3   Training the Fuzzy Cognitive Map by Using Nonlinear Hebbian Learning

Up to now, the results of the simulation qualitatively describe the state of the climate system. However, in order to establish action strategies that allow for planning and support decision processes in terms of environmental policy, it is necessary to know what might be the changes of the weights in order to obtain a desired state for the system. To do this, we used the nonlinear Hebbian learning (NHL) algorithm proposed by Papageorgiou [14] to adjust the weights of the map. This algorithm was used because it is referred by Papakostas [15] in a comparative study as the one which exhibits a satisfactory behavior in control systems, having at the same time a low algorithmic complexity when it is compared with similar algorithms [15].

Two restrictions for the algorithm were used. First: Do not change the edges values that describe relationships between concepts that operate over natural processes. For example, it cannot modify the dependency of climate change with respect to the increase in the atmospheric CO_2 concentration, since doing so it would be "modifying" a natural process. Therefore, the algorithm was restricted for to only change the weights of the edges starting out from the node industrialization. The values of the other edges were kept equal after each iteration. Second: the weight of these edges could not be less than zero, since the causal relations were positively established by experts. Therefore, the value of the concepts were kept > 0.

The algorithm operates by using targets for the values of the concepts to be adjusted. For this, only targets over variables climate change (1) and biodiversity loss (7) were used. This is because we were interested in reducing the concepts that are operating over the proposed boundary that were simulated by the model. The targets were established to perform a reduction in this two variables by 0.2 under its initial value. The values of the other concepts, in the state vector, were allowed to variate freely. Then, the algorithm operates as follows:

Step 1: Given the initial vector (A^0) and the initial adjacent matrix of weights W^0.
Step 2: For each iteration k.
Calculate A^k by using Eq (1).
Update W^k for edges coming out Industrialization by using the equation:

$$w_{ij}^{(t+1)} = w_{ij}^{(t)} + \eta A_j^{(t)} \left(A_i^{(t)} - sgn(w_{ij}^{(t)} A_j^{(t)} w_{ij}^{(t)}) \right) \qquad (5)$$

Where η is the learning rate, and $sgn(\cdot)$ function takes the sign of the argument. Calculate error associated with climate change and biodiversity loss, defined as the difference between the current value of each concept and its established target value. Stop when error $\leq$ acceptable difference (0.05).
Step 3: Return the final weights.

The final weights of the concepts are shown in Table 4. The system converges into a final vector:
0.49, 0.14, 0.17, 0.20, 0.49, 0.22, 0.67, 0.13, 0.45, 0.29, 0.06, 0.29, 0.2, 0.4 after 24 iterations. We found the nodes climate change and biodiversity loss, 0.2 and 0.23 below its original values, respectively. This coincides with the established targets of 0.2 units below for each one.

A first remarkable result is that the greater reduction appears in the relation industrialization - CO_2 atmospheric concentration, that goes from 0.7 to 0.45. This can be interpreted from the point of view of the possible action policies as the most important processes that must be taken into account.

In general, we can arrange our results in four classes, depending on the reduction of the weights. Relations with minimal reduction (0.05), industrialization—fresh water use, industrialization—chemical pollution, and industrialization—biodiversity loss). Relations with reduction of 0.1, industrialization—stratospheric ozone depletion and industrialization—agriculture. Relations with a reduction of 0.2, industrialization—Land use, and concepts with weight reduction of 0.25 industrialization—CO_2 atmospheric concentration. These results can be useful to establish different levels of action. However, it is important to say that declaring different levels of action do not imply neglect other processes while serving one. It is necessary to take into account the fact that all the systems are interrelated, as well as the nonlinear processes that could appear in their interactions. Nevertheless, these results allow us to establish hierarchies for a policy design.

5 Conclusions and Further Work

The methodology presented for evaluation and simulation of the climate system showed qualitatively consistent results, both with climate system state, as well as with the expected scenarios. Moreover, the adjustments to the weights obtained through the implementation of the algorithm (where we clearly observed different levels of adjustment) allow the design and assessment of environmental policies, helping at the same time, to their planning.

Given that the system is highly sensitive to changes in the strength of interactions between subsystems, and considering the feedback processes (Sect. 4.1), it is clear that more research must be developed to increase the description accuracy of the model in the first case, and to analyze possible irreversible degradation processes within the feedback loops. However, since the system allows to asses and differentiate the importance between these relationships, further research can be prioritized. In that sense, this model can also help to the elaboration of the research agenda.

The methodological approach presented (based on [11] clearly allows the construction of stronger collective cognitive maps when compared with the capacity of descriptions of the individual ones. As cognitive maps highly depend on the expert's opinion, this methodology also allows us to evaluate the strength and descriptive capacity of the model. The performed simulation was capable to identify the feedback processes, and when linguistic variables were used and aggregated to produce an overall linguistic weight for each edge with the associated defuzzification by using center of gravity, the resulting model matched with the current description of the climate system referred by Rockström [18]. Although the algorithm for the adjustment of the weights was restricted, the adjustments where the principal reduction occurs in the relation industrialization—CO_2 atmospheric concentration, were plausible in the context of the current reports on climate change. This, together with the results of the simulations, supports the idea that the developed model can be used for the planning, implementation, and evaluation of policies. A possible further work, in order to analyze the adjustments performed, could implement migration or evolutionary algorithms to adjust the weights [21] and evaluate the performance of each type from the point of view of the stakeholders.

Acknowledgments The present work was developed with the support of the Programa de Investigación en Cambio Climático (PINCC) of the Universidad Nacional Autónoma de México (UNAM).

References

1. Cakmak, E., Dudu, H., Eruygur, O., Ger, M., Onurlu, S., Tonguc, Ö.: Visions for the future of water in Seyhan basin, Turkey: a backcasting application international environmental modelling and software society (iEMSs). In: Swayne, D.A., Yang, W., Voinov, A.A., Rizzoli, A., Filatova, T. (eds.) International Congress on Environmental Modelling and Software Modelling for Environments Sake, Fifth Biennial Meeting, Ottawa, Canada. http://www.iemss.org/iemss2010/index.php?n=Main.Proceedings (2010)
2. Eden, C.: On the nature of cognitive maps. J. Manage. Stud. **29**, 261265 (1992)
3. Eden, C., Ackermann, F., Cropper, S.: The analysis of cause maps. J. Manage. Stud. **29**, 309–324 (1992)
4. Foley, JA.: Boundaries for a healthy planet. Sci. Am. **302**, 54–57 (2010)
5. IPCC Intergovernamental Panel on Climate Change: Cambio climtico 2007: Informe de sntesis. Contribucin de los Grupos de trabajo I, II y III al Cuarto Informe de evaluacin del Grupo Intergubernamental de Expertos sobre el Cambio Climtico. IPCC, Suiza (2007)
6. IPCC: Summary for policy makers. In: Field, C.B., Barros, V.R., Dokken, D.J., Mach, K.J., Mastrandrea, M.D., Bilir, T.E., Chatterjee, M., Ebi, K.L., Estrada, Y.O., Genova, R.C., Girma, B., Kissel, E.S., Levy, A.N., MacCracken, S., Mastrandrea, P.R., White, L.L. (eds.) Climate Change 2014: Impacts, Adaptation, and Vulnerability. PartA: Global and Sectoral Aspects. Contribution of Working Group II to the Fifth Assessment Report of the Intergovernmental Panel on Climate Change, pp. 1–32. Cambridge University Press, Cambridge, United Kingdom and NewYork, NY, USA (2014)
7. Kosko, B.: Fuzzy cognitive maps. Man Mach. Stud. **24**, 65–75 (1986)
8. Kosko, B.: Neural Networks and Fuzzy Systems: A Dynamical Systems Approach to Machine Intelligence. Prentice Hall, New Jersey (1991)
9. Mace, G., Masundire, H., Baillie, J., Ricketts, T., Brooks, T.: Chapter 4: Biodiversity in "Ecosystems and Human Well-being: Current State and Trends". Island Press, USA (2005)
10. Ozesmi, U., Ozesmi, S.L.: Ecological models based on peoples knowledge: a multi- step fuzzy cognitive mapping approach. Ecol. Model. **176**, 4364 (2004)
11. Papageorgiou, E., Kontogianni, A.: Using fuzzy cognitive mapping in environmental decision making and management: a methodological primer and an application. In: Young, S. (ed.) International Perspectives on Global Environmental Change. ISBN: 978-953-307-815-1, InTech. http://www.intechopen.com/books/international-perspectives-on-global-environmental-change/using-fuzzy-cognitive-mapping-in-environmental-decision-making-and-management-a-methodological-prime (2012)
12. Papageorgiou, E., Stylios, C.D.: Fuzzy cognitive maps. In: Pedrycz, W., Skowron, A., Kreinovich, V. (eds.) Handbook of Granular Computing, Chapter 34, pp. 755–775. Wiley, New York (2008)
13. Papageorgiou, Elpiniki I., Stylios, Chrysostomos D., Groumpos, Peter P.: Fuzzy Cognitive Map Learning Based on Nonlinear Hebbian Rule. In: Gedeon, Tamás (Tom) Domonkos, Fung, Lance Chun Che (eds.) AI 2003. LNCS (LNAI), vol. 2903, pp. 256–268. Springer, Heidelberg (2003)
14. Papageorgiou, E.I., Stylios, C.D., Groumpos, P.P.: Unsupervised learning techniques for fine-tuning fuzzy cognitive map casual links. Int. J. Hum. Comput. Stud. **64**(8), 727–743 (2006)
15. Papakostas, G.A., Polydoros, A.S., Koulouriotis, D.E., Tourassis, V.D.: Training fuzzy cognitive maps by using Hebbian learning algorithms: a comparative study. In: IEEE International Conference on Fuzzy Systems, pp. 851–858 (2011)
16. Paz-Ortiz, A.: Uso de mapas cognitivos para el estudio de la estabilidad en sistemas climáticos terrestres. Master degree thesis, National Autonomous University of Mexico, Mexico (2011)
17. Gay-García, C., Paz-Ortiz, I.: Stability analysis of climate system using fuzzy cognitive maps. In: Obaidat, M.S., Filipe, J., Kacprzyk, J., Pina, N. (eds.) Simulation and Modeling Methodologies, Technologies and Applications International Conference, SIMULTECH 2012, Rome, Italy, 28–31 July 2012

18. Rockström, J., Steffen, W., Noone, K., Persson, A., Chapin, F.S., Lambin, E.F., Lenton, T.M., Scheffer, M., Folke, C., Schellnhuber, H.J., Nykvist, B., Wit, C.A., Hughes, T., Van der Leeuw, S., Rodhe, H., Srlin, H., Snyder, P.K., Costanza, R., Svedin, U., Falkenmark, M., Karlberg, L., Corell, R.W., Fabry, V.J., Hansen, J., Walker, B., Liverman, D., Richardson, K., Crutzen, P., Foley, J.A.: A safe operating space for humanity. Nature **461**, 472–475 (2009)
19. Stylios, C.D., Georgopoulos, V.C., Groumpos, P.P.: The use of fuzzy cognitive maps in modeling systems. In: 5th IEEE Mediterranean Conference on Control and Systems MED5 paper, vol. 67, p. 7 (1997)
20. Soler, L., Kok, K., Camara, G., Veldkamp, A.: Using fuzzy cognitive maps to describe current system dynamics and develop land cover scenarios: a case study in the Brazilian Amazon. J. Land Use Sci. **iFirst**, 127 (2011)
21. Vaščák, J.: http://neuron-ai.tuke.sk/vascak/publications/p-11.pdf (2012)
22. Zadeh: http://www-bisc.cs.berkeley.edu/zadeh/papers/1986-CWW.pdf (1986)

The Optimization of a Surgical Clinical Pathway

Roberto Aringhieri and Davide Duma

Abstract A Clinical Pathway (CP) can be conceived as an algorithm based on a flow chart that details all decisions and treatments related to a patient with a given pathology. CPs can be considered an operational tool in the clinical treatment of diseases, from a patient-focused point of view. Although it has been shown their benefits in clinical practices, little attention has been dedicated to study how CP can optimize the use of resources. We focus our attention on the analysis of a surgical CP from a patient-centred point of view in order to optimize the most critical resources of a surgical CP, and to evaluate the impact of the optimization with respect to a set of patient- and facility-centred indices.

Keywords Surgical clinical pathway · Operating room management · Simulation · Optimization

1 Introduction

The current development of the health care systems is aimed to recognize the central role of the patient as opposed to the one of the health care providers. In this context, Clinical Pathways (CPs) shift the attention from a single health benefit to the health care chain that starts to resolve the illness episode. They can be defined as "health-care structured multidisciplinary plans that describe spatial and temporal sequences of activities to be performed, based on the scientific and technical knowledge and the organizational, professional and technological available resources" [1].

R. Aringhieri · D. Duma (✉)
Department of Computer Science, Università degli Studi di Torino,
Corso Svizzera 185, 10149 Torino, Italy
e-mail: davide.duma@unito.it

R. Aringhieri
e-mail: roberto.aringhieri@unito.it

© Springer International Publishing Switzerland 2015
M.S. Obaidat et al. (eds.), *Simulation and Modeling Methodologies,
Technologies and Applications*, Advances in Intelligent Systems
and Computing 402, DOI 10.1007/978-3-319-26470-7_16

A CP can be conceived as an algorithm based on a flow chart that details all decisions, treatments, and reports related to a patient with a given pathology, with a logic based on sequential stages [2]. A CP is therefore "the path" that a patient suffering from a disease walks in the National Health System. This pathway can be analysed at a single, local level of care (a single hospital, or a single region) or globally, taking into account every level of health care from education and prevention, to diagnosis of diseases, treatment, and recovery. CPs are specifically tailored to stimulate continuity and coordination among the treatments given to the patient through different disciplines and clinical environments. For this reason, they can be considered an operational tool in the clinical treatment of diseases, from a patient-focused point of view [3].

The aim of a care pathway is to enhance the quality of care by improving patient outcomes, promoting patient safety, increasing patient satisfaction, and optimizing the use of resources as stated by the European Pathway Association. Moreover, while many papers show that, appropriately implemented, CPs have the potential to increase patient outcome, reduce patient length of stay, and limit variability in care, thereby yielding cost savings [4], little attention has been dedicated to study how CP can optimize the use of resources as reported in [5] to which the reader can refer to deepen this topic.

In this chapter, we focus our attention on the analysis of a surgical CP from a patient-centred point of view. The main concern of this work is the introduction of some optimization modules in the management of the most critical resources in a surgical CP, that is, the stay beds and the operating rooms, and to evaluate their impact with respect to a set of patient- and facility-centred indices. Many optimization solutions have been proposed for the operating room management as reported in [6, 7].

The aims are to reduce the waiting list according to a prioritized admission system, to operate patients within a given time limit depending on their level of urgency, to improve the utilization of the above critical resources, and to minimize the number of cancellations.

As reported in [8], health care optimization problems are challenging, often requiring the adoption of unconventional solution methodologies. The solution approach proposed herein belongs to this family. Our approach is a hybrid simulation and optimization model. Simulation is used in order to generate a real situation with respect to the inherent stochasticity of the problem while optimization is used to take the best decisions in different points of the surgical CP. Accordingly to [9], we consider the operative decisions concerning the advanced scheduling and allocation scheduling of patients. Furthermore, we consider the real time management of the operating room planning.

The chapter is organized as follows. The problem is depicted in Sect. 2 while the integrated simulation and optimization model is discussed in Sect. 3. Model validation and its main results are discussed in Sect. 4 and Sect. 5, respectively. Section 6 closes the chapter.

2 Problem Statement

We consider the problem of managing a single surgical clinical pathway taking into account the optimization problems arising when dealing with the decision levels concerning the advanced and the allocation scheduling [9]. In order to guarantee the execution of such decisions, we deal with the Real Time Management (RTM) of an operating room planning. RTM consists in a sort of centralized surveillance system whose main task is to supervise the execution of the planning and, in the case of delays, to take a decision regarding the patient cancellation or the overtime assignment.

The definition of the surgical CP is inspired to that presented and analysed in [10] for the thyroid surgical treatment. The reader can refer to this paper for further details. From a management point of view, a surgical CP can be seen as made up of three phases.

The first phase concerns the *pre-admission phase* and it is related to all the activities regarding the patients before the admission as depicted in Fig. 1.

In this phase, a relevant information is the Diagnosis Related Groups (DRG). A DRG defines a general time limit before which the patient should be operated on. In our context, an *Urgency Related Group* (URG) is assigned to each patient belonging to the same DRG: the URG states a more accurate time limit. In other word, URG allows to define a partition of the patients in such a way to prioritize their surgery. The optimization problem arising in this phase—the advanced scheduling problem—is that of selecting patients from the (usually long) waiting list and to assign them to an *OR session* (i.e. an operating room on a given day) in such a way to satisfy several operative constraints (number of beds available during the patient stay, total time available for the OR session, and so on). Our objective is to maximize the utilization of the operating rooms in each day in such a way to guarantee that each patient is operated within the time limit defined by the URG. This problem is well known in literature as Surgical Case Assignment Problem (SCAP) [11].

The *hospital phase* is concerned with all the activities involving the admitted patient stay except for those related to the operating theatre as depicted in Fig. 2a. The relevant information in this phase is the Length Of Stay (LOS) of each patient,

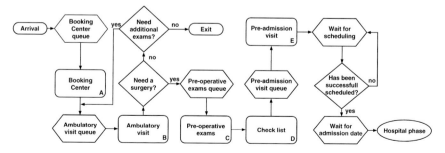

Fig. 1 Pre-admission phase flowchart

(a)   **(b)**

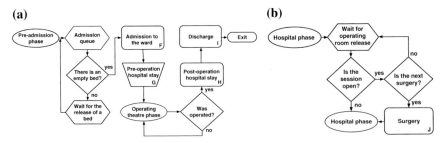

Fig. 2 Flowcharts of the hospital and operating theatre phases

that is, the number of days required before the discharge. The optimization problem arising in this phase—the allocation scheduling problem—is that of finding a sequence of patients to decide the order in which they are operated on. The objective is to minimize the risk of cancellation according to their assigned position in the sequence taking into account a patient-centred point of view (considering waiting time, class of urgency, possible previous referrals) with respect to the available operating time.

Figure 2b depicts the *operating theatre phase* which is a component of the hospital phase—as highlighted in Fig. 2a. Due to its importance in a surgical CP, it requires to be treated separately. Patients assigned to a given OR session will be operated on following the sequence previously defined unless delays impose to define a new sequence. Patient not operated on will be rescheduled. The optimization problem arising in this phase is the real time management of the operating room planning. When the *Estimated Operating Time* (EOT) differs from the *Real Operating Time* (ROT), we could have delays. If such delays become significant, that is, exceeding the total operating time allowed, the real time management should deal with the following possible decision to be taken every time a patient finish its surgery:

- to use some overtime reducing the total amount weekly available;
- to cancel 1 or more patients and to reschedule them, if possible;
- to change the sequence of the remaining patients in order to minimize the number of cancellations.

The first two choices are generally non-trivial and alternatives requiring to consider several aspects. For instance, the decision of postponing a patient could violate the so-called Maximum Time Before Treatment (MTBT) determined by its URG. Further, it determines an increased patient stay lowering the patient satisfaction and, by consequence, the quality of the service. These decisions have to take into account the inherent uncertainty. On the other side, overtime is a scarce resource. So, it seems crucial to establish some criteria driving the decisions to use the overtime to avoid cancellations. Table 1 introduces the notation of the problem used hereafter in the chapter.

Table 1 Notation

N: number of OR sessions	S_j: duration of j-th OR session
d_j: operating day (from Mon to Fri) of the j-th OR session	k: index of the day, $k = 1, \ldots, 7$
B_k: number of beds available on the k-th day of the week	Ω: weekly overtime available
I: set of patients in the pre-admission waiting list	L: set of scheduled patients
$L^{(j)}$: set of patients scheduled into the j-th OR session	M_i: MTBT of patient i
t_i: waiting time of the i-th patient	ℓ_i: LOS of patient i
e_i: EOT of patient i	r_i: ROT of patient i

3 The Hybrid Simulation and Optimization Model

This section discusses the hybrid simulation optimization model proposed in this chapter. Simulation allows to deal with the inherent stochasticity of the problem while optimization allows to deal with the three problems arising in the three phases depicted in Sect. 2.

3.1 The Simulation Framework

The simulation framework is based on a DES methodology. It is a straightforward implementation of the surgical CP depicted in Figs. 1 and 2. Simulation makes the whole model capable to deal with the inherent stochasticity of the problem. The literature usually considers three main sources of uncertainty, that is, the arrival of patients, the variability of patient length of stays, and the variability of patient operating times (see, e.g. [12–14]).

Note that the hybrid model is implemented using AnyLogic 6.9 [15]: its Enterprise Library is exploited for the implementation of the DES simulation framework whilst the optimization modules are implemented from scratch in Java.

The main parameters of the simulation model, and their distribution, are depicted in the following.

3.1.1 Parameters

We report the parameters of the simulation framework and their setting for the model validation and the quantitative analysis. In brackets is given the unit of measure.
Flow and Patient Characteristics:

r_0: patient interarrival rate [patients/minutes],
R_0: initial length of the pre-admission waiting list [patients],
p_1: patient probability to require a surgical treatment during the ambulatory visit (see Fig. 1),

p_2: patient probability to do not require a surgical treatment but requiring further exams during the ambulatory visit (see Fig. 1),

$p_A, \ldots, p_G$: urgency class A,…,G patient probability.

Duration of Activities:

$T^{min,avg,mod}_{A,\ldots,F,I}$: minimum, average, and modal for the service time of A,…,F,I [minutes] (see Figs. 1 and 2b),

$\ell_{min, max, mod; A,\ldots,G}$: minimum, maximum, and modal LOS for patients belonging to the urgency class A,…,G [days],

$\bar{e}_{A,\ldots,G}$: average EOT for the surgery of a patient belonging to the urgency class A,…,G [minutes],

e_{max}: maximum duration of a surgery [minutes],

$\sigma_{A,\ldots,G}$: EOT standard deviation [minutes],

σ: ROT standard deviation for each patient [minutes],

τ: tolerance time within which the surgical team operates a patient at the end of OR session without resorting to the overtime [minutes].

Table 2 shows the distributions used to generate the required time for the execution of the activities A,…,J. Table 3 reports the values assigned to the parameters for the model validation and for the quantitative analysis.

The parameters k and ϑ for the Gamma distributions (see Table 2) were obtained from the empirical data reported [10], requiring that the expected and the modal values of these distributions coincide with the empirical values reported in that paper. Further, we compute the value of the survival function on the maximum time for the execution of activities (always reported in the paper), obtaining a value less than 10 %. The choice to use a lognormal distribution derives from the literature (see, e.g. [16–18]).

Table 2 Distribution of the activity durations

Activities	Durations	Parameters
A,…, F, I	$T^{A,\ldots,F,I}_{min} + T,$ $T \sim \text{Gamma}(k, \vartheta)$	$k = T^{A,\ldots,F,I}_{avg} - T^{A,\ldots,F,I}_{mod},$ $\vartheta = \dfrac{T^{A,\ldots,F,I}_{avg} - T^{A,\ldots,F,I}_{min}}{T^{A,\ldots,F,I}_{avg} - T^{A,\ldots,F,I}_{mod}}$
H (LOS)	$\lfloor \text{Triangular}(l_{min; A,\ldots,G}, l_{max; A,\ldots,G}, l_{mod; A,\ldots,G}) + \frac{1}{2} \rfloor$	
J (EOT)	$\min\left\{ \max\left\{ \lfloor \frac{T}{u} + \frac{1}{2} \rfloor u, 0 \right\}, e_{max} \right\},$ $T \sim \text{Lognormal}(\mu, s^2)$	$\mu = \log \epsilon_{A,\ldots,G} - \frac{1}{2} \log\left(\frac{\sigma^2_{A,\ldots,G}}{\epsilon^2_{A,\ldots,G}} + 1 \right),$ $s = \sqrt{\log\left(\frac{\sigma^2_{A,\ldots,G}}{\epsilon^2_{A,\ldots,G}} + 1 \right)}$
J (ROT)	$\min\left\{ \max\left\{ 0, T \right\}, e_{max} \right\},$ $T \sim \text{Gaussian}(\text{EOT}, \sigma^2)$	

Table 3 Parameters used in the simulation framework

Parameters	Unit of measure	Validation	Quantitative analysis
r_0	Patients/minutes	$5.8 \cdot 10^{-3}$	$2.0 \cdot 10^{-2}$
R_0	Patients	140	420
p_1, p_2		0.2, 0.1	0.2, 0.1
$p_A, \cdots, p_G$		0.0245, 0.1401, 0.4136, 0.1785 0.1140, 0.0749, 0.0544	0.0245, 0.1401, 0.4136, 0.1785 0.1140, 0.0749, 0.0544
$T_{min}^{A,...,F,I}$	Minutes	5, 25, 25, 25, 40, 25, 35	5, 25, 25, 25, 40, 25, 35
$T_{avg}^{A,...,F,I}$	Minutes	7.5, 31.5, 31, 28, 62.5, 32, 41	7.5, 31.5, 31, 28, 62.5, 32, 41
$T_{mod}^{A,...,F,I}$	Minutes	6, 30, 26, 25, 50, 30, 40	6, 30, 26, 25, 50, 30, 40
$\ell_{min; A,...,G}$	Days	2, 1, 1, 1, 1, 1, 1	2, 1, 1, 1, 1, 1, 1
$\ell_{max; A,...,G}$	Days	29, 16, 7, 9, 5, 5, 5	29, 16, 7, 9, 5, 5, 5
$\ell_{mod; A,...,G}$	Days	3, 2, 2, 2, 2, 2, 2	3, 2, 2, 2, 2, 2, 2
e_{max}	Minutes	360	420
$\bar{e}_{A,...,G}$	Minutes	145, 171, 149, 153, 171, 164, 166	145, 171, 149, 153, 171, 164, 166
$\sigma_{A,...,G}$	Minutes	85, 85, 66, 60, 61, 51, 60	85, 85, 66, 60, 61, 51, 60
σ	Minutes	0	30
τ	Minutes	30	10
Ω	Minutes	0	300
u	Minutes	30	30
$B_1, \ldots, B_7$	Beds	18, 18, 18, 18, 18, 18, 18	50, 50, 50, 50, 50, 35, 35
$M_{URG\,A,...,URG\,G}$	Days	8, 15, 30, 60, 90, 120, 180	8, 15, 30, 60, 90, 120, 180

3.2 The Advanced Scheduling Problem

Inspired by [19], we proposed a simple metaheuristic based on a greedy construction of an initial solution and then a local search to improve that solution. Note that we should take into account the fact that the resources available can be reduced since patients admitted the previous week are already in the hospital phase, usually waiting for the discharge but also for their surgery.

3.2.1 Constructive Greedy Algorithm

The algorithm associates to each patient $i \in I$ the following values:

$$w_i = \frac{t_i + \min_{1 \leq j \leq N} d_j}{M_i}, \tag{1}$$

$$\tilde{w}_i^k = \frac{t_i + \min_{1 \leq j \leq N} d_j + \pi(k)}{M_i} = w_i + \frac{\pi(k)}{M_i}, \tag{2}$$

where $\pi(k)$ measures the distance of the current day k to the next Friday, that is, the day in which the next week schedule is decided: for instance, at the moment of determining a solution for the advance scheduling problem $\pi(k)$ is equal to 7. The value w_i measures the ratio of the time elapsed before the surgery and the MTBT associated to the URG of the patient $i \in I$ whilst $\tilde{w}_i$ is the same meaning but referred to the next week.

Patients are ordered by decreasing value of w_i in such a way to promote the scheduling of those patients which are close to their MTBT. Then, each patient is considered for the scheduling. A patient will be inserted in the current schedule if there exists an OR session available in such a way so as to satisfy the operative constraints regarding the bed occupation and the operating time S_j.

Among different possible OR sessions, the algorithm tries to schedule the patient first in a day k such that $k + \ell_i \leq 5$. If it is not possible, the algorithm tries the insertion in a day k such that $k + \ell_i > 5$. The rationale here is to avoid the use of the weekend stay beds which are usually a limited resource. This rule can be overridden when $\tilde{w}_i \geq 1$ assigning the patient to the first day $k = 1$, if possible, or to the second day $k = 2$, and so on. In this case, we would like to reduce the probability to do not satisfy the URG requirements in case of cancellation.

Finally, if a patient cannot be scheduled, the algorithm will consider the next patient. The algorithm terminates when all patients are considered for the insertion in the schedule.

3.2.2 Improvement Local Search Algorithm

The Local Search tries to improve the solution computed by the greedy exchanging pairs of patients already scheduled in such a way to cluster them in a reduced number of OR sessions and, by consequence, to allow the insertion of new patients previously not scheduled. On a given schedule, let j^* be the OR session having the maximum operating time yet available, that is, the one having the minimal utilization. The local search algorithm follows these criteria to select a new solution when exploring the neighbourhood:

- the new solution will be that providing the maximal increase of the time yet available of j^*;
- otherwise, if the two schedules are equivalent in j^*, the algorithm will consider the second less-utilized OR session, and so on;
- otherwise, if the two schedules are equivalent in all OR sessions, the algorithm selects those solutions having OR sessions less utilized at the end of the week.

3.3 The Allocation Scheduling Problem

In our settings, the allocation scheduling problem consists in establishing the order in which patients $i \in L^{(j)}$ will be operated on in such a way to minimize the inefficiency due to possible cancellations.

Considering a given schedule, there is a set of patients for which it is better to avoid the cancellation of their surgical operation, that is, those patients whose $\tilde{w}_i^k$ is greater than or equal to 1 and those patients whose surgery was already postponed. To deal with these special cases, let us introduce the following values:

$$W_i = \begin{cases} \tilde{w}_i^k & \text{if } \tilde{w}_i^k > 1 \\ 0 & \text{otherwise} \end{cases}, \tag{3}$$

and let $D_i > 0$ be the number of days elapsed after a cancellation, 0 otherwise. Finally, we define the value

$$s_i = \alpha_1 W_i + \alpha_2 D_i + \alpha_3 e_i \tag{4}$$

for each $i \in L^{(j)}$ where α_1, α_2, and α_3 are parameters. Setting

$$\alpha_1 \gg \alpha_2 \gg \alpha_3 = \begin{cases} 1 & \text{case (A)} \\ -1 & \text{case (B)} \end{cases},$$

the sequencing of patients $i \in L^{(j)}$ is simply obtained by ordering them by the decreasing order of the associated s_i.

Within such ordering, the use of α imposes three levels of priorities when breaking ties. First, we schedule patients close to their MTBT. Then we schedule those whose surgery was previously postponed in such a way to foster those waiting for more days after the cancellation. Finally, when the first two components of s_i, that is, $\alpha_1 W_i$ and $\alpha_2 D_i$, yield to the same value for two different patients, we break ties by ordering them following a LPT or a SPT policy (with respect to EOT) in the case (A) and in the case (B), respectively.

3.4 Operating Room Real Time Management

During the execution of the operating room plan, it could happen that the EOT differs from ROT. When $r_i > e_i$, for a patient $i \in L^{(j)}$, the whole plan could be delayed. When the overall delay could determine the exceeding of the jth OR session duration S_j, RTM should deal with the problem of postponing a surgery or using a part of the available overtime. Such a decision poses the problem of evaluating the impact of consuming overtime or to have a cancellation.

Let us consider the jth OR session on day $k = d_j$ having duration S_j and a list $L^{(j)}$ of scheduled and sequenced patients. We suppose that $m < |L^{(j)}|$ patients are already operated on. Let ρ_m be the effective time elapsed to operate on the m patients, that is,

$$\rho_m = \sum_{i=i_1,\dots,i_m} r_i. \tag{5}$$

Let us introduce the following parameter:

$$\beta_{km}^j = 1 + \frac{N_k}{N} - \frac{\Omega_{km}^j}{\Omega} \qquad (6)$$

where N_k is the number of OR sessions from the day after k and Ω_{km}^j is the remaining overtime after the surgery of patient i_m after elapsing ρ_m.

The value β_{km}^j would measure the overtime still available with respect to the number of OR sessions to be still performed. Actually, β_{km}^j is close to 1 when the overtime has been used proportionally; it is between 0 and 1 or it is greater than 1 when it is underused or overused, respectively. Because N_k is equal to 0, we remark that the last day of the week is always less than or equal to 1 hence promoting the use of overtime.

RTM starts every time a surgery ends and $\rho_m > \sum_{i=i_1,\dots,i_m} e_i$. It consists of three procedures.

Sequencing Check.   The sequencing of the remaining patients is checked in such a way to ensure that (i) all the remaining patients having $\tilde{w}_i^k > 1$ are scheduled prior to the other patients and (ii) those having $\tilde{w}_i^k > 1$ are ordered by decreasing value of $\tilde{w}_i^k$; if such patients run out the available operating time S_j, the remaining patients maintain the same ordering, otherwise the algorithm selects a number of patients to fill S_j following a rule similar to the best fit rule for the bin packing problem [20];

Patient Postponing.   Let i_{m+1} be the next patient in the schedule. Then, if

$$e_{i_{m+1}} > S_j - \rho_m ,$$

the patient i_{m+1} could incur in a cancellation. Therefore, RTM checks if

$$\beta_{km}^j \left(\frac{e_{i_{m+1}} + \rho_m}{S_j} \right) \leq 1 \qquad (7)$$

and if (7) is satisfied, the required overtime is assigned to patient i_{m+1}.

Rescheduling.   At the end of the day, all the postponed surgeries must be rescheduled. Our algorithm tries to insert a surgery with EOT e_i in the jth OR session planned in the next days in such a way to minimize the difference between S_j and the sum of the EOTs of the already assigned patients plus e_i. If an insertion is not possible, the patient will be scheduled next week.

Table 4 Model validation: comparison with real measures

	Bed utilization (%)	OR session utilization (%)
Real measures	51.10	77.33
Simulation model	49.10	80.82
Difference	2.00	3.49

4 Model Validation

The validation of a simulation model requires a quite complex analysis. In our case, we are interested in the logical correctness of the simulation model representing the surgical CP. On the other side, we are not interested in the replication of a real system.

To this purpose, we adapted our simulation model to represent the inspiring case, that is, reported in [10]. In that paper, the proposed model dealt with two patient flows having similar EOT but different LOS. Note that the LOS of the second flow is roughly the double of the first one while the number of patients in the first flow is roughly the double of the second flow. Since our model deals with only one patient flow, we adapted our patient flow generator in such a way to have, on average, the same number of patients having the LOS of the first flow which is the most numerous. In this validation scenario, we have $N = 7$ OR sessions having the same duration equal to 360 minutes. Two OR sessions are scheduled on from Tuesday to Thursday and one on Friday. The other parameters are reported in Sect. 3.1. Furthermore, we turn off all the optimization during the three phases.

Let us introduce the following performance indices: u_{bed} is the bed utilization whilst u_{OR} is the OR session utilization.

Table 4 reports the results of the comparisons with respect to the measures reported in [10]. Our results are the average value over those obtained by running the hybrid model 30 times with different starting conditions. Each of these computational experiments runs for a time horizon of 2 years but collects data only in the second year.

The differences in the two performance indices can be accounted to the different composition of the patient flow as depicted above. For instances, the gap of 3.49 % for u_{OR} expressed in minutes corresponds to the execution of one surgery having average duration. On the basis of these considerations, the comparison is satisfactory with respect to our objective, that is, the validation of the logical correctness of our simulation model.

5 Impact of the Optimization

In this section we would like to evaluate the impact of the optimization modules integrated in the simulation model. Section 5.1 describes how the computational experiments are carried out reporting the possible configurations of the optimization modules, the performance indices, and the evaluation scenario. Section 5.2 reports the results of the computational tests made on a given scenario. Finally, Sect. 5.3 reports about the impact of such optimization on bed levelling.

As reported for the model validation, the results reported in the following sections are the average value among those obtained by running the hybrid model 30 times on a given configuration and, each time, starting from a different initial condition. Finally, we remark that all the simulation parameters are reported in Sect. 3.1.

5.1 Test Configurations, Performance Indices, and Scenarios

The optimization modules depicted in Sects. 3.2–3.4 can be combined in different ways, each one determining a different final overall performance.

In order to evaluate their actual impact, we define a *baseline configuration* with respect to the three phases as follows:

Phase 1: advanced scheduling performed by a first fit algorithm, that is, (i) consider patients by decreasing order of w_i, (ii) place a patient in the first OR session available from Monday to Friday;

Phase 2: patients are sequenced in a random order;

Phase 3: overtime is assigned *a priori* uniformly to all OR sessions;

Phase 3: surgeries are rescheduled only at the end of the day using a first fit algorithm.

Besides the baseline configuration, we define further configurations to evaluate the impact of the optimization modules. Each configuration is defined with respect to the baseline configuration.

- **Phase 1**:

  option 1: computing w_i w.r.t Monday instead of the previous Friday (in the simulation model, Friday is the day in which the advance scheduling is performed);

  option 2: adopting the greedy depicted in Sect. 3.2.1 (instead of the first fit algorithm);

  option 3: adopting the local search depicted in Sect. 3.2.2;

- **Phase 2**:

  LPT/SPT: use LPT or SPT rules in sequencing (case (A) or (B) in Sect. 3.3), respectively;

- **Phase 3**:

  option A: adopting RTM after each surgery;

  option B: adopting algorithm depicted in Sect. 3.4 for rescheduling patients at the end of the day (instead of the first fit algorithm).

Table 5 Patient-centred and facility-centred indices

Index	Definition
Patient-centred	
C	Number of cancellations
f_{MTBT}	Percentage of patients operated within the MTBT
l_{avg}	Average length (number of patients) of the waiting list
t_{avg}	Average waiting time spent in the waiting list
w_{avg}	Average value of patient's w_i at the time of their surgery
w_{max}	Maximum value of patient's w_i at the time of their surgery
Facility-centred	
u_{bed}	Bed utilization
u_{OR}	OR session utilization

Table 6 Scenario

	OR 1	OR 2	OR 3	OR 4	OR 5
Mon	300	360	420	420	420
Tue	300	360	420	420	
Wed	300	360	420		
Thu	300	360	420	420	420
Fri	300	360	420	420	

Table 5 reports the two types of indices adopted to evaluate the impact of the optimization modules. We define a set of patient-centred indices in such a way to evaluate the performance from a patient point of view. We also define a set of facility-centred indices in such a way to evaluate them against to the patient-centred ones.

It is quite evident that different indices can affect each other. For instance, the increase of the number of cancellations can affect the bed utilization and, in its turn, can reduce the percentage of patients operated within the MTBT. Our aim is to identify a test configuration which increases the patient-centred indices without deteriorating the facility-centred ones.

Table 6 depicts the scenario in which we evaluate the optimization solutions on different operating contexts.

5.2 Quantitative Analysis

This section reports the analysis on the evaluation scenario. First, the impact of each optimization modules is evaluated through the quantitative analysis. Based on these results, two further configurations are studied. The results of such analysis are summarized in Table 7 which reports the value of the performance indices for each test

Table 7 Performance indices for each test configurations

| Option(s) | | | | | | | Performance indices | | | | | | | |
id	1	2	3	seq.	A	B	C	f_{MTBT}	l_{avg}	t_{avg}	u_{bed}	u_{OR}	w_{avg}	w_{max}
(0)	Baseline configuration						234 (2348)	32.63 %	338	55	63.61 %	89.88 %	1.17	4.05
(1)	✓						235 (2347)	31.85 %	346	56	60.20 %	89.78 %	1.11	3.29
(2)	✓	✓					226 (2340)	25.97 %	360	58	60.57 %	89.32 %	1.16	3.27
(3)			✓				252 (2346)	35.95 %	324	52	60.35 %	89.60 %	1.12	3.61
(4)	✓		✓				246 (2349)	35.34 %	330	53	60.26 %	89.78 %	1.06	3.41
(5)	✓	✓	✓				230 (2338)	27.17 %	355	58	60.79 %	89.55 %	1.17	3.10
(6)				LPT			236 (2367)	47.85 %	292	48	60.52 %	90.76 %	1.03	3.79
(7)				SPT			240 (2261)	12.14 %	452	72	58.57 %	86.35 %	1.51	4.91
(8)					✓		197 (2384)	74.62 %	213	35	59.26 %	91.34 %	0.80	2.64
(9)						✓	236 (2315)	30.70 %	339	55	72.56 %	88.79 %	1.18	3.79
(10)					✓	✓	222 (2372)	72.97 %	223	37	64.16 %	90.65 %	0.83	2.68
(11)			✓	LPT	✓		239 (2389)	79.91 %	192	32	60.27 %	91.75 %	0.73	2.62
(12)	✓		✓	LPT	✓		248 (2390)	85.53 %	207	34	60.58 %	91.78 %	0.71	1.87

configurations denoted by the value in the first column "id". Note that the column reporting the number of cancellations also reports, in brackets, the total number of patients operated on. All the analysis are compared with the baseline configuration.

Regarding the impact of the advanced scheduling optimization module, we can observe a lower waiting time in the waiting list and an improvement of the performance indices related to MTBT in test configurations (3) and (4). On the other side, the minimal number of cancellations is obtained with configuration (2) but, at the same time, the percentage of patients operated on before their MTBT decreases consistently. Note that the use of local search allows to insert more patients determining the improvement measured in (3) and (4).

Regarding the impact of the allocation schedule optimization module, we can observe a significant better performances when LPT policy is adopted. Figure 3 shows the trend of I_{avg} under the baseline, (6) and (7) configurations.

Regarding the impact of RTM, we observe a remarkable improvement of all the performance indices (see configurations (8) and in particular f_{MTBT}). On the other side, we observe the negligible impact of the algorithm for the rescheduling postponed patients at the end of the day (see configurations (9) and (10)).

Figures 4 and 5 show, respectively, the trend of I_{avg} and w_{avg} under the baseline and (8) configurations. Note that it is positive when $w_{avg} < 1$ which means that, on average, all the patients are operated on before their MTBT.

Finally, configurations (11) and (12) report about the combination of the different best options. We note a further improvement of the performance indices except for that related to the number of cancellations if compared with configuration (8). This is due to the fact that local search allows to insert more patients in the advanced scheduling thus reducing the waiting time in the waiting list but increasing the probability of incurring in a cancellation. Figure 6 shows the trend of w_{avg} under the baseline, (11) and (12) configurations. While baseline configuration shows a value of w_{avg} always greater than 1, we remark that both configurations (11) and (12) tend to be less than 1. Further, configuration (12) seems more stable and powerful in reducing these indices.

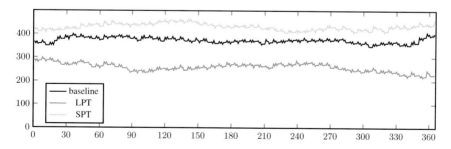

Fig. 3 Trend of I_{avg} (data referred to the second year, days on x-axis, patients on y-axis).

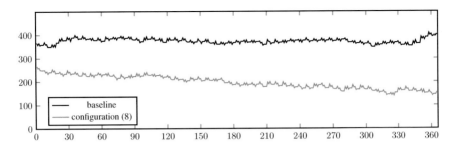

Fig. 4 Trend of I_{avg} (data referred to the second year, days on x-axis, patients on y-axis).

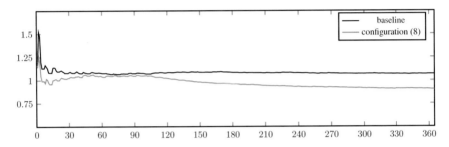

Fig. 5 Trend of w_{avg} (data referred to the second year, days on x-axis, patients on y-axis).

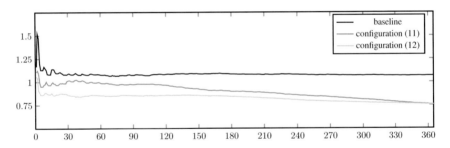

Fig. 6 Trend of w_{avg} (data referred to the second year, days on x-axis, patients on y-axis).

5.3 Bed Levelling

Among many different performance criteria [6], the evaluation of the ward stay bed levelling seems to be one of the more challenging [21]. A planning leading to a smooth—without peaks—stay bed occupancy, will determine a smooth workload in the ward and, at the end, an improved quality of care provided to patients.

Figure 7 depicts the bed occupancy during the week reporting both the average (Fig. 7a) and 95 % percentile (Fig. 7b) values. The results for the baseline and configuration (12) show a peak on Friday determined by an increased bed occupancy of about 50 % with respect to Monday. This behaviour seems to be not affected by the

(a)   **(b)**

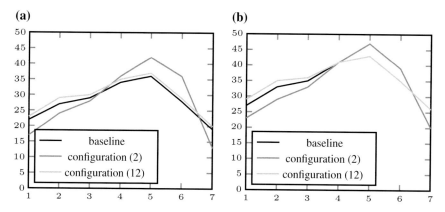

Fig. 7 Bed occupancy during the week (average and 95 % percentile values)

optimization of the clinical pathway since the baseline configuration and configuration (12) are really similar.

On the other side, the behaviour of the configuration (2) shows how planning decisions can affect the bed levelling during the week. In the case of configuration (2), the advance scheduling would compute a solution limiting the use of the weekend stay beds since they are limited in number. This decision largely affects the bed levelling as shown both in Fig. 7a and 7b where bed occupancy is doubled, approximately. These results confirms those available in the literature leading to the need of ad hoc optimization methods for bed levelling as in [22].

6 Conclusions

In this paper we proposed a hybrid simulation and optimization model for the analysis of a surgical CP from a patient-centred point of view: simulation is used to generate a real situation with respect to the inherent stochasticity of the problem while optimization is used to take the best decisions in different points of the surgical CP.

The quantitative analysis discussed in Sect. 5 shows the positive impact of the optimization in the management of the surgical CP. In particular, the most effective optimization module is the operating room real time management determining a general improvement of all the performance indices with respect to a baseline configuration of the surgical CP.

Comparing the baseline configuration with configurations (11) and (12) we can observe a great improvement of the performance indices related to the waiting list in terms of its length and the waiting time. This allows to double (at least) the percentage of the patients operated on before their MTBT time limit. These improvements can determine a general improvement of the quality of service from a patient-centred point of view without deteriorating the facility-centred performance indices.

The quantitative analysis confirms the trade-off between the number of cancellations and the number of operated patients (or, equivalently, the OR session utilization) as discussed in [23]. From this point of view, the proposed hybrid model could help the hospital management in the evaluation of this trade-off. Furthermore, the quantitative analysis confirms also the challenging problem of determining a planning leading to a smooth stay bed occupancy.

Acknowledgments The authors acknowledge support from the Italian Ministry of Education, University and Research (MIUR), under the grant n. RBFR08IKSB, "Firb – Futuro in Ricerca 2008". This paper extends the preliminary work reported in [24].

References

1. Campbell, H., Bradshaw, N., Porteous, M.: Integrated care pathways. Br. Med. J. **316** (1998)
2. De Bleser, L., Depreitere, R., De Waele, K., Vanhaecht, K., Vlayen, J., Sermeus, W.: Defining pathways. Journal of Nursing Management 14 (2006)
3. Panella, M., Marchisio, S., Stanislao, F.: Reducing clinical variations with clinical pathways: do pathways work? Int. J. Qual. Health Care **15**, 509–521 (2003)
4. Rotter, T., Kinsman, L., James, E., Machotta, A., Gothe, H., Willis, J., Snow, P., Kugler, J.: Clinical pathways: effects on professional practice, patient outcomes, length of stay and hospital costs (review). Cochrane Libr. **7** (2010)
5. Aringhieri, R., Addis, B., Tànfani, E., Testi, A.: Clinical pathways: inInsights from a multidisciplinary literature survey. In: Proceedings ORAHS 2012. (2012). ISBN 978-90-365-3396-6
6. Cardoen, B., Demeulemeester, E., Beliën, J.: Operating room planning and scheduling: a literature review. Eur. J. Oper. Res. **201**, 921–932 (2010)
7. Guerriero, F., Guido, R.: Operational research in the management of the operating theatre: a survey. Health Care Manage. Sci. **14**, 89–114 (2011)
8. Aringhieri, R., Tànfani, E., Testi, A.: Operations research for health care delivery. Comput. Oper. Res. **40**, 2165–2166 (2013)
9. Magerlein, J., Martin, J.: Surgical demand scheduling: a review. Health Serv. Res. **13**, 418–433 (1978)
10. Ozcan, Y., Tànfani, E., Testi, A.: A simulation-based modeling framework to deal woth clinical pathways. In: Jain, S., Creasey, R., Himmelspach, J., White, K., Fu, M. (eds.) Proceedings of the 2011 Winter Simulation Conference, pp. 1190–1201 (2011)
11. Tànfani, E., Testi, A.: A pre-assignment heuristic algorithm for the master surgical schedule problem (MSSP). Ann. Oper. Res. **178**, 105–119 (2010)
12. Herring, W., Herrmann, J.: A stochastic dynamic program for the single-day surgery scheduling problem. IIE Trans. Healthc. Syst. Eng. **4**, 213–225 (2011)
13. Lamiri, M., Xie, X., Dolgui, A., Grimaud, F.: Optimization methods for a stochastic surgery planning problem. Int. J. Prod. Econ. **120**, 400–410 (2009)
14. Tànfani, E., Testi, A., Alvarez, R.: Operating room planning considering stochastic surgery durations. Int. J. Health Manage. Inf. **1**, 167–183 (2010)
15. Borshchev, A.: The Big Book of Simulation Modeling. Multimethod Modeling with Any Logic, vol. 6. (2013). ISBN 978-0-9895731-7-7
16. Strum, D., May, J., Vargas, L.: Modeling the uncertainty of surgical procedure times: comparison of lognormal and normal models. Anesthesiology **92**, 1160–1167 (2000)
17. May, J., Strum, D., Vargas, L.: Fitting the lognormal distribution to surgical procedure times. Dec. Sci. **31**, 129–148 (2000)
18. Spangler, W., Strum, D., Vargas, L., Jerrold, H.: Estimating procedure times for surgeries by determining location parameters for the lognormal model. Health Care Manage. Sci. **7**, 97–104 (2004)

19. Aringhieri, R., Landa, P., Soriano, P., Tànfani, E., Testi, A.: A two level metaheuristic for the operating room scheduling and assignment problem. Comput. Oper. Res. **54**, 21–34 (2015). Available online 6 September 2014
20. Martello, S., Toth, P.: Knapsack Problems: Algorithms and Computer Implementations. Wiley-Interscience Series in Discrete Mathematics and Optimization. Wiley, New York (1990)
21. Beliën, J., Demeulemeester, E., Cardoen, B.: Building cyclic master surgery schedules with levelled resulting bed occupancy: a case study. Eur. J. Oper. Res. **176**, 1185–1204 (2007)
22. Aringhieri, R., Landa, P., Tanfani, E.: Assigning surgery cases to operating rooms: a VNS approach for leveling ward beds occupancies. In: 3rd International Conference on Variable Neighborhood Search. Electronic Notes in Discrete Mathematics, vol. 47, pp. 173–180 (2015) To appear
23. Beaulieu, I., Gendreau, M., Soriano, P.: Operating rooms scheduling under uncertainty. In Tànfani, E., Testi, A. (eds.) Advanced Decision Making Methods Applied to Health Care. International Series in Operations Research & Management Science, vol. 173, pp. 13–32. Springer, Milan (2012)
24. Aringhieri, R., Duma, D.: A hybrid model for the analysis of a surgical pathway. In: Proceedings of the 4th International Conference on Simulation and Modeling Methodologies, Technologies and Applications (HA-2014), pp. 889–900 (2014). ISBN 978-989-758-038-3. Winner of the Best Paper Award

Managing Emergent Patient Flow to Inpatient Wards: A Discrete Event Simulation Approach

Paolo Landa, Michele Sonnessa, Elena Tànfani and Angela Testi

Abstract In recent years, a growing proportion of patients flowing through inpatient hospital wards come from Emergency Departments (EDs). Because of ED overcrowding and the reduction of hospital beds, it is becoming crucial to improve the management of emergent patient flows to be admitted into inpatient wards. This study evaluates the impact and potential of introducing the so-called Bed Management function in a large city's health district. Thanks to the collaboration with the Local Health Authority of the Liguria region, an observational analysis was conducted based on data collected over a 1-year period to develop a discrete event simulation model. The model has been utilised to evaluate several bed management strategies. Two scenarios at a tactical level, i.e. the opening of a discharge room and blocking elective arrivals, have also been simulated. The effects of such scenarios have been compared with respect to a set of performance metrics, such as waiting times, misallocated patients, trolleys in EDs, and inpatient bed occupancy rates.

Keywords Health care services · Bed management · Discrete event simulation

P. Landa (✉) · M. Sonnessa · E. Tànfani · A. Testi
Department of Economics and Business Studies, University of Genova,
Via Vivaldi 5, 16126 Genova, Italy
e-mail: paolo.landa@unige.it

M. Sonnessa
e-mail: michele.sonnessa@edu.unige.it

E. Tànfani
e-mail: etanfani@economia.unige.it

A. Testi
e-mail: testi@economia.unige.it

P. Landa
Agenzia Regionale Sanitaria Liguria, Genova, Italy

© Springer International Publishing Switzerland 2015
M.S. Obaidat et al. (eds.), *Simulation and Modeling Methodologies,
Technologies and Applications*, Advances in Intelligent Systems
and Computing 402, DOI 10.1007/978-3-319-26470-7_17

1 Introduction and Problem Addressed

Emergency Department (ED) overcrowding is a critical issue in many health care systems [6]. In general, two approaches to face ED overcrowding are suggested. The first approach is devoted to reducing the number of patients inappropriately visiting the EDs, while the second is directed at facilitating early discharges from inpatient wards to smooth emergent admissions in some peak periods during the day [3].

Recently, hospital decision makers have primarily been focusing on the second intervention. This approach entails recognising that the so-called "ED problem" is actually a "system problem", and that to manage inpatient bed capacity, hospitals must consider all patient flows competing for such a resource. Increasing capacity in the ED without managing transferrals of patients into inpatient wards can worsen the problem [13]. In addition, a high percentage of patients visiting EDs are transferred after ED treatment to inpatient hospital wards. Hospitals are thus faced with an increasing proportion of their inpatient work coming from the fluctuating and unpredictable demand of emergency admissions and a decreasing portion coming from planned elective patients.

In the past, the problem had already been addressed extensively in many health systems, primarily in North America and the UK [1, 2, 5, 13]. To manage such a situation, greater coordination and communication among different health care providers and facilities involved in the pathway flows of patients are required. A solution suggested is the introduction of the so-called Bed Manager (BM) function.

The BM it is not a new concept. Twenty years ago, Green and Armstrong [7] defined the BM as a way of "keeping a balance between flexibility for admitting emergency patients and high bed occupancy, which is an indicator of good hospital management". Its main task is to report at given interval time slots during the day the volume, census, and occupancy rates of the available ward stay beds to synchronise expected discharges, i.e. bed supply, with expected admissions from the ED, i.e. bed demand [8].

The BM function has been demonstrated to be effective. For instance, Howell et al. [11] reported an increase of the ED throughput, substantially due to the reduction of approximately 25 % (approximately 1 hour and a half) of the time spent inside the ED. This effect was still larger in transferring patients from the ED to intensive care units [12].

The Health Foundation provides an extensive collection of empirical research studies aimed at improving emergent patient flows across inpatient wards [10]. From analysis of the literature, it appears that the scope, responsibility, and role of the bed manager are not clearly defined, and it is actually rather different in practical applications: medical, nursing, or managerial staff can perform the bed manager function, and even tasks or levels may be different [16]. The non-existence of standards of bed management practice is taken to reflect a lack of systematic attention paid to this role [4]. In the operative scenario of the Italian Health System investigated in this study, bed management has been introduced in hospitals only recently, and no systematic results are available to date [19].

2   Materials and Methods

The analysis of patient flows and clinical pathways are key issues in the recent operational research literature (see, for instance, [20, 21]). Other methods from industrial and business process modelling have also been employed, such as lean health care or six sigma [24].

Bagust et al. [3] applied a stochastic simulation model to determine what is the optimal level of spare capacity in presence of flows, by their nature stochastic and difficult to predict, resulting in 85 % of bed occupancy at most. Harper et al. [9] utilised a DES model not only to manage but also to plan bed capacity, with particular regard to the trade-off between bed occupancy and refusals. Schmidt et al. [18] implemented a DES model to assess a decision support system for bed management in a context where there is inherent uncertainty in the length of stay and number of ED patients to be admitted.

This study benefits from a collaboration with the Local Health Authority (LHA) of the Liguria region aimed at studying the effects of bed management strategies without increasing the inpatient bed capacity. To study the interrelation of the flows of emergent and elective admissions into inpatient departments in the observed case study, a Discrete Event Simulation (DES) model has been developed. In a preliminary version of the present work, it has been demonstrated how the model can help in supporting bed management decisions [14].

The complex nature of emergent and elective patient flow and the unpredictability involved makes discrete event simulation an ideal tool for evaluating the effects of alternative bed management on both ED and inpatient ward performance. Indeed, the advantage of employing a DES model, as also described in [15], is primarily related to its ability to provide deep analysis of the dynamic flows of patients throughout different time windows. It is not sufficient to consider the average distribution pattern because capacity and demand may match on average, and it may look as though the system ought to flow smoothly. However, even when capacity and demand match on average, the degree of variation in the timing of patient arrivals (demand) and the ability of beds to absorb that demand can result in admission delays and cancellations.

In this paper, simulation is utilised to evaluate scenarios pertaining to two primary areas of intervention. First, from the ED point of view, the model has been utilised to compare different bed management rules to manage the allocation of patients coming from the ED to the wards. Second, from an inpatient ward perspective, the possibility to smooth elective flows (admissions and discharges), in some period when peaks in the emergency demand are expected, has been modelled and evaluated.

The effects of both interventions are compared by means of a set of performance indexes. They are chosen to take into consideration both the hospital point of view (bed occupancy, additional beds and trolleys in the ED) and the patient point of view (misallocations, postponements of elective admissions already scheduled, and excessive waits in the ED for patients to be admitted).

The paper is organised as follows. In Sect. 3, a description of the simulation model structure and performance metrics employed are given. In Sect. 4, the case study is introduced together with the data collection and analysis performed with the collaboration of the Local Health Authority (LHA) of the Liguria region. In Sect. 5, the results of the scenario analysis tests are given. Finally, conclusions are reported.

3 Discrete Event Simulation Model Development

To study the patient flows between emergency departments and inpatient hospital wards, as well as their interrelation and synchronisation, a discrete event simulation model has been developed using the simulation software environment Witness 2013 [22]. Elective outpatient flows are also considered and also patient transferrals among the wards.

The overview of the system, the main components and the patient flows involved in the model are reported in Fig. 1. The system is made up of one ED and a given set of hospital inpatients wards with a given bed capacity. To simplify the analysis, the hospital wards are divided into four groups (in the case study reported in Sect. 4, they have been grouped as: medical, surgery, orthopaedics, and other wards). Both emergent and elective patients flow across the system. Emergent patients exiting the ED with a decision to admit should be allocated to a given ward and occupy a free bed. Elective patients are previously registered on a waiting list, and the date of admission is a priori determined.

For both emergent and elective patient arrivals, a detailed shift pattern is created to simulate the different arrival distributions.

For each emergent patient who arrives in the system, four attributes are generated. The attribute "to_admit" decides whether after the ED primary care and diagnosis, the patient must be admitted into a hospital ward or discharged. The attribute "ward" defines the proper ward destination of a patient who will be admitted. The "Time_in_ED" is the service time of the patient within the ED, while the "Length of Stay" (LOS) represents the service time of the hospitalised patients in the ward assigned to them. All these attributes are generated utilising different distributions depending on the slot time of arrival of the patients.

When an emergent patient has been in the ED, the generated "Time_in_ED", he/she becomes "ready" and is processed by a machine that verifies whether he/she should be admitted into a hospital ward. In the first case, he/she is pushed into one of the buffers "To_be_admitted", depending on the ward they are assigned to.

Obviously, emergent patients can be admitted into a ward, if and only if, there is a free bed available.

The "bed manager" machine works at a priori determined time windows during the day. It applies the defined BM strategy and manages the flow of emergent patients to be admitted into the inpatients hospital wards.

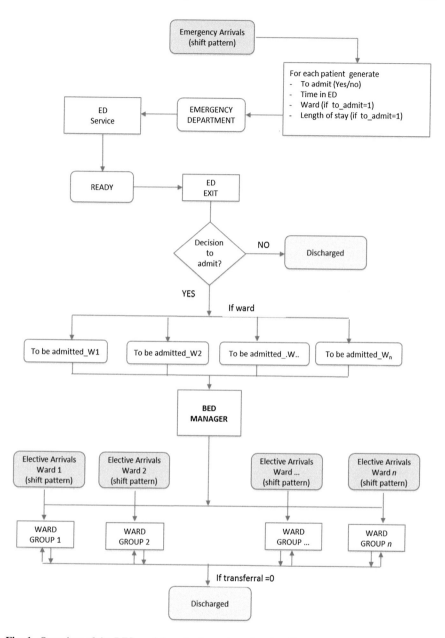

Fig. 1 Overview of the DES model: main elements and patient flows

After being admitted into the inpatient wards, elective or emergent patients stay in the ward for a given period of time, on the basis of their LOS. During their stay, patients can also be transferred from one hospital ward to another.

The effects of alternative bed management scenarios should be assessed by a set of performance metrics that must take into account the patient, the ED, and the hospital perspective. In this paper, the following five metrics have been used:

1. *Misallocation Index*: percentage of patients admitted into a ward different from the one assigned to them at the ED decision to admit. In a baseline scenario, patients can be placed in the first available bed, but this is not always appropriate. It is important to, for instance, limit the number of medical ward beds occupied by surgical patients and vice versa [1]. Wolstenholme et al. [23] observed that the most common strategy to face ED queues is to move patients to surgical wards until the "right" bed becomes free and that it can, if not properly handled, produce an increase in system congestion.
2. *Average Number of Patients Waiting to be Admitted*: this index is a proxy of additional beds or trolleys required in the ED.
3. *Average Waiting Time Before Admission in Inpatient Wards for emergent patients*. Note that national guidelines give a maximum time of 4 h for the UK [17] and 8 h in Italy [19].
4. *Percentage of Delayed Elective Admissions* due to unavailability of a free bed in the day a priori fixed for patient admission.
5. *Bed Utilisation Rate of Stay Beds* for each ward.

Some of them are designed to consider the patient's point of view, i.e. measuring patient discomfort and the risk of reduced outcome in the health care delivery process related to allocation of patients to a wrong specialty (misallocation index, waiting time before admissions, and number of patients delayed). Other metrics instead take the health facility point of view and measure the ED overcrowding with respect to patients with a decision to admit and the production efficiency of the inpatient wards (number of patients waiting to be transferred to inpatient beds, requiring additional beds and trolleys in the ED and, finally, the inpatient stay beds occupancy rate).

4 Case Study Data Collection and Analysis

The case study is developed with the collaboration of the Local Health Authority of Liguria (Agenzia Regionale Sanitaria), which provided the data referring to a large teaching hospital in the city of Genova. In particular, data have been collected for a 1-year period (January–December 2012) from two different databases, tracking, respectively, the flow of patient across the emergency department and the hospital inpatient beds. From the first database, we obtained the day and time of arrival, the exit decision from the ED (whether to be admitted into a hospital ward), the day and time of discharge form the ED, and in the case a patient is discharged with a decision to admit, the inpatient ward assigned. The second database collects, for both elective and emergent patients admitted into inpatient beds, the day and time

Table 1 Number of patients admitted

Number of ED accesses	84,781
% discharged	(75.3 %)
% transferred to inpatient wards	(24.7 %)
Number of elective admissions	45,638
Elective	24,696
Coming from ED	20,942

of admission, the ward of admission, the possible transferrals to other wards during the stay and the date and time of discharge from the hospital.

The overall number of patients entering the ED was 84,781. Approximately one out of four (24.7 %), i.e. 20,942, were subsequently admitted to an inpatient ward. In the same period, 24,696 elective patients were also admitted to the inpatient wards, Table 1.

The total number of inpatient beds available in the hospital and distributed among 79 wards was 1,256. The wards have been grouped into four groups, i.e. medicine, surgery, orthopaedics, and other wards.

In Table 2, the number of beds available in each group of inpatient wards and the distribution of emergent and elective patient among the wards is reported.

In Figs. 2 and 3, the relevant patient flows during the observation period are represented. In particular, Fig. 2 presents the ED flows for each day of the data collection period. Two series of data are reported, representing the total number of patients entering the ED and the number that are admitted in inpatients beds after the ED diagnosis, respectively.

In Fig. 3, the flow of emergent patients admitted into the hospital wards is compared along with the flow of elective patients. It can be easily observed that the flow of elective patients is more relevant and fluctuating with respect to the ED one.

To collect necessary data to build the model, a deep analysis has been devoted to estimate the interarrival and service rates for both emergent and elective patients. They are, apparently, very variable. Indeed, there is a large difference in arrivals between weekdays and weekends, working and resting times. Their variance, however, can be reduced when particular time slots are considered. With the help of the medical staff, 42 time slots have been chosen (Table 3).

In Fig. 4, for each slot, the yearly average consistency of patient flows during the week is presented. In particular, for each time slot, the number of patient discharged

Table 2 Distribution of beds, emergent, and elective admissions among wards

Ward group	Number of ward beds	% emergent admissions	% elective admissions
Medicine	332	62.47	14.15
Surgery	157	4.80	16.00
Orthopaedics	80	5.93	3.64
Other	569	26.80	66.21
Total	1141	100.00	100.00

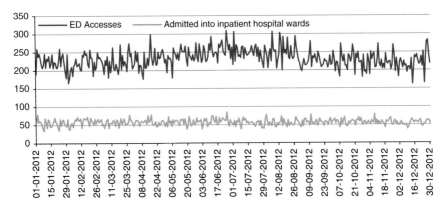

Fig. 2 Total ED arrivals and emergent patients admitted into inpatients wards

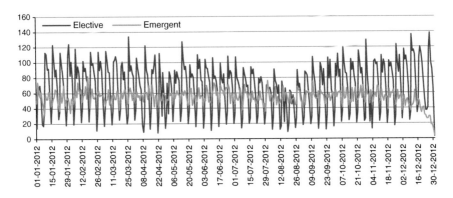

Fig. 3 Elective and emergent admissions in inpatient wards

Table 3 Time slot pattern

Time slot	Monday	Tuesday	Wednesday	Thursday	Friday	Saturday	Sunday
0–7	1	7	13	19	25	31	37
7–10	2	8	14	20	26	32	38
10–13	3	9	15	21	27	33	39
13–16	4	10	16	22	28	34	40
16–19	5	11	17	23	29	35	41
19–24	6	12	18	24	30	36	42

and admitted is reported. The plot of patients admitted is given as the sum of elective and emergent patients.

These flows can be used as a proxy for each day and time slot of the "supply" of beds, i.e. the number of patients discharged that are leaving a bed free for a new

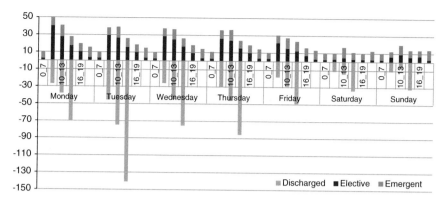

Fig. 4 Distribution of entrances and exits from inpatient wards

admission, and the "demand" of beds, i.e. the number of elective and emergent patients admitted.

Looking at the internal distribution of arrival flows it is clear how the elective patients are admitted, mostly at the beginning of the week, while the emergent ones are more equally distributed. This phenomenon is quite reasonable and must be considered to plan the bed allocation effectively. Discharges are distributed in three slots (i.e. 7_10, 10_13, 13_16). Most of the discharges are finalised in the slot (13–16), immediately after the lunch and after the doctor visits.

In Figs. 5, 6, 7 and 8, the same analysis has been conducted specifically for each ward group. In this case, patients transferred among inpatient beds have also been considered. Generally, it seems that admissions are primarily concentrated in the early morning slots, while discharges in the afternoon ones. The internal flows (transferrals in and transferrals out) have a different impact depending on the ward group considered. These flows are relevant, both as in and out flows, in the medicine ward group (Fig. 5), primarily referred to as transferrals out in the surgery and

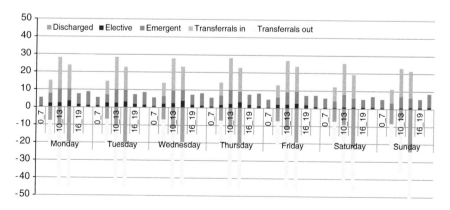

Fig. 5 Distribution of entrances, exits, and transferrals (medicine)

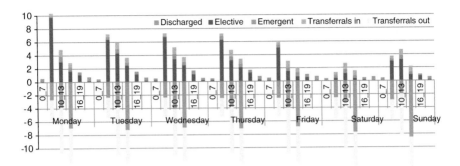

Fig. 6 Distribution of entrances, exits, and transferrals (surgery)

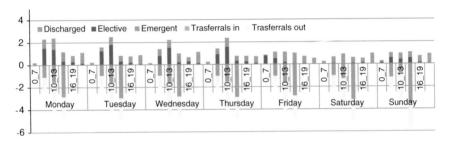

Fig. 7 Distribution of entrances, exits, and transferrals (orthopaedics)

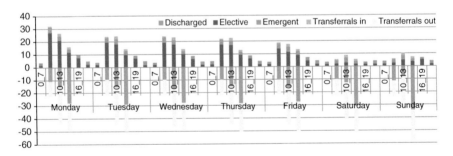

Fig. 8 Distribution of entrances, exits, and transferrals (others)

other wards (Figs. 6 and 8) and generally not present in the orthopaedics specialties (Fig. 7).

After analysis of the patient flows utilised to assess the interarrival and transferral rates, the service times have been estimated (Figs. 9, 10 and 11).

Figure 9 compares the time (in hours) spent in the ED by patients discharged and those transferred to inpatient beds. It demonstrates how on average the length of

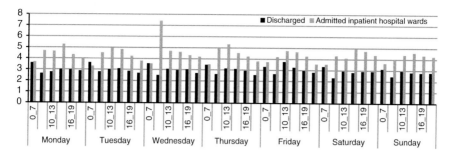

Fig. 9 Time in ED

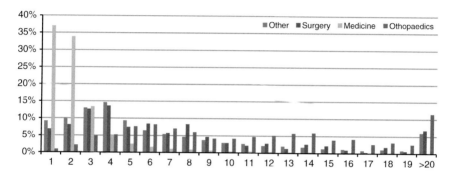

Fig. 10 LOS distributions (in days) in the inpatient wards (elective patients)

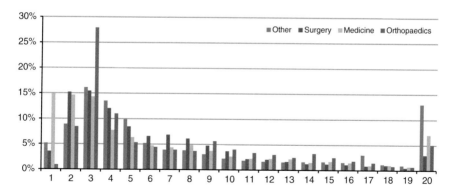

Fig. 11 LOS distributions (in days) in the inpatient wards (emergent patients)

stay of patient with serious pathology requiring a subsequent admission is longer than others.

Figures 10 and 11 present the distribution of the length of stay (LOS) for each ward, for elective and emergent patients, respectively. On the x-axis, the length of

P. Landa et al.

stay in days is represented, while on the y-axis the percentage of patients is reported for each ward where they have been admitted.

The largest number of emergent patients exits the ED with a decision to be admitted into one of the medicine wards, and their stay is quite short (the largest number of patients stays in the hospital for no more than 3 days, which is probably the time to be diagnosed). The elective patients are more equally distributed over different wards. The average length of stay is greater in this case.

5 Scenario Analysis and Results

Before performing the scenario analysis, the developed discrete event simulation model has been validated. The parametric tests and numerical description of the validation process is reported in [14]. Successively, the model was employed to analyse the effects of different bed manager rules (i.e. decision about the ward where an emergency patient should be admitted) and hospital organisation (i.e. slots and frequencies of patients discharge and elective admissions). The resulting scenarios are reported in Table 4. Two series of analysis have been conducted. First, starting from the baseline scenario (Scenario 0), which mimics current hospital practice without the BM function, the first three scenarios (Scenarios 1–3) introduce growing level of complexity in the BM function (BM decision postponement and forecast component). Second, starting from the best detected BM rule (Scenario 3) Scenarios 4 and 5 simulate at a tactical level two different organisational changes in the inpatient wards: (i) blocking the elective arrivals in some time slots where peaks in the emergent arrivals are forecasted; (ii) opening a discharge room to make beds available from the morning.

In the baseline scenario (Scenario 0), before admitting an emergent patient to an inpatient ward, the model just verifies whether there is a free bed available in the ward assigned by the ED clinicians after the ED treatment and diagnosis. If no bed is available in the ward assigned to the patient, the model checks bed availability in

Table 4 Scenario analyses

| Scenario | BM strategies | | Organisational changes | |
	BM decision postponement	BM lookahead	Elective admission blocking	New discharge policy
Baseline	–	–	–	–
Scenario 1	x	–	–	–
Scenario 2	–	x	–	–
Scenario 3	x	x	–	–
Scenario 4	x	x	x	–
Scenario 5	x	x	–	x

other wards and pushes the patient to the ward where a free bed is found. In this case, the patient becomes a misallocated patient.

In Scenario 1, the ability to postpone the assignment of a patient to a "wrong" ward is introduced. This allows verifying whether after a short time (usually half an hour), beds in the right ward assigned to the patient will be available. The number of times the BM can postpone its decision and also the period between the BM re-evaluations are input parameters of the simulation model. In the tests herein presented, these parameters are set to 4 times and 30 min, respectively. This finding means that the BM can postpone the admission decision for 2 h at most in an effort to reduce patient misallocations.

Afterwards, the BM ability to forecast (Scenarios 2 and 3) is introduced starting from Scenario 0 and Scenario 1, respectively. In these configurations, the BM component takes the decision to admit a patient into a given ward; it considers not only the number of ward beds available at each moment but also the expected number of elective patient arrivals, transferrals, and discharges during the day.

For each scenario, in Table 5, the number of patients misallocated, the average number of patients waiting to be admitted in the wards, the misallocation indexes, the average waiting time before admission, the bed utilisation rate, and the percentage of delayed elective patient admissions are reported.

First, the impact of the BM decision postponement is evaluated by comparing Scenario 0 with Scenario 1 and Scenario 2 with Scenario 3. The misallocations, as expected, are lower in both the cases. The total number of misallocated patients decreases from 619 to 239 and from 2270 to 368. Of course, if a greater priority is giving to allocate the patients in the "right" beds, we would expect that waiting time before admissions would increase. This is what appears in Table 5, where times for Scenario 1 are greater than for Scenario 0 and for Scenario 3 than for Scenario 2. Coherently, also average time and average total number of patients waiting (additional trolleys) are increasing. In conclusion, as expected there is a trade-off between the right assignment and the wait time, and the model can assess how much you have to wait longer for a better probability to be assigned to the correct ward.

Second, by comparing Scenario 2 with Scenario 0 and Scenario 3 with Scenario 1, the impact of the lookahead rule is assessed. This rule has a major impact on reducing the number of delayed elective admissions, which are always better (except for ward 3), thus penalising the emergent admissions. In fact, this benefit causes a large increase in the misallocation index and in waiting times, as well. By including both the decision postponements and the lookahead in the BM rules (Scenario 3), the number of patients misallocated and the number of elective patients delayed are reduced by approximately 40 %. This scenario produces a good balance between the benefits obtained by emergent and elective patients and is considered as the baseline scenario in the second scenario analysis directed to simulate organisational changes in the inpatient wards aimed at modifying the flow of elective admissions and discharges, respectively. In Table 6, the performance metrics of the second scenario analyses (Scenarios 3–5) are reported.

Table 5 Performance measures—first series of scenario analyses (scenarios 0–3)

	Scenario 0 (baseline)	Scenario 1	Scenario 2	Scenario 3
Total number of elective admissions	24,669	24,749	24,660	24,748
Total number of emergent admissions	20,974	20,946	20,941	20,939
Total # of patients misallocated	619	239	2,270	368
Average number of ED patients waiting to be admitted	4.13	4.67	4.13	4.60
Misallocation index				
W1	2.88	1.34	8.68	1.74
W2	0.00	0.00	3.79	0.00
W3	0.40	0.00	2.33	0.00
W4	4.20	1.13	19.00	2.50
Avg. waiting time before admission in inpatient wards				
W1	1.80	2.56	2.58	2.99
W2	1.58	1.60	1.67	1.67
W3	1.56	1.75	1.74	1.68
W4	1.71	2.23	2.06	2.54
% of delayed elective admissions				
W1	2.09	3.58	0.23	2.65
W2	0.00	0.00	0.59	0.00
W3	0.45	0.00	1.25	0.00
W4	3.83	3.77	0.25	1.29
Bed utilisation rate				
W1	92.35	92.52	92.42	92.32
W2	67.21	66.31	69.66	68.06
W3	73.49	70.02	75.93	71.25
W4	92.69	92.49	90.71	92.75

Scenario 4 assumes to block the elective admissions when a bottleneck in the flow of emergent patients is expected. This scenario tries to manage the inpatient demand flow to smooth the flow of emergent patients coming from the ED. Scenario 5 simulates the possibility to transfer patients to be discharged during the day in a dedicated room, referred as the discharge room. This simulation enables the freeing of some beds in advance early in the morning, thereby facilitating the admissions of emergent patients in the morning time slots.

From Table 6, it appears that blocking elective admission and using the discharge rooms have different benefits. Blocking elective admission appears to be preferable to increase the number of patients admitted in inpatient beds, without benefit for wait times, but limiting the additional trolleys in the ED. Instead, the discharge room allows for an increase in the bed utilisation rate. As to the misallocation index, both strategies can improve the index.

Table 6 Performance measures—second series of scenario analyses (scenarios 3–5)

	Scenario 3 (baseline)	Scenario 4	Scenario 5
Total number of elective admissions	24,681	24,586	24,719
Total number of emergent admissions	20,939	21,009	20,977
Total # of patients misallocated	368	189	229
Avg. number of ED patients waiting to be admitted	4.60	4.20	3.82
Misallocation index			
W1	1.74	1.25	1.05
W2	0.00	0.00	0.10
W3	0.00	0.08	0.24
W4	2.50	0.44	1.55
Avg. waiting time before admission			
W1	2.99	2.83	2.53
W2	1.67	1.70	1.60
W3	1.68	1.68	1.68
W4	2.54	2.29	2.34
Bed utilisation rate			
W1	92.32	91.68	92.38
W2	68.06	66.73	67.73
W3	71.25	70.01	72.23
W4	92.75	89.28	91.45
% of delayed elective admissions			
W1	2.67	1.56	1.67
W2	0.00	0.00	0.00
W3	0.00	0.00	0.00
W4	1.29	0.44	0.41
Bed utilisation rate			
W1	92.32	91.68	92.38
W2	68.06	66.73	67.73
W3	71.25	70.01	72.23
W4	92.75	89.28	91.45

Note, however, that additional considerations should be taken to complete the scenario analyses because costs also should be estimated to assess the impact of proposed variation, not only benefits. For instance, blocking elective admissions on some days generates a lengthening of waiting lists for elective patients (social cost) or even cancellations (revenue loss for the hospital).

Even Scenario 5 entails costs because it is necessary to set up a dedicated room (organisational costs) and/or requires higher staff costs to manage the discharges in different and enlarged time periods.

There are, of course, many other possible scenarios, not considered in this study, that imply strategic decisions. For example, facilitating the discharge of elderly patients by improving the continuity of care and linking the discharge coordinator with social services; extending the hours of community nurse teams by increasing home assistance and transferrals to the intermediate care ward, i.e. a lower intensity ward [4].

6   Conclusions

This chapter presents an analysis of patient flows between the emergency department and the inpatient stay wards of the same hospital. A discrete event simulation model has been developed to evaluate the impact of introducing the so-called bed management function within the hospital organisation.

The so-called "bed management" is a business function that allows increasing the efficient use of beds. The primary aim is to improve bed capacity allocation of emergent patients to inpatient hospital wards. As a result, the synchronisation of emergent and elective patients can be obtained.

The model herein developed could be employed as a decision support tool for assessing the effects of different bed management strategies, both from a patient and a hospital point of view. Thanks to the collaboration with the Local Health Authority of the Liguria region, the model has been applied to the San Martino University hospital, situated in the city of Genova (Italy), utilising data collected over a 1-year period.

Bed manager decisions affect how other hospital resources, such as operating theatres and elective inpatients wards, perform because all hospital services are dependent on bed availability. In turn, other hospital departments' inefficient performance, primarily lengthening hospital stays, can impact upon ED crowding. This study has recognised this effect in line with the literature findings.

Through the use of such tool, hospital management can assess the direct and indirect impact of different BM rules over the ED and inpatient ward flows. Moreover, the definition of specific measures introduced can help to better evaluate BM strategy. Indeed, the analysis of the simulation results gives relevant information about the effectiveness of the different strategies.

In particular, the introduction of the discharge room and elective patient blocking, are both able to increase performance for the hospital facility and also as for the benefit of patients. Note that these changes entail different tactical and organisational decisions to be taken from the hospital management.

Of course, the model could also include other patient characteristics and patient flows versus other hospital facilities and can easily be adapted to simulate other case studies, changing the system constraints and the organisational models of the ED and hospital wards.

In addition, the future direction of the research will be devoted to verifying the impact of an alternative assignment of beds among the ward groups on the system performance, integrating the developed DES model with an optimization model.

Acknowledgments The authors acknowledge support from the Italian Ministry of Education, University and Research (MIUR), under the grant FIRB n. RBFR081KSB. Data have been made available thanks to the collaboration between ARS Liguria (Dr. Francesco Quaglia and Domenico Gallo) and the Department of Economics and Business, University of Genova.

References

1. Audit Commission: Bed Management. HMSO, London (1993)
2. Audit Commission: Lying in Wait: the Use of Medical Beds, in Acute Hospitals. HMSO, London (1992)
3. Bagust, A., Place, M., Posnett, J.: Dynamics of bed use in accommodating emergency admissions: stochastic simulation model. Br. Med. J. **319**(7203), 155–158 (1999)
4. Boaden, R., Proudlove, N., Wilson, M.: An exploratory study of bed management. J. Manag. Med. **13**(4), 234–250 (1999)
5. Evans, B., Potvin, C., Johnson, G., Henderson, N., Yuen, I., Smith, T., Metham, S., Taylor, S., Sniekers, D.: Enhancing patient flow in acute care hospital: successful strategies at the Juravinski Hospital. Healthc. Q. **14**(3), 66–75 (2011)
6. Forero, R., McCarthy, S., Hillman, K.: Access block and emergency department overcrowding. Crit. Care **15**(216), 1–6 (2011)
7. Green, J., Armstrong, D.: The views of service providers. In: Morrell, D., Green, J., Armstrong, D., Bartholomew, J., Gelder, F., Jenkins, C., Jankowski, R., Mandalia, S., Britten, N., Shaw, A., Savill, R. (eds.) Five Essays on Emergency Pathways, Institute, for the Kings Fund Commission on the Future of Acute Services in London. Kings Fund, London (1994)
8. Haraden, C., Resar, R., Horton, S.S., Henderson, D., Dempsey, C., Appleby, D.: Capacity management: breakthrough strategies for improving patient flow. Front. Health Serv. Manag. **20**(4), 3–15 (2004)
9. Harper, P.R., Shahani, A.K.: Modelling for the planning and management of bed capacities in hospitals. J. Oper. Res. Soc. **53**, 11–18 (2002)
10. Health Foundation: Improving patient flow across organisations and pathways. http://www.health.org.uk/public/cms/75/76/313/4196/Improving%20patient%20flow.pdf?realName=TxPs1T.pdf (2013)
11. Howell, E., Bessman, E., Kravet, S., Kolodner, K., Marshall, R., Wright, S.: Active bed management by hospitalists and emergent department throughput. Ann. Intern. Med. **149**, 804–810 (2008)
12. Howell, E., Bessman, E., Marshall, R., Wright, S.: Hospitalist bed management effecting throughput from the emergency department to the intensive care unit. J. Crit. Care **7**(2), 184–189 (2010)
13. IHI, Institute for Healthcare Improvement: Optimizing patient flow: moving patients smoothly through acute care settings. Innovation series (2003)
14. Landa, P., Sonnessa, M., Tanfani, E., Testi, A.: A discrete event simulation model to support bed management. In: Proceedings of the 4th International Conference on Simulation and Modeling Methodologies, Technologies and Applications (HA-2014), pp. 901–912 (2014)
15. Norouzzadeh, S., Garber, J., Longacre, M., Akbar, S., Riebling, N., Clark, R.: A modular simulation study to improve patient flow to inpatient units in the emergency department. J. Hosp. Adm. **3**(6), 205–215 (2014)
16. Proudlove, N., Boaden, R., Jorgensen, J.: Developing bed managers: the why and the how. J. Nurs. Manag. **15**, 34–42 (2007)
17. Proudlove, N., Gordon, K., Boaden, R.: Can good bed management solve the overcrowding in A&E? Emerg. Med. J. **20**(2), 149–155 (2003)

18. Schmidt, R., Geisler, S., Spreckelsen, C.: Decision support for hospital bed management using adaptable individual length of stay estimations and shared resources. BMC Med. Inform. Decis. Making **13**(3), 54 (2013)
19. Simeu: Standard organizzativi delle strutture di emergenza e urgenza. Organizational standards of A&E facilities. http://www.comesemergenza.it/files/Standard_FIMEUC-SIMEU_2011.pdf. Accessed 24 June 2014 (2011)
20. Vanberkel, P.T., Boucherie, R.J., Hans, E.W., Hurink, J.L., Litvak, N.: A survey of health care models that encompass multiple departments. Int. J. Health Manag. Inf. **1**(1), 37–69 (2010)
21. Vissers, J., Beech, R.: Health Operations Management: Patient Flow Logistics in Health Care. Routledge, Oxon (2005)
22. Witness: User guide. Lanner Group, London, UK (2013)
23. Wolstenholme, E., Monk, D., McKelvie, D., Arnold, S.: Coping but not coping in health and social care: masking the reality of running organisations beyond safe design capacity. Syst. Dyn. Rev. **23**(4), 371 (2007)
24. Young, T., Brailsford, S., Connell, C., Davies, R., Harper, P., Klein, J.K.: Using industrial processes to improve patient care. Br. Med. J. **328**, 162–164 (2004)

Author Index

© Springer International Publishing Switzerland 2015
M.S. Obaidat et al. (eds.), *Simulation and Modeling Methodologies,*
Technologies and Applications, Advances in Intelligent Systems
and Computing 402, DOI 10.1007/978-3-319-26470-7